Edited by
Charles Q. Meng and Ann E. Sluder

Ectoparasites

Titles of the Series "Drug Discovery in Infectious Diseases"

Selzer, P.M. (ed.)

Antiparasitic and Antibacterial Drug Discovery

From Molecular Targets to Drug Candidates

2009

Print ISBN: 978-3-527-32327-2, also available in digital formats

Becker, K. (ed.)

Apicomplexan Parasites

Molecular Approaches toward Targeted Drug Development

2011

Print ISBN: 978-3-527-32731-7, also available in digital formats

Caffrey, C.R. (ed.)

Parasitic Helminths

Targets, Screens, Drugs and Vaccines

2012

Print ISBN: 978-3-527-33059-1, also available in digital formats

Jäger, T., Koch, O., Flohé, L. (eds.)

Trypanosomatid Diseases

Molecular Routes to Drug Discovery

2013

Print ISBN: 978-3-527-33255-7, also available in digital formats

Doerig, C., Späth, G., Wiese, M.

Protein Phosphorylation in Parasites

Novel Targets for Antiparasitic Intervention

2013

Print-ISBN: 978-3-527-33235-9, also available in digital formats

Unden, G., Thines, E., Schüffler, A. (eds)

Host – Pathogen Interaction

Microbial Metabolism, Pathogenicity and Antiinfectives

2016

Print-ISBN: 978-3-527-33745-3, also available in digital formats

Müller, S., Cerdan, R., Radulescu, O. (eds.)

Comprehensive Analysis of Parasite Biology

From Metabolism to Drug Discovery

2016

Print-ISBN: 978-3-527-33904-4, also available in digital formats

Edited by
Charles Q. Meng and Ann E. Sluder

Ectoparasites

Drug Discovery Against Moving Targets

WILEY-VCH

Editors

Charles Q. Meng
Boehringer Ingelheim Animal Health
3239 Satellite Boulevard
Duluth, GA 30096
United States

Ann E. Sluder
Vaccine and Immunotherapy Center
Massachusetts General Hospital
149 13th St.
Charlestown, MA 02129
United States

Series Editor

Paul M. Selzer
Boehringer-Ingelheim Animal Health
Binger Straße 173
55216 Ingelheim am Rhein
Germany

Cover
Hungry *Ixodes ricinus* females gathered on a small tree seedling questing for a host in front of a molecular cartoon of a ligand gated chloride channel (LGCC). The photo was taken by Jan Erhart in March 2012 in an oak wood in South Bohemia, Czech Rep. Courtesy of Jan Erhart and Petr Kopáček, Institute of Parasitology, BC CAS, Czech Rep. The schematic representation of the CysLGCC sectional view was taken from figure 12.4, chapter 12 by Tina Weber & Paul M. Selzer.

Library of Congress Card No.: applied for

British Library Cataloguing-in-Publication Data
A catalogue record for this book is available from the British Library.

Bibliographic information published by the Deutsche Nationalbibliothek
The Deutsche Nationalbibliothek lists this publication in the Deutsche Nationalbibliografie; detailed bibliographic data are available on the Internet at <http://dnb.d-nb.de>.

Print ISBN: 978-3-527-34168-9
ePDF ISBN: 978-3-527-80290-6
ePub ISBN: 978-3-527-80289-0
Mobi ISBN: 978-3-527-80291-3
oBook ISBN: 978-3-527-80288-3

Cover Design Adam Design, Weinheim, Germany
Typesetting SPi Global Private Limited, Chennai, India
Printing and Binding C.O.S. Printers Pte Ltd, Singapore

Printed on acid-free paper

10 9 8 7 6 5 4 3 2 1

Contents

List of Contributors

Eric A. Benner
Stine-Haskell Research Center
DuPont Crop Protection
1090 Elkton Rd.
Newark, DE 19714
USA

*John M. Clark**
University of Massachusetts
Department of Veterinary &
Animal Science
Amherst, MA 01003
USA
jclark@vasci.umass.edu

*Jeffrey N. Clark**
JNC Consulting Services
38 Wild Rose Lane
Pittsboro, NC 27312
USA
depotrd33@yahoo.com

Pat N. Confalone
Stine-Haskell Research Center
DuPont Crop Protection
1090 Elkton Rd.
Newark, DE 19714
USA

Daniel Cordova
Stine-Haskell Research Center
DuPont Crop Protection
1090 Elkton Rd.
Newark, DE 19714
USA

Andrew A. DeRosa
Zoetis, Veterinary Medicine Research
and Development
333 Portage St. Kalamazoo
MI 49007
USA

Robert F. Dietrich
Stine-Haskell Research Center
DuPont Crop Protection
1090 Elkton Rd.
Newark, DE 19714
USA

Christian Epe
Boehringer Ingelheim Animal Health
3239 Satellite Blvd
Duluth, GA 30096
USA

*Corresponding author.

Brandon R. Gould
Stine-Haskell Research Center
DuPont Crop Protection
1090 Elkton Rd.
Newark, DE 19714
USA

Ondřej Hajdušek
Czech Academy of Sciences
Institute of Parasitology
Biology Centre
České Budějovice
Czech Republic

Eric J. Hartline
Stine-Haskell Research Center
DuPont Crop Protection
1090 Elkton Rd.
Newark, DE 19714
USA

Gail S. Jones
DuPont Crop Protection
Stine-Haskell Research Center
1090 Elkton Rd.
Newark, DE 19714
USA

Ronald Kaminsky*
paraC Consulting for Parasitology and Drug Discovery
79685 Haeg-Ehrsberg
Germany
para.C@gmx.de

John B. Kinney
DuPont Crop Protection
Stine-Haskell Research Center
1090 Elkton Rd.
Newark, DE 19714
USA

Petr Kopáček*
Czech Academy of Sciences
Institute of Parasitology
Biology Centre
Czech Academy of Sciences
České Budějovice
Czech Republic
kopajz@paru.cas.cz

George P. Lahm*
Stine-Haskell Research Center
DuPont Crop Protection
1090 Elkton Rd.
Newark, DE 19714
USA
george.lahm@fmc.com

Diane Larsen
Boehringer Ingelheim Animal Health
3239 Satellite Blvd
Duluth, GA 30096
USA

Laura Letendre*
Boehringer Ingelheim Animal Health
631 Route 1 South
North Brunswick, NJ 08902
USA
laura.letendre@boehringer-ingelheim.com

Alan Long*
Boehringer Ingelheim Animal Health
3239 Satellite Blvd
Duluth, GA 30096
USA
Al.Long@merial.com

Jeffrey K. Long
Stine-Haskell Research Center
DuPont Crop Protection
1090 Elkton Rd.
Newark, DE 19714
USA

Léonore Lovis
Elanco Animal Health
Mattenstrasse 24A
4058 Basel
Switzerland

Michael Mahaffey
Stine-Haskell Research Center
DuPont Crop Protection
1090 Elkton Rd.
Newark, DE 19714
USA

Richard G. McDowell
Stine-Haskell Research Center
DuPont Crop Protection
1090 Elkton Rd.
Newark, DE 19714
USA

Tom L. McTier
Zoetis, Veterinary Medicine Research and Development
333 Portage St. Kalamazoo
MI 49007
USA

*Ard M. Nijhof**
Freie Universität Berlin
Institute for Parasitology and Tropical Veterinary Medicine
Robert-von-Ostertag-Str. 7–13
14163 Berlin
Germany
ArdMenzo.Nijhof@fu-berlin.de

Sandra Noack
Boehringer Ingelheim Animal Health
Binger Str. 173
55216 Ingelheim
Germany

Thomas F. Pahutski
Stine-Haskell Research Center
DuPont Crop Protection
1090 Elkton Rd.
Newark, DE 19714
USA

Cedric J. Pearce
Mycosynthetix, Inc.
505 Meadowland Drive
Suite 103
Hillsborough, NC 27278
USA

Jan Perner
Czech Academy of Sciences
Institute of Parasitology
Biology Centre
Czech Academy of Sciences
České Budějovice
Czech Republic

Daniel F. Rhoades
Stine-Haskell Research Center
DuPont Crop Protection
1090 Elkton Rd.
Newark, DE 19714
USA

*Heinz Sager**
Elanco Animal Health
Mattenstrasse 24A
4058 Basel
Switzerland
sager_heinz@elanco.com

Theo P. M. Schetters*
ProtActivity R&D
Cuijk
The Netherlands

and

Université de Montpellier
Laboratoire de Biologie Cellulaire
et Moléculaire
Montpellier
France
th.schetters@protactivity.com

Sandra Schorderet-Weber*
Sablons 30
2000 Neuchâtel
Switzerland
sandra.schorderet@hotmail.com

Mark E. Schroeder
Stine-Haskell Research Center
DuPont Crop Protection
1090 Elkton Rd.
Newark, DE 19714
USA

Paul M. Selzer*
Boehringer Ingelheim Animal Health
Binger Str. 173
55216 Ingelheim
Germany
paul.selzer@boehringer-ingelheim.com

Rafael Shapiro
Stine-Haskell Research Center
DuPont Crop Protection
1090 Elkton Rd.
Newark, DE 19714
USA

Wesley L. Shoop
Stine-Haskell Research Center
DuPont Crop Protection
1090 Elkton Rd.
Newark, DE 19714
USA

Radek Šíma
Czech Academy of Sciences
Institute of Parasitology
Biology Centre
Czech Academy of Sciences
České Budějovice
Czech Republic

Ben K. Smith
Stine-Haskell Research Center
DuPont Crop Protection
1090 Elkton Rd.
Newark, DE 19714
USA

Daniel Sojka
Czech Academy of Sciences
Institute of Parasitology
Biology Centre
Czech Academy of Sciences
České Budějovice
Czech Republic

Mark Soll
Boehringer Ingelheim Animal Health
3239 Satellite Blvd
Duluth, GA 30096
USA

Katharine R. Tyson
AgBiome
Research Triangle Park, NC 27709
USA

Molly E. Waddell
Stine-Haskell Research Center
DuPont Crop Protection
1090 Elkton Rd.
Newark, DE 19714
USA

Ty Wagerle
Stine-Haskell Research Center
DuPont Crop Protection
1090 Elkton Rd.
Newark, DE 19714
USA

Tina Weber
Merck KGaA
Frankfurter Str. 250
64293 Darmstadt
Germany

Debra J. Woods*
Zoetis, Veterinary Medicine Research
and Development
333 Portage St. Kalamazoo
MI 49007
USA
debra.j.woods@zoetis.com

Ming Xu
Stine-Haskell Research Center
DuPont Crop Protection
1090 Elkton Rd.
Newark, DE 19714
USA

Foreword

The attempts of humans to control the influence of ectoparasites on the health of themselves and their associated animals have been documented throughout recorded time. Within the past 100 years, we have witnessed major gains for ectoparasite control with the use of synthetic insecticides; but through time, we have found that these gains are episodic, primarily because of environmental issues and selection of drug resistance in arthropod populations. Therefore, the constant discovery of novel and safe drugs for ectoparasite control is a modern need. Volume 8 of the series *Drug Discovery in Infectious Diseases* provides a valuable snapshot of the timeline in the battle to control ectoparasites. The contributing authors have provided current perspectives on control of ectoparasites and transmission of agents of disease, strategies for discovery and development of drugs, and the development and potential uses of isoxazolines.

Ectoparasites have impacts on human and animal health by both direct and indirect mechanisms, and the reduction of these different impacts can be achieved by approaches that are not dependent on pesticides. The control program for the New World screwworm using the area-wide release of sterile males has been highly effective in controlling the direct impact of obligatory myiasis in North and Central America. Area-wide programs to control the indirect effects of ectoparasites, such as using vaccines for protection against agents of vector-borne diseases like yellow fever, and controlling onchocerciasis by targeting the microfilarial populations of humans also have been effective. However, the success of these programs is based on very specific parameters that lead to narrow applications, which leaves the need for broader spectrum control methods as a top priority.

The need for drug discovery for use in the control of ectoparasites of humans and animals will continue to be a major factor in the preservation of human and animal health. The One Health approach considers the facts that these entities cannot be separated and will only become more important due to global changes in the environment, as well as human population growth and movement. The majority of vector-borne human diseases have zoonotic cycles which can be affected by the effective use of ectoparasite control. Even for anthroponoses such as malaria and visceral leishmaniasis, zoonotic blood sources maintain many species of potential vectors of pathogens that are drivers of major causes of death in humans. Ectoparasites do truly represent a moving target for control efforts relative to population density and susceptibility. The timely and rational use of extant and novel drugs against these moving targets and upon a changing global stage can provide leverage for humans in our race against ectoparasites, as long as the discovery and development of new and effective drugs can maintain the pace.

April 2018

Lane Foil
Professor of Entomology
Louisiana State University

Preface

Infestation by ectoparasites has plagued humans, figuratively and literally, since ancient times; for example, lice are listed among the Biblical plagues visited upon Egypt (*Exodus 8:17, KJV*) and fleas transmitting bubonic plague have had devastating impacts on numerous civilizations over the centuries. Strategies for battling ectoparasites have an equally deep history, as evidenced by mummified lice found in ancient Egyptian combs and by perforated necklace beads that doubled as personal flea traps in medieval Europe. Although human ectoparasite infestations are less prevalent in modern developed countries due to dramatically improved living and hygiene conditions, infestation on domesticated animals remains a major challenge, causing nuisance in companion animals and livestock as well as lowering livestock productivity. Ectoparasites can move between animals and from animals to humans, potentially transmitting various diseases in the process. Ectoparasite control strategies must therefore contend with the ability of the target to move, often quite quickly, as anyone who has ever wanted to kill a flea can attest. This eighth volume in the Drug Discovery for Infectious Diseases series reviews strategies and models for discovery and development of ectoparasiticidal treatments for use in both human and animal health. The challenges presented by moving targets are a common theme throughout, ranging from the market requirement for a rapid speed of kill to the design of effective containment strategies in whole-organism drug screening assays.

The first section of the volume, Strategies & Resistance, presents various perspectives on what is needed to achieve effective therapeutic control of ectoparasite infestations. The section begins with a comparison by Woods *et al.* of therapeutic strategies against moving target ectoparasites with those against the less-mobile endoparasites. Weber *et al.* review strategies for preventing disease transmission by ectoparasite vectors, for which speed of kill is an important consideration. Schetters reviews promising progress toward development of vaccines against ticks. The emergence of drug resistance threatens the utility of ectoparasiticides, especially for cattle tick and human head lice. Sager *et al.* and Lovis *et al.* discuss the threat, reality, and monitoring of drug resistance in cattle tick, particularly relevant for Southern Hemisphere markets such as Brazil and Australia. Clark reviews new developments in the control of human lice.

The second section focuses on laboratory screens and *in vivo* models for discovery of new treatments against ectoparasites. Compared to human diseases, the molecular targets of parasites, especially ectoparasites, are much less clear, and few can be utilized for screening. The chapter by Kopáček considers the challenges in identifying candidate small-molecule drug targets in ticks. Currently, discovery of new treatments against ectoparasites relies heavily on phenotypic-based screening against whole organisms such as fleas and ticks. Chapters by Clark and Pearce and by Nijhof and Tyson discuss the design and implementation of various whole-organism assays to detect different aspects of the desired treatments, for example, the flea ingestion assay to detect the ability of a compound to work through ingestion rather than through contact. Compared to drug discovery for humans, a major advantage of drug discovery for animal health is that a new investigative drug can be tested in the target host much sooner in the latter. This might seem to make testing in rodent models less critical. However, testing in rodent models remains an important step in drug discovery for animal health, because these models require much less quantity of a compound and save valuable animals of the target species, as discussed in depth by Weber *et al.* Of course, testing in the target host species is an essential aspect of late-stage development of a new drug for animal health, and in the concluding chapter of this section Clark reviews protocols for controlled laboratory testing in host species and provides numerous examples of how these testing strategies have been applied in successful ectoparasiticide development programs.

Drugs effective against ectoparasites comprise only a few chemical classes, the pyrethroids, the phenylpyrazoles, and the macrocyclic lactones being the major ones. On average a new class appears about every 20 years. The isoxazolines are the most recent addition to the roster. The last section of this volume is devoted exclusively to this fascinating new class of ectoparasiticides, which has attracted tremendous interest in the animal health and crop protection industries. Weber and Selzer first discuss the new mode of action that underlies the rapid speed of kill by the isoxazolines. Chapters by Lahm *et al.* and by Letendre *et al.* detail the complete drug discovery and development process for afoxolaner, the first commercial product launched from this class. The development of sarolaner, reviewed by Woods and McTier, gives another story from a different setting. The final chapter by Long presents a comprehensive overview of the entire isoxazoline chemical class to date.

We thank Dr. Paul M. Selzer, the series editor, and the various representatives of Wiley for the opportunity to shepherd this volume, and for their guidance and support. We also thank the authors who have generously contributed their time and expertise. The combined result of their efforts is a volume designed to be of both interest and utility to those scientists in academia and industry willing to undertake the discovery of drugs aimed at moving targets.

April 2018

Charles Q. Meng
Duluth, Georgia
Ann E. Sluder
Boston, MA

Part One
Strategies & Resistance

Ectoparasites: Drug Discovery Against Moving Targets, First Edition.
Edited by Charles Q. Meng and Ann E. Sluder

1
Comparison of Anti-ectoparasite and Anti-endoparasite Therapies and Control Strategies

Debra J. Woods, Tom L. McTier, and Andrew A. DeRosa*

Abstract

In this chapter, we consider the similarities and differences between management of ecto- and endoparasites. We discuss the general approaches of prevention and control of ecto- and endoparasites (historic and current chemotherapies, environmental management/host management), while considering the different challenges faced relating to lifecycle, host distribution, genetics, and selection pressure.

Introduction

The Merriam Webster dictionary defines a parasite as an organism living in, with, or on another organism. "Parasitism" refers to the intimate association between the parasite and host, whereby the parasite obtains part or all of its nutrition or needs from the host and results in an overall negative effect on the host. Simply, ectoparasites live on the outside of the animal and endoparasites on the inside. Microparasites (bacteria, viruses, protozoa) establish infections where it is hard to quantify numbers of infectious agents present, so numbers of infected hosts are quantified, rather than numbers of parasites within each host. Microparasites are small and have rapid generation times relative to their hosts. Macroparasites (nematodes, flies, ticks, etc.) are larger and can be counted; so the unit of study is the individual parasite, not the infected host. Macroparasites are also small and have rapid generation times, but there is less of a difference than between microparasites and host. Epiparasites are an interesting class of parasites whereby a parasite parasitizes a parasite in a host–parasite interaction referred to as hyperparasitism (as referred to in the well-known poem by Jonathan Swift: "a flea has smaller fleas that on him prey, And these have smaller still to bite 'em: And so proceed ad infinitum"). Examples of this are the larvae of the tapeworm,

*Corresponding author.

Ectoparasites: Drug Discovery Against Moving Targets, First Edition.
Edited by Charles Q. Meng and Ann E. Sluder

Dipylidium caninum, which infect fleas (*Ctenocephalides species*) and biting lice (*Trichodectes canis*). When a dog ingests a parasitized flea/louse when grooming, the tapeworm develops into an adult in the dog's intestine.

Fleas, ticks, and flies are the most visible and treated ectoparasites, but lice and mites also affect health and wellness. Infestation with ectoparasites causes many pathogenic effects, including tissue damage and blood loss due to feeding; hypersensitivity responses following exposure to ectoparasite antigens; secondary infections; and, most importantly disease transmission. Ectoparasites have evolved to fill many niches, but may be considered in terms of their host association. Many mites and lice live almost completely in permanent association with their host and, as such, have fairly low mobility and are open to risk of desiccation and death without the protection of their host. Other parasites, such as fleas, ticks, and flies, are more mobile and relatively resistant to damaging factors when off the host. As a result, the first category of organisms, mites and lice, often has a commensal relationship with the host as opposed to a parasitic interaction. The latter are able to find new hosts relatively easily, so are less impacted by death of a host and therefore likely to impose greater harm to the host. Most medically important ectoparasites have short generation times, large numbers of offspring, and very high rates of population growth [1].

Roundworms are the major infective internal parasite in both humans and animals, although cestodes (tapeworms) and trematodes (flukes) also have a significant impact on health. Helminth infections cause significant long-term, chronic debilitating disease and even death. In humans, it is estimated that around 125 000 deaths occur every year, and these are mainly due to infections with the hookworms, *Ancylostoma duodenale* and *Necator americanus*, or the roundworm, *Ascaris lumbricoides* [2]. In companion animals, endoparasite infections are primarily a disease of younger animals, with peak occurrence in dogs less than 6 months old and cats under 18 months old [3], with prevalence ranging from 5% to 70% worldwide [4]. Clinically, symptoms can vary from zero to critical (emaciation, anemia, death) and the zoonotic risks associated with some helminths are an additional concern. The economic impact of helminth infections on livestock, especially ruminant, production is well recognized [5, 6]; in pigs, it has been shown that the presence of endoparasites induces a reduction in body weight [7]. The mechanisms for the impact of helminths on production include direct tissue damage and diminished function of the affected organs; diversion of energy and protein resources of the host from production toward defense and immune mechanisms and reduced feed intake. In companion animals, there are similar adverse effects on health; unfortunately, roundworm infection is common, due to the ubiquity of infective stage larvae in the environment, and concerns are elevated due to zoonotic health risks.

Approaches for Ectoparasite and Endoparasite Control

Treatment of parasites results in removal of an existing infection, whereas prevention is a process by which infection is deterred. For dog and cat ectoparasite infections, experts generally recommend prophylaxis (year round in some climates) over therapeutic treatment, to effectively manage control of the lifecycle, as well as to reduce the risk of disease transmission from ectoparasite vectors [8, 9]. The benefit from regular preventative treatment is particularly recognized for the control of fleas due to the nature of their lifecycle; an adult flea infestation is only a very small part of the population, which includes immature stages present in the pet's environment. It is critical to control these stages, either by the use of products that target these early lifecycle stages or by regular use of products that eliminate adult fleas on the animal, which will progressively lead to the reduction of environmental lifecycle stages. CAPC (Companion Animal Parasite Council) goes as far as to recommend "avoiding initial infestation altogether by placing pets on life-long prevention programs is the best option for pets and their owners" [8]. Transmission of diseases (i.e., *Rickettsia rickettsia* and *Borrelia burgdorferi*) by vectors, especially ticks, in dogs and cats is a major concern, and reducing the ability of a vector to attach and/or feed with an effective ectoparasite control program will reduce the risk of disease transmission. Tick-borne diseases in dogs and cats are becoming increasingly important, with several tick species responsible for the continued spread of multiple diseases. Among the other more important diseases are babesiosis, hepatazoonosis, Ehrlichiosis, anaplasmosis, cytauxzoonosis (cats), and tick paralysis. Although control of internal parasites is the primary concern for horses, ectoparasites can also impact the welfare of horses, either through dermatological effects or nuisance bites, which affect the ability of horses to thrive. The primary ectoparasites of horses are houseflies, stable flies, mosquitoes, and horse and deer flies; ticks, lice, and mites also parasitize horses. The major problem is a limited supply of effective, licensed products for horses [10], combined with the challenges of managing ectoparasite species that are able to live for extensive periods off the animal, requiring frequent treatment. Fly repellents tend to have a very short duration of efficacy, if any, and need frequent reapplication. Taylor's 2001 review [11] highlighted how few pharmaceutical agents are available for treating horse ectoparasites and this situation has improved little in the intervening years.

For livestock, as for companion animals, ectoparasite control is dependent on the parasite lifecycle – do they spend their whole life on the host, like lice; or only spend time on the animal to feed, as for some species of mites, which then return to protected spaces in the environment? For the former, treating just the animal will suffice; for the latter, the environment must also be treated. In a 1992 review [12], Byford *et al.* gave an authoritative overview of the commercial and health impact of ectoparasite infestation in the United States, focusing on the horn fly, *Haematobia irritans*, commercially the most important and widespread pest in cattle in the southern United States. Although a complicated condition, the overall implication was that the damaging effect on production and performance of cattle results from an alteration of the total energy balance following ectoparasite infestation. This is a major problem, considering the widespread resistance of horn flies to pyrethroids, probably accelerated by the use of pyrethroid-impregnated ear tags [13].

Humans are as susceptible to ectoparasite infestation as animals are, and are often affected by the same pests; for example, close contact with pets can result in infestation with fleas, ticks, lice, and mites and, although more common in animals, humans can also suffer from myiasis, especially in tropical regions. Scabies and head lice [14], as well as being socially embarrassing, can cause significant health problems. Resistance is a major issue, with multiple resistance mechanisms identified in different populations of head lice, including kdr (knockdown resistance) mutations of the sodium channel and oxidative metabolism resistance mechanisms (see chapter 6 by J. M. Clark in this volume). Although head lice are the most prevalent parasites causing pediculosis, body louse prevalence is also increasing, which heightens the public health threat due to risk of transmission of a number of diseases, including typhus (*Rickettsia prowazekii*), louse-borne relapsing fever (*B. recurrentis*), and quintana (trench) fever (*Bartonella quintana*). Tungiasis occurs in tropical and subtropical regions and is caused by the tiny flea, *Tunga penetrans*, the chigoe flea or jigger, which embeds itself under the stratum corneum and can lead to dangerous complications from secondary infections.

However, the biggest impact on human health globally is from ectoparasite vectors. Malaria, caused by the protozoan parasite *Plasmodia* spp., is commonly transmitted by infected female *Anopheles* spp. mosquitoes and, in 2015, there were approximately 214 million malaria cases and an estimated 438 000 malaria deaths [15]. Ticks are becoming increasingly important as a cause of significant disease in humans, as well as their pets. Examples of disease common to both pets and humans include the bacterial Lyme disease (*B. burgdorferi*), transmitted by the deer tick, *Ixodes scapularis* (*Ixodes ricinus* in the European Union); Rocky Mountain spotted fever (*R. rickettsia*), transmitted by *Dermacentor variabilis*; and ehrlichiosis (*Ehrlichia chaffeensis*), transmitted by the lone star tick, *Amblyomma americanum* and *I. scapularis*. The protozoal disease babesiosis is caused by infection with *Babesia microti* or *Babesia equi*, transmitted by *I. scapularis* and *Ixodes pacificus*. Viral diseases can also be transmitted by ticks, for example, tick-borne encephalitis (TBE) (caused by the flavivirus, TBE virus), transmitted by *Ixodes* spp. and there are even toxins, such as the tick paralysis toxin transmitted by *Dermacentor* spp. in the United States and *Ixodes holocyclus* in Australia. Ticks and mosquitoes may cause significant disease, but fleas have also had a major effect on human history. The vector for bubonic plague, *Xenopsylla cheopis*, transmits the bacterium *Yersinia pestis* when it feeds and this was thought to be the cause of the Black Death, which killed an estimated 50 million people in the fourteenth century [16].

For helminth infections, prevention is managed by disrupting the lifecycle of the parasite, which, in humans, is usually achievable by good sanitation and hygiene; but in animals, this is often less feasible. For livestock, experts recommend combining anthelmintic control with minimizing exposure to reinfection; while in companion animals with exposure to the external environment. Where contamination of the environment with infective larvae is extensive, prevention usually requires a strict treatment regimen, combined with regular egg production monitoring. A unique situation exists with heartworm, where a very high degree (up to 100%) of efficacy is required to control this potentially life-threatening disease of dogs and cats. Fortunately, regular dosing (1-month and 6-month products) with a macrocyclic lactone (ML)-based anthelmintic prevents development of the larval-stage heartworms. Heartworm larvae are very sensitive to ML products and until recently efficacy was thought to be 100% for the various products. However, more recent evidence of heartworm resistance to MLs has been detected in some areas of the United States (Mississippi Delta) and is a cause for concern. The American Heartworm Society [17] generates guidelines for canine and feline heartworm prevention, which it updates regularly based on the latest scientific understanding of the disease; the most recent revision was in 2014. For horses, as mentioned earlier, internal parasites are a major concern, especially as few new drugs are being approved for horses. In the face of increasing anthelmintic resistance [18], more sustainable methods for helminth control are being sought.

Ectoparasiticides

There are many mechanisms of action utilized in the management of ectoparasites in animals and humans, most older ectoparasiticides being historically leveraged from the crop protection industry. Numerous agricultural pests and veterinary ectoparasites are insects and acarines; and agrochemicals with activity against crop pests also frequently work against animal health ectoparasites. Add to this the fact that the market for Animal Health ectoparasiticides is significantly smaller than the market for agricultural pesticides, and it makes commercial sense to leverage the learnings and assets for animal health utility. Ivermectin is a major exception, being discovered by a pharmaceutical company animal health group (Merck Sharp & Dohme), and was first used on animals and later for agriculture and human medicine.

A primary driver for the development of these multiple therapies is the development of resistance. Resistance is a shift in susceptibility to a drug [19] and is recognized as a failure of drugs to control parasitism. Resistance is often measured as survival of parasites following a treatment that would be expected to be effective, or as a reduction in the protection period that a persistent treatment provides. Resistance development is multifactorial and involves parasite genetic factors (dominance of resistance alleles, gene frequency, fitness of resistant parasites, linkage disequilibrium, etc.); the host–parasite interaction (immunogenicity, pathogenicity, levels of refugia, etc.); biological factors (breeding patterns, numbers of offspring, generation time, behaviors that impact gene flow and opportunities for selection – migration, refugia, host range, etc.); and the parasite management system (method of application, frequency and timing of treatments, life cycle stage treated, selection threshold, etc.).

Insecticide resistance was first documented in 1908 by Melander [20] who noticed significant levels of survival of the San Jose scale insect, *Quadraspidiotus perniciosus* (Comstock), after exposure to lime-sulfur. In a 1984 review, Forgash [21] described the emergence of 428 resistant insect and acarine species in the following years, with 61% having medical/veterinary importance, and numbers still growing. By 2014, the cumulative increase in species resistant to insecticides was 586 [22]. Significantly, the numbers of resistant species started to increase dramatically after the introduction of synthetic organic insecticides (i.e., DDT (dichlorodiphenyltrichloroethane), cyclodienes, and organophosphates) in the 1940s. These products had better efficacy and broader spectrum of activity and consequently were used more extensively and repetitively, a practice that likely resulted in the observed resistance.

The Insecticide Resistance Action Committee (IRAC) was set up in 1984 to provide a coordinated response from industry to delay resistance development in insect and mite pests [23]. Its primary objective is to ensure long-term efficacy of insecticides and acaricides, thereby enabling sustainable agriculture and improved public health. One of the tools used is a mode of action classification scheme [22], which classifies pesticides based on the target site of action or mode of action. This can then be utilized, along with guidance on resistance management, to support alternation or rotation-based resistance management programs. The current classification includes 25 different mechanisms of action. Although focused on the crop protection industry, these classifications are also valid for effective management of human and animal health insect and acarine infestations.

Methods of ectoparasiticide use vary depending on the parasite and the host. For animal health, convenience is a major driver of route of administration [24]. Treating livestock is a very costly and resource-intensive process, so farmers have traditionally sought methods that allow whole-herd administration, such as foggers, dusts, sprays, dips, and so on, primarily incorporating formulations of organophosphates and synthetic pyrethroids to control ticks, mites, lice, and blowflies. Dosing frequency is highly dependent on the persistence of ectoparasiticide on the skin, hair, or wool of the animal, not just the lifecycle of the parasite; duration of efficacy is therefore generally longer in sheep, as the persistence on wool is higher. Cattle ear tags, primarily formulated with pyrethroids and some organophosphates, are still used for management of biting flies; when attached to the ear, insecticide is released from the formulation and dissolves in the sebum, spreading over the whole body, likely by grooming, ear/tail flapping, and contact between animals. In the past 30–40 years, agents and formulations with systemic efficacy have been developed and have enabled easy pour-on delivery (i.e., avermectins/milbemycins, synthetic pyrethroids, and some organophosphates); and even parenteral delivery for control of some ectoparasites, primarily endectocides (i.e., avermectins/milbemycins).

Companion animal ectoparasiticide products have progressed significantly in the past 30 years. Historically, dusting powders, baths, and aerosol sprays and impregnated insecticidal collars, with organophosphates, carbamates, and synthetic pyrethroids as the active agents, were the only available control measures for fleas, ticks, mites, and lice. Efficacy was variable and often with short duration and there were higher risks associated with toxicity both for the owner applying the product and for the animal, than with products developed in recent years. Spot-on application increased in popularity in the 1990s, with formulations incorporating the nicotinergic flea agent, imidacloprid, the GABA-gated chloride channel antagonist for flea and tick control, fipronil, and the chloride channel agonist, selamectin, for control of endo- and ectoparasites. More recently, oral ectoparasiticides such as spinosad [25, 26], a putative nicotinic acetylcholine receptor agonist; and the GABA-gated chloride channel blocking isoxazolines [27–29] have become available on the market (see chapter 15, by Woods and McTier, in this volume). The latter are potent insecticidal and acaricidal molecules which have provided safe, oral chewable tablets for the treatment and control of fleas and ticks in dogs. Interestingly, insecticidal collars, such as Scalibor® (Merck) and Seresto® (Bayer) that provide both repellency and direct killing of parasites have made significant sales in recent years, due partly to an increased concern about the spread of *Leishmania* into northern Europe [30], as well as convenient prevention of flea and tick infestations. Domestic dogs are the primary reservoirs for human visceral leishmaniasis, caused by the zoonotic protozoa *Leishmania infantum*. Control of the sand fly vectors, *Phlebotomine* spp., is the primary approach to managing disease transmission and collars impregnated with pyrethroids, such as Merck's Scalibor, are able to deliver an extended duration of prevention.

As outlined, a key motivator for managing ectoparasite infestation in humans is reduction of the risk of vector-borne disease transmission [31], although we should not underestimate the potential for significant morbidity from other ectoparasites in susceptible populations [32]. Treatments are generally topical [33]. For head lice treatment, pyrethroids are the main pediculicides [14]; scabies is treated with topical scabicides (pyrethroids, lindane, malathion, crotamiton, benzyl benzoate) and off-label oral ivermectin [34]; in tungiasis, the flea (*T. penetrans*) is removed physically. The incidence of flea- and tick-borne diseases is thought to be greater than is recognized by doctors and health authorities, and hence diagnosis and treatment are often delayed as they are not initially considered when attempting to determine the cause of the illness [35]. Control of fleas on pets and in the environment is the best approach for preventing disease transmission, as discussed earlier. Similarly, preventing exposure to ticks is recommended to prevent transmission of diseases such as Lyme disease; for example, with the use of an insect repellent, either DEET (*N,N*-diethyl-*meta*-toluamide) or a pyrethroid spray. In areas where TBE is prevalent (central and eastern Europe and northern Asia), the World Health Organization (WHO) recommends immunization with the TBE vaccine, which has been shown to be highly effective [36], and is on the WHO list of essential medicines [37]. There is increasing interest in utilizing oral ivermectin for control of ectoparasites, especially in poor countries where populations are infected with multiple parasites and ivermectin is already used in antifilarial control programs, although spectrum gaps have been identified [38], so this is not a universal solution.

Endoparasiticides

There are fewer classes of endoparasiticides than ectoparasiticides due in part to a reduced emphasis on the discovery of endoparasiticide agents by the agrochemical industry, although there are still examples of nematicidal molecules being discovered by crop protection companies and leveraged for development as animal health anthelmintics; emodepside being one example. In 1990, Meiji Seika Kaisha patented PF1022A, a novel cyclooctadepsipeptide anthelmintic (European patent 0382173A2) [39]. Fujisawa Pharmaceutical Co. (Japan) then filed another patent which included the bis-*para*-morphonyl derivative of PF1022A, named emodepside; which was licensed by Bayer Animal Health and developed as an anthelmintic for dogs and cats; marketed as Profender®, in combination with praziquantel [40].

Similar to the situation with ectoparasiticides, resistance is a strong driver for identification of novel endoparasiticides. However, for the major commercially important host species (cattle and dogs) nematodes have been slow to develop resistance to the endectocidal avermectins and milbemycins, so investment in novel endoparasitic drug classes has been limited. As a result, only three new drug classes have been marketed in the past 30 years [41]. Endoparasiticide resistance is, however, now being reported in cattle gastrointestinal nematodes [42–44] and heartworm (*Dirofilaria immitis*) resistance to MLs (avermectins and milbemycins) in dogs is now acknowledged to have emerged in the Mississippi Delta in the United States [45, 46]. This has no doubt stimulated investment in Animal Health endoparasiticide research. For example, more than 800 anthelmintic families were filed in animal health company patents during the past 10 years; at least 175 of these describing novel compounds. Along with increased investment in anthelmintic discovery for human filarial diseases, this is very encouraging for future management of endoparasitic diseases in animals and humans.

In animal health, anthelmintics are used therapeutically to treat existing infections or clinical outbreaks or prophylactically where treatment timing is dependent on the disease epidemiology. When viewed across all hosts, anthelmintics are primarily administered orally: as drenches in livestock, tablets for dogs and cats, and pastes for horses; but parenteral dosing, by injection or with pour-on formulations, is also widely used in cattle (and to a limited degree in dogs), to reduce time and resources needed to treat the animals. Due to increasing resistance, which is widespread and serious in sheep, multiple drug classes are used, both alone and in combination (in some cases, with multiple active agents); with the newer aminoacetonitrile derivative and spiroindole drugs being utilized increasingly in sheep to control infections in areas where all other drug classes no longer work. There is a strong advocacy among experts for more sustainable approaches to resistance management [47], and this will be discussed in more detail in the section titled "Endoparasite Challenges" Section 2.2.

Challenges for Ecto- and Endoparasite Control

Ectoparasite Challenges

Unfortunately, selective breeding for "improved" traits in livestock and companion animals has generally increased susceptibility to parasites; for example, some breeds of dog (Dalmation, American Bulldog, and American Pit Bull Terrier) appear to be more susceptible to *Demodex canis* [48]. This is exacerbated by intensive production practices for livestock and the increasing zoonosis concerns with the growth of pets being viewed as family members; in a US 2015 Harris Poll [49], 95% of pet owners considered their pets to be members of the family (up 7% since 2007).

With the inherent variability in lifecycles, climate, and hosts, it is difficult to make broad recommendations on management of ectoparasite infestations. It can be challenging to control parasites, such as ticks and flies, that only spend part of their lifecycle on the host. Some parasite infections are seasonal (e.g., tick infections are common in spring and autumn and louse/mite infections more common in autumn and winter), allowing seasonally targeted treatments. With changes in climate, there are increasing predictions and observations of the spread of diseases from warmer to previously more temperate climates, due to movement of the vectors, as for sand flies in Europe [30]. There is also no doubt that increased international trade and travel is leading to reemergence of ectoparasite diseases; epidemiological studies support that ectoparasite diseases and their vectors are hyperendemic in the developing world [50].

At what point is intervention optimal? This is a surprisingly difficult question to answer. Logically, it makes sense to intervene before welfare is impacted, but this is not always well understood and can sometimes be difficult to measure. For example, with hypersensitivity, as in flea allergy dermatitis [51], once sensitization has occurred, recurrence of signs can be initiated by just a small number of bites, although the threshold of sensitivity varies between individual dogs [52], so preventing flea infestation with monthly treatments, either topical or oral, is recommended to break the lifecycle. There are established guidelines for companion animal parasite management [8, 9, 17], which include guidance for ectoparasites. As highlighted in the section titled "Ectoparasite Challenges", experts generally recommend prophylaxis over therapeutic treatment, to effectively manage control of the lifecycle, and to reduce the risk of vector disease transmission. For livestock, it is a continuous battle to maintain the efficacy of parasiticides [53]. There is a dichotomy between the desire of farmers for easy application and fewer interventions, and the prevention of resistance by minimizing selection pressure and maintaining refugia, thereby ensuring the population is constantly refreshed with unexposed, susceptible parasites [54]. It is no coincidence that resistance developed more rapidly in the single host tick, *Rhipicephalus* (*Boophilus*) *microplus*, than in multihost ticks [55]. Understanding population dynamics is a valuable tool, but can be challenging. Experts advocate the development of more sustainable, integrated pest management programs [1, 56], incorporating strategic, directed treatments, environmental control, disease management, and resistant breeds. However, this would require significant changes in management practices in the industry.

In humans, head lice infestations are a significant issue in developed, as well as developing countries [57–59], with evidence that prevalence is increasing around the world [60]. It is therefore surprising that monitoring and reporting are not standard practice in many countries [60]. It is clear that epidemics spring up frequently in populations of children, where, if left untreated, the infestations spread rapidly. Unfortunately, resistance to topical ectoparasiticides is widespread [61], with many plant-based products now being used, although safety and efficacy have not been well established. A key recognition from mathematical modeling is that synchronized treatment of "potentially" infected individuals (relatives, classmates, and other close contacts) should interrupt transmission [62], with systematic treatments being another key to successful eradication of infections.

Vaccines have long been championed as the solution to parasiticide resistance; however, despite decades of investment and research into host–parasite interactions and evaluation of many putative vaccine antigens, the number of marketed ectoparasite and helminth parasite vaccines is disappointingly limited [63]. Ectoparasite vaccines are particularly challenging, as the parasites live either on the surface of the host or even off the host. Bm86 is the only ectoparasite recombinant vaccine and works by immunizing cattle with a "hidden" tick gut antigen; antibodies generated against Bm86 rupture the gut wall of the tick and give good levels of protection against tick infection, although repeated immunizations are required to maintain antibody levels [64]. The lower levels of efficacy when compared to drug treatment, requiring parallel drug treatment, led to poor sales and removal from the market. Interestingly, a new formulation, developed by the U.S. Department of Agriculture (USDA) Agricultural Research Services (ARS), USDA Veterinary Services, and Zoetis, was recently given a conditional license for management of *R. microplus* infestations in both permanent and temporary quarantine zones in Texas [65].

Environmental concerns have led in the past to removal of pesticides from the market. Everyone is familiar with DDT, once considered the solution to all our pest problems. Its use to control ectoparasite vectors undoubtedly saved many lives before tolerance/resistance started to emerge. Unfortunately, it is also now known to be a persistent organic pollutant, which is readily adsorbed to soil/sediment and is resistant to environmental degradation; and add to this a high lipophilicity, which leads to bioaccumulation in the food chain and impacts on wildlife. There was a widespread ban implemented in most countries between 1970 and 1990, although a limited supply of DDT is still used for vector control, by indoor residual spraying (spraying the inside walls of homes made of mud or wood) [66]. Environmental impact now has to be evaluated for every new antiparasitic drug; ectoparasiticide, anthelmintic, or endectocide. This is generally not an issue for the newer oral companion animal ectoparasiticides, where environmental exposure is very limited, but can require a considerable program of work for livestock ectoparasiticides. Some ectoparasites, such as red mites on chickens, spend the majority of their lifecycle in crevices in buildings, only leaving to feed on the host for a short period at night. In this case, spraying the buildings with pesticide is the most effective control method; however, the products are considered biocides and require an extensive environmental program for approval.

Endoparasite Challenges

In this section we focus on the management of helminth parasites in livestock where drug resistance is a major challenge for both the control and prevention of endoparasite infections.

Both internal and external parasites of grazing livestock are ubiquitous and therefore all grazing livestock should be considered an at-risk or an exposed population to infection and infestation [67]. Even light-to-moderate infections and infestations negatively impact the welfare, thrift, and production efficiency of grazing animals. Approaches to minimize parasite infections and infestations include husbandry practices founded on an understanding of the epidemiology of the organism, chemotherapeutic interventions to prevent or remove the effect of the organism on the host, or a combination of both. Prior to the 1960s the use of chemotherapeutic agents to achieve modern-day expectations for animal welfare and production efficiency was virtually nonexistent. Livestock producers were essentially dependent on the genetic resistance or tolerance of the host to sustain body growth and reproduction, often at a high cost to the welfare and production efficiency of the animal. Furthermore, little was known about the epidemiology and biology of internal and external parasites, which could assist management decisions to moderate parasite infections. Even with the current knowledge base of parasite epidemiology and host genetics and breeding techniques, improved husbandry and hygiene practices and genetic selection as stand-alone methods for parasite control are far from achieving the level of animal welfare and production efficiency expected from modern livestock producers and society [68].

Since the 1960s, global beef production has more than doubled and carcass weights have increased by approximately 30% [69]. Improvements in animal welfare gained from effective parasite control by the three main classes of anthelmintics (benzimidazole, imidazothiazole/tetrahydropyrimidines, and MLs) have contributed to the efficiency of livestock production. The endectocidal characteristic of the MLs has also enabled livestock producers to depart from the once common use of plunge dipping or bath treatments for external parasite control; now limited primarily to tropical and subtropical regions of the world. Livestock managed under effective internal and external parasite control programs founded on chemotherapeutic control are more efficient converters of feedstuff to meat, enabling more efficient utilization of land and feed resources.

Inherent with the administration of any anthelmintic is the genetic selection of the subpopulation of organisms that are genetically tolerant or resistant to the active ingredient. Anthelmintic resistance, at least to the major classes of compounds, is conferred by multiple alleles and therefore constitutes a small percentage of a naive parasite population. As selection pressure is increased on a parasite population, the proportion of resistant parasites increases until they are the dominant genotype in the parasite population. In addition to the frequency of exposure/selection pressure, underdosing (exposing parasites to subtherapeutic levels of a drug) will also increase the resistant population by further selecting parasites that are genetically tolerant to the active pharmaceutical ingredient (API). This phenomenon has been observed for all classes of anthelmintics and will likely be the case if other classes of anthelmintic compounds become commercially available [42].

The greatest prevalence of anthelmintic resistance in livestock has been observed in the sheep industry where frequent anthelmintic administrations were common practice in internal parasite control programs. The frequency of administration was driven primarily by high mortality and clinical morbidity associated with parasitic gastroenteritis and anemia due in large part to the hematophagous parasite *Haemonchus contortus. H. contortus* is often the most prevalent internal parasite in sheep due to its high fecundity (egg shedding) relative to other species. The practice of frequent anthelmintic application and resistance development in sheep has often been inaccurately extrapolated to characterize the use pattern of anthelmintics and endectocides in cattle parasite management programs. This is an unfortunate and mistaken characterization of the industry as a whole. There are likely cases of misuse, overuse, or erroneous application of anthelmintics; but without empirical evidence demonstrating common practice, such statements are mere generalizations. It would not be possible to have 30 years of effective use of these compounds in the cattle industry if frequent indiscriminate or misuse was common practice in the industry. A brief review of the scientific literature will demonstrate that anthelmintic resistance is now, however, emerging in all species of livestock that are exposed to the current classes of compounds [42, 67, 70]. Industry and scientific leaders in conjunction with veterinarians and producers are working toward solutions to maintain the longevity of existing anthelmintics. For example, there is a slow shift in some market segments away from pour-on formulations to injectable formulations to ensure proper dose rate and application and the goal is to continue this trend. The introduction and proper use of combination products or concurrent use of anthelmintics with disparate modes of action [71] are also being introduced to producers along with education on use patterns.

Targeted selective treatments have long been advocated as a refugia-based approach to resistance management [47]. A number of biomarkers have been proposed, including Famacha® for haemonchosis [72] and measures of health and/or performance [73–75]. However, even the advocates recognize that it can be a challenge to convince farmers of the value of these approaches, when weighed against the additional time, energy, and costs required to implement [76, 77].

The availability of all of the existing classes of anthelmintics is vital to maintaining the current level of health and welfare of livestock in modern production systems, with a need to introduce new classes to support and maintain these levels, alongside helminth management programs that include anthelmintic resistance management as a variable.

Perspectives on Current and Future Strategies for Ecto- and Endoparasite Control

It is evident that antiparasitic agents greatly enhance the welfare and subsequently, for livestock, the production efficiency of the host. Unfortunately, these advantages diminish over time if parasiticide-susceptible parasite populations are not maintained within the environment where the host–parasite interaction occurs. When consistent genetic selection pressure by an antiparasitic agent is maintained on a parasite population, a threshold is reached where the parasite population is no longer susceptible to the antiparasitic agent and its benefits are no longer observed. Such is the case for small ruminants and horses, and the situation is now emerging in cattle. Researchers in more recent years have been evaluating mechanisms to manage resistance development and keep products effective for longer; but there is a mismatch between the reality of dosing regimens for animals and for humans in the developing world and the reality of what is required for "best practice" for management of resistance development.

Challenges of Bringing New Antiparasitic Drugs to the Market

In earlier sections we highlighted the importance of agrochemical pesticide development for leveraging substrate for animal health application, particularly for ectoparasiticide drugs, with the isoxazoline class highlighting the value of this resource; as well as the importance of animal health drug development as a source of human health antiparasitic medicines. We have previously shown figures highlighting the consolidation of the animal health industry over time [78, 79]. As a consequence of acquisitions and mergers, the overall resources available for antiparasitic discovery have reduced considerably over the past 25 years. We have updated the figure for this chapter (Figure 1.1) to show that the recent acquisition of Novartis by Elanco, and the acquisition of Merial by Boehringer Ingelheim Vetmedica have only intensified this consolidation. Although antiparasitic drugs are core to the success of animal health companies, these changes have resulted in fewer players, reduced competition, and potentially less opportunity for the discovery of novel antiparasitic molecules. A reduction in resources available for research and development (R&D) will impact availability of scientists and funds to discover and develop new products.

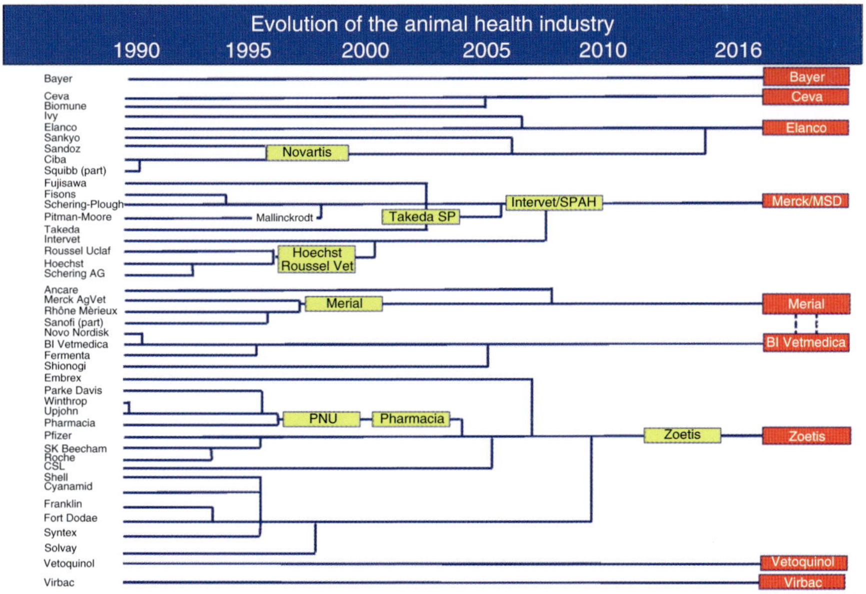

Figure 1.1 Consolidation of animal health companies 1990–2016.

As discussed in previous reviews [41, 78, 79], identifying molecules that kill parasites *in vitro* is the easiest part of the R&D process. A major challenge is delivering the drug with the optimal pharmacokinetic profile for efficacy, via the preferred, convenient route of administration. The drug has to be safe, both to the animal and the handler; with additional regulatory hurdles for human food safety (and environmental safety) for livestock products. Added to this are increasing regulatory pressures on new and even on existing products. For example, European Medicines Agency (EMEA) has concerns about persistent, bioaccumulative and toxic (PBT) or very persistent and very bioaccumulative (vPvB) substances in veterinary medicinal products [80]. Finally, the product has to be manufactured to high levels of quality, both for the API and the formulated product. All of these challenges increase the time and resources required to bring a product to the market. Companies then have to balance the chance of success and R&D costs against the potential value and return on investment (ROI) long term. If this is low, or negative, the product will not be developed. For human health the business model is different, with nongovernmental (nonprofit) organizations (NGOs) increasingly investing in neglected parasite diseases, including helminths. Although they still face all the challenges of the animal health industry, there is no expectation of ROI, but still a proven route to success is partnering with animal health companies. Historically, there are good examples of pharmaceutical companies partnering with the WHO to develop products for human health use. Mectizan® is a great example in which Merck led a collaboration with WHO in the late 1980s, running field studies to demonstrate the efficacy and safety of ivermectin for the treatment of onchocerciasis (river blindness). They have continued to commit to the Mectizan Donation Program [81] to provide ivermectin to treat both onchocerciasis and lymphatic filariasis (elephantiasis) in Africa. DNDi (Drugs for Neglected Diseases Initiative) has also recently partnered with Bayer Animal Health to evaluate the anthelmintic emodepside for macrofilaricidal activity against onchocerciasis [82].

Prevention and Control – Balancing the "Ideal" Against the "Real World"

In an ideal world, we would manage parasite infections and infestations with good hygiene and husbandry but, with the realities of the conditions of the developing world for humans and the intensive farming required to feed the world's population, this is never going to be possible, even for infections that could be managed this way. Parasiticides are therefore a fact of life, and it is to everyone's advantage to extend the lifespan of existing and new drugs. There is a clash, however, between treatment regimens that minimize resistance development and the desire for convenience in dosing and duration of efficacy, both for humans in the developing world, where access to treatments may be limited and challenging to reach, and for animals where reducing handling is a major driver. Farmers and pet owners now expect treatments and preventatives that meet their needs, be they chewable monthly flea and tick products for dogs (Simparica®,[1] NexGard®, and Bravecto®) or long-acting products (heartworm preventatives for dogs (ProHeart® 6 and 12) and anthelmintics for cattle (LongRange®)). Indications are that the market is progressing even further down the path of convenience, to meet the demands of their customers. So, how do we balance the demands of the customer with responsible parasite management?

Integrated parasite management (IPM) has been championed for many years as a responsible tactic to minimize the impact of parasite resistance [83, 84]. IPM aims to improve host resistance by combining multiple approaches, both chemical and nonchemical (targeted use of parasiticides, improved monitoring of resistance and infection levels, and incorporation of nonchemical control methods, e.g., fungi). There are examples of initiatives to utilize agroecological approaches in developing countries [85]. However, there is reluctance among end users in the developed world to accept the increased costs and resources to implement such schemes and the reality that there will likely be some loss in production and a level of parasitism present.

One approach being used successfully in sheep for helminth control is development of combination products. Historically there have been concerns about combinations increasing parasite selection, but modeling has shown that combining a new drug (with low resistance frequency and very high efficacy) with another class of anthelmintic will delay development of resistance to the new drug [71, 86]. The modeling showed that resistance to even a new active drug can develop rapidly if it is used in an inappropriate manner. However, although the benefit of the anthelmintic in slowing resistance development to the new entity is influenced by the level of resistance to the older drug, Leathwick's model [71] predicted that even at 50% efficacy of the older drug, the development of resistance to the novel drug should be slowed in a combination and vice versa. Nonetheless, best practice would be to use the combination while the older drug still has relatively high efficacy and resistance genes are still infrequent. The model also illustrated that a large percentage of the population must remain unexposed to the treatment – as refugia decreased, resistance developed more rapidly, reducing the benefit of the combination; encouraging management strategies such as rotational grazing. Another important observation was that resistance was still delayed even when resistance to one of the drugs was functionally dominant, as long as a high level of refugia was maintained. This is likely to be due to fully overlapping generations and small proportions of populations exposed to each treatment.

For insecticides too, modeling shows that mixtures are effective at delaying resistance (even better than alternation), as long as a proportion of the population is not exposed to the treatment [87]. In order for these mixtures to be effective for delaying resistance, the initial resistance frequencies should be low, the agents should be close to 100% effective against treated susceptible homozygotes, and the combination components should be nearly equal in persistence.

There have been some efforts to identify and validate nutritional supplements for sustainable control of gastrointestinal nematodes in livestock [77]; both targeting direct anthelmintic effects and the indirect effect of supplementary feeding improving an animal's resilience against gut nematode infections (nutraceuticals).

[1] SIMPARICA is a trademark or registered trademark of Zoetis Services LLC in the United States and other countries.

Tannin-rich plants (TRP), for example, may have a direct effect on reduction of larvae establishment in the host, as well as benefit for the host from the nutrients in the fodder [88]. This approach however depends on supporting the level of resilience and resistance against gastrointestinal nematode infections, which varies among ruminant species and also among and within breeds.

Biological control is well established for control of agricultural pests, where a range of control methods are used, including introduction of pathogens (bacteria, fungi, viruses, etc.), predators (insect larvae (ladybugs), entomopathogenic nematodes, predatory mites), and parasitoids (wasps and flies). More recently, RNA interference (RNAi) is also being evaluated. Transgenic plants offer the opportunity to express pathogens/toxins and this has been incredibly successful, especially with the use of *Bacillus thuringiensis* (Bt) toxin, which has replaced chemical insecticide use for many crops. This strategy is now being applied to RNAi, the latest tool for pest management [89, 90], with an RNAi-enhanced corn engineered to contain RNA devised to kill rootworms in development by Monsanto. RNAi sprays are also being developed, and could be on the market by 2020 [89]. However, there are concerns about effects on biodiversity and a need to evaluate potential levels of risk posed to nontarget species by biological control strategies [91]. Although there are studies evaluating fungi for veterinary control of livestock gastrointestinal nematodes [92–94], biological control strategies have yet to make an impact on the management of veterinary parasites.

The reality for veterinary and human parasite control is that there will continue to be an expectation for ectoparasiticides that rapidly clear existing infestations and prevent reinfestation for extended periods of 1 month and longer. For helminth control in livestock, the aim is to keep the challenge to young livestock at a minimum rate by both periodic and strategic deworming. For companion animals, the recommendation is year-round broad-spectrum parasite control with efficacy against roundworms, hookworms, and whipworms [8].

It is clear that parasites, both internal and external, have a major impact on the health and well-being of humans, both directly and through their effect on companion animals (with associated zoonotic diseases) and livestock (influencing the efficiency of food production). Building our understanding of the biology of the responsible organisms will help in the development of new drugs, vaccines, and control strategies.

References

1. Wall, R. (2007) Ectoparasites: future challenges in a changing world. *Vet. Parasitol.*, **148**, 62–74.
2. Stepek, G., Buttle, D.J., Duce, I.R., and Behnke, J.M. (2006) Human gastrointestinal nematode infections: are new control methods required? *Int. J. Exp. Pathol.*, **87**, 325–341.
3. Gates, M.C. and Nolan, T.J. (2009) Endoparasite prevalence and recurrence across different age groups of dogs and cats. *Vet. Parasitol.*, **166**, 153–158.
4. Stull, J.W., Carr, A.P., Chomel, B.B., Berghaus, R.D., and Hird, D.W. (2007) Small animal deworming protocols, client education, and veterinarian perception of

zoonotic parasites in western Canada. *Can. Vet. J.*, **48**, 269–276.

5. Charlier, J., van der Voort, M., Kenyon, F., Skuce, P., and Vercruysse, J. (2014) Chasing helminths and their economic impact on farmed ruminants. *Trends Parasitol.*, **30**, 361–367.
6. Mavrot, F., Hertzberg, H., and Torgerson, P. (2015) Effect of gastro-intestinal nematode infection on sheep performance: a systematic review and meta-analysis. *Parasit. Vectors*, **8**, 557.
7. Kipper, M., Andretta, I., Monteiro, S.G., Lovatto, P.A., and Lehnen, C.R. (2011) Meta-analysis of the effects of endoparasites on pig performance. *Vet. Parasitol.*, **181**, 316–320.
8. https://www.capcvet.org/guidelines (accessed 23 June 2016).
9. http://www.esccap.org (accessed 23 June 2016).
10. Little, J. (1999) Control of ectoparasites in horses. *In Practice*, **21**, 418–424.
11. Taylor, M.A. (2001) Recent developments in ectoparasiticides. *Vet. J.*, **161**, 253–268.
12. Byford, R.L., Craig, M.E., and Crosby, B.L. (1992) A review of ectoparasites and their effect on cattle production. *J. Anim. Sci.*, **70**, 597–602.
13. Byford, R.L., Craig, M.E., DeRouen, S.M., Kimball, M.D., Morrison, D.G., Wyatt, W.E., and Foil, L.D. (1999) Influence of permethrin, diazinon and ivermectin treatments on insecticide resistance in the horn fly (Diptera: Muscidae). *Int. J. Parasitol.*, **29**, 125–135.
14. Burgess, I.F. (2004) Human lice and their control. *Annu. Rev. Entomol.*, **49**, 457–481.
15. http://www.who.int/features/factfiles/malaria/en (accessed 23 June 2016).
16. http://www.who.int/mediacentre/factsheets/fs267/en (accessed 23 June 2016).
17. https://www.heartwormsociety.org (accessed 23 June 2016).
18. Matthews, J.B. (2014) Anthelmintic resistance in equine nematodes. *Int. J. Parasitol. Drugs Drug Resist.*, **4**, 310–315.
19. Sangster, N.C. and Dobson, R.J. (2002) Anthelmintic resistance, in *The Biology of Nematodes* (ed. D.L. Lee), Harwood, Newark, NJ, pp. 531–567.
20. Melander, A.L. (1914) Can insects become resistant to sprays? *J. Econ. Entomol.*, **7**, 167–173.
21. Forgash, A.J. (1984) History, evolution and consequences of insecticide resistance. *Pestic. Biochem. Physiol.*, **22**, 178–186.
22. Sparks, T.C. and Nauen, R. (2015) IRAC: mode of action classification and insecticide resistance management. *Pestic. Biochem. Physiol.*, **121**, 122–128.
23. http://www.irac-online.org (accessed 23 June 2016)
24. Taylor, M.A., Coop, R.L., and Wall, R.L. (2016) *Veterinary Parasitology*, 4th edn, John Wiley & Sons, Ltd, Chichester, pp. 323–328.
25. Sparks, T.C., Thompson, G.D., Larson, L.L., Kirst, H.A., Jantz, O.K., Worden, T.V., Hertlein, M.B., and Busacca, J.D. (1995) Biological characteristics of the spinosyns: a new naturally derived insect control agents. Proceedings of the 1995 Beltwide Cotton Production Conference, National Cotton Council, Memphis, TN, Vol. 2, pp. 903–907.
26. Blagburn, B.L., Young, D.R., Moran, C., Meyer, J.A., Leigh-Heffron, A., Paarlberg, T., Zimmermann, A.G., Mowrey, D., Wiseman, S., and Snyder, D.E. (2010) Effects of orally administered spinosad (Comfortis) in dogs on adult and immature stages of the cat flea (*Ctenocephalides felis*). *Vet. Parasitol.*, **168**, 312–317.
27. McTier, T.L., Chubb, N., Curtis, M.P., Hedges, L., Inskeep, G.A., Knauer, C.S., Menon, S., Mills, B., Pullins, A., Zinser, E., Woods, D.J., and Meeus, P. (2016) Discovery of sarolaner: a novel, orally administered, broad-spectrum, isoxazoline ectoparasiticide for dogs. *Vet. Parasitol.*, **222**, 3–11.
28. Shoop, W.L., Hartline, E.J., Gould, B.R., Waddell, M.E., McDowell, R.G., Kinney, J.B., Lahm, G.P., Long, J.K., Xu, M., Wagerle, T., Jones, G.S., Dietrich, R.F., Cordova, D., Schroeder, M.E., Rhoades, D.F., Benner, E.A., and Confalone, P.N. (2014) Discovery and mode of action of afoxolaner: a new isoxazoline parasiticide for dogs. *Vet. Parasitol.*, **201**, 179–189.

29. Rohdich, N., Roepke, R.K., and Zschiesche, E. (2014) A randomized, blinded, controlled and multi-centered field study comparing the efficacy and safety of Bravecto (fluralaner) against Frontline (fipronil) in flea- and tick-infested dogs. *Parasit. Vectors*, **7**, 83.

30. Maia, C. and Cardoso, L. (2015) Spread of *Leishmania infantum* in Europe with dog travelling. *Vet. Parasitol.*, **213**, 2–11.

31. Barnes, A.M. (1962) Ectoparasite control in public health. Proceedings of the 1st Vertebrate Pest Conference, Paper, Vol. 19, pp. 250–270.

32. Muehlen, M., Heukelbach, J., Wilcke, T., Winter, B., Mehlhorn, H., and Feldmeier, H. (2003) Investigations on the biology, epidemiology, pathology and control of *Tunga penetrans* in Brazil II: prevalence, parasite load and topographic distribution of lesions in the population of a traditional fishing village. *Parasitol. Res.*, **90**, 449–455.

33. Heukelbach, J. and Feldmeier, H. (2004) Ectoparasites – the underestimated realm. *Lancet*, **363**, 889–891.

34. Bockarie, M.J., Alexander, N.D., Kazura, J.W., Bockarie, F., Griffin, L., and Alpers, M.P. (2000) Treatment with ivermectin reduces the high prevalence of scabies in a village in Papua New Guinea. *Acta Trop.*, **75**, 127–130.

35. Bitam, I., Dittmar, K., Parola, P., Whiting, M.F., and Raoult, D. (2010) Fleas and flea-borne diseases. *Int. J. Infect. Dis.*, **14**, e667–e676.

36. WHO (2011) Vaccines against tick-borne encephalitis: WHO position paper. WHO Weekly Epidemiological Record No. 24, Vol. 86, pp. 241–256.

37. http://www.who.int/medicines/publications/essentialmedicines/EML2015_8-May-15.pdf (accessed 24 June 2016).

38. Derua, Y.A., Kisinza, W.N., and Simonsen, P.E. (2015) Differential effect of human ivermectin treatment on blood feeding *Anopheles gambiae* and *Culex quinquefasciatus*. *Parasit. Vectors*, **8**, 130.

39. Sasaki, T., Takagi, T., Yaguchi, T., Miyadoh, S., Okada, T., and Koyama, M. (1992) A new anthelmintic cyclodepsipeptide, PF1022A. *J. Antibiot.*, **45**, 692–697.

40. Harder, A. and von Samson-Himmelstjerna, G. (2002) Cyclooctadepsipeptides – a new class of anthelmintically active compounds. *Parasitol. Res.*, **88**, 481–488.

41. Woods, D.J., Vaillancourt, V.A., Wendt, J.A., and Meeus, P.F. (2011) Discovery and development of veterinary antiparasitic drugs: past, present and future. *Future Med. Chem.*, **3**, 887–896.

42. Gasbarre, L.C. (2014) Anthelmintic resistance in cattle nematodes in the US. *Vet. Parasitol.*, **204**, 3–11.

43. Geurden, T., Chartier, C., Fanke, J., Frangipane di Regalbono, A., Traversa, D., von Samson-Himmelstjerna, G., Demeler, J., Vanimisetti, H.B., Bartram, D.J., and Denwood, M.J. (2015) Anthelmintic resistance to ivermectin and moxidectin in gastrointestinal nematodes of cattle in Europe. *Int. J. Parasitol. Drugs Drug Resist.*, **5**, 163–171.

44. Ramos, F., Portella, L.P., de Souza Rodrigues, F., Zamperete Reginato, C., Potter, L., Skrebsky Cezar, A., Sangioni, L.A., and Flores Vogel, F.S. (2016) Anthelmintic resistance in gastrointestinal nematodes of beef cattle in the state of Rio Grande do Sul, Brazil. *Int. J. Parasitol. Drugs Drug Resist.*, **6**, 93–101.

45. https://www.avma.org/News/JAVMANews/Pages/131101f.aspx (accessed 23 June 2016).

46. Pulaski, C.N., Malone, J.B., Bourguinat, C., Prichard, R., Geary, T., Ward, D., Klei, T.R., Guidr, T., Smith, G., Delcambre, B., Bova, J., Pepping, J., Carmichael, J., Schenker, R., and Pariaut, R. (2014) Establishment of macrocyclic lactone resistant *Dirofilaria immitis* isolates in experimentally infected laboratory dogs. *Parasit. Vectors*, **7**, 494.

47. Kenyon, F., Greer, A.W., Coles, G.C., Cringoli, G., Papadopoulos, E., Cabaret, J., Berrag, B., Varady, M., Van Wyk, J.A., Thomas, E., Vercruysse, J., and Jackson, F. (2009) The role of targeted selective treatments in the development of refugia-based approaches to the control of gastrointestinal nematodes of small ruminants. *Vet. Parasitol.*, **164**, 3–11.

48. http://www.hsvma.org/assets/pdfs/guide-to-congenital-and-heritable-disorders.pdf (accessed 26 June 2016).

49. http://www.theharrispoll.com/health-and-life/Pets-are-Members-of-the-Family.html (accessed 26 June 2016).
50. Diax, J.H. (2006) The epidemiology, diagnosis, management, and prevention of ectoparasitic diseases in travelers. *J. Travel Med.*, **13**, 100–111.
51. Dryden, M.W. and Rust, M.K. (1994) The cat flea: biology, ecology and control. *Vet. Parasitol.*, **52**, 1–19.
52. Carlotti, D.N. and Jacobs, D.E. (2000) Review article: Therapy, control and prevention of flea allergy dermatitis in dogs and cats. *Vet. Dermatol.*, **11**, 83–98.
53. Levot, G.W. (1995) Resistance and the control of sheep ectoparasites. *Int. J. Parasitol.*, **25**, 1355–1362.
54. Nolan, J. (1987) New approaches in the development and management of drugs used in ectoparasite control. *Vet. Parasitol.*, **25**, 135–145.
55. Reck, J., Klafke, G., Webster, A., Dall'Agnol, B., Scheffer, R., Souza, U.A., Corassini, V.B., Vargas, R., dos Santos, J.S., and Martins, S. (2014) First report of fluazuron resistance in *Rhipicephalus microplus*: a field tick population resistant to six classes of acaricides. *Vet. Parasitol.*, **201**, 128–136.
56. Peter, R.J., Van den Bossche, P., Penzhorn, B.L., and Sharp, B. (2005) Tick, fly, and mosquito control – lessons from the past, solutions for the future. *Vet. Parasitol.*, **132**, 205–215.
57. Feldmeier, H. (2012) *Pediculosis capitis*: new insights into epidemiology, diagnosis and treatment. *Eur. J. Clin. Microbiol. Infect. Dis.*, **31**, 2105–2110.
58. Moosazadeh, M., Afshari, M., Keianian, H., Nezammahalleh, A., and Enayati, A.A. (2015) Prevalence of head lice infestation and its associated factors among primary school students in Iran: a systematic review and meta-analysis. *Osong Public Health Res. Perspect.*, **6**, 346–356.
59. Borges, R. and Mende, J. (2002) Epidemiological aspects of head lice in children attending day care centres, urban and rural schools in Uberlandia, Central Brazil. *Mem. Inst. Oswaldo Cruz*, **97**, 189–192.
60. Falagas, M.E., Matthaiou, D.K., Rafailidis, P.I., Panos, G., and Pappas, G. (2008) Worldwide prevalence of head lice. *Emerg. Infect. Dis.*, **14**, 1493–1494.
61. Burgess, I.F., Brunton, E.R., and Burgess, N.A. (2010) Clinical trial showing superiority of a coconut and anise spray over permethrin 0.43% lotion for head louse infestation. *Eur. J. Pediatr.*, **169**, 55–62.
62. Laguna, M.F. and Risau-Gusman, S. (2011) Of lice and math: using models to understand and control populations of head lice. *PLoS One*, **6**, e21848.
63. Meeusen, E.N.T., Walker, J., Peters, A., Pastoret, P.-P., and Jungersen, G. (2007) Current status of veterinary vaccines. *Clin. Microbiol. Rev.*, **20**, 489–510.
64. Willadsen, P. (2004) Anti-tick vaccines. *Parasitology*, **129**, S367–S387.
65. http://texasfarmbureau.org/new-tool-fever-tick-eradication (accessed 26 June 2016).
66. UNEP/POPS/DDT-EG.3/3 (2010) Stockholm Convention on Persistent Organic Pollutants. Report of the Third Expert Group Meeting on DDT.
67. Rinaldi, L., Hendrickx, G., Cringoli, G., Biggeri, A., Ducheyne, E., Catelan, D., Morgan, E., Williams, D., Charlier, J., von Samson-Himmelstjerna, G., and Vercruysse, J. (2015) Mapping and modelling helminth infections in ruminants in Europe: experience from GLOWORM. *Geospat. Health*, **9**, 257–259.
68. Sutherland, I. and Leathwick, D. (2011) Anthelmintic resistance in nematode parasites of cattle: a global issue? *Trends Parasitol.*, **27**, 176–181.
69. FAO (2010) Food and Agriculture Organization of the United Nations Statistical Databases, http://faostat.fao.org (accessed 26 June 2016).
70. Papadopoulos, E., Gallidis, E., and Ptochos, S. (2012) Anthelmintic resistance in sheep in Europe: a selected review. *Vet. Parasitol.*, **189**, 85–88.
71. Leathwick, D.M. (2012) Modelling the benefits of a new class of anthelmintic in combination. *Vet. Parasitol.*, **186**, 93–100.
72. Van Wyk, J.A. and Bath, G.F. (2002) The FAMACHA© system for managing haemonchosis in sheep and goats by clinically identifying individual animals for treatment. *Vet. Res.*, **33**, 509–529.

73. Greer, A.W., Kenyon, F., Bartley, D.J., Jackson, E.B., Gordon, Y., Donnan, A.A., McBean, D.W., and Jackson, F. (2009) Development and field evaluation of a decision support model for anthelmintic treatments as part of a targeted selective treatment (TST) regime in lambs. *Vet. Parasitol.*, **164**, 12–20.

74. Höglund, J., Morrison, D.A., Charlier, J., Dimander, S.O., and Larsson, A. (2009) Assessing the feasibility of targeted selective treatments for gastrointestinal nematodes in first-season grazing cattle based on mid-season daily weight gains. *Vet. Parasitol.*, **164**, 80–86.

75. Greer, A.W., McAnulty, R.W., and Gibbs S.J. (2010) Performance-based targeted selective anthelmintic treatment regime for grazing dairy calves. Proceedings of the 4th Australasian Dairy Science Symposium, pp. 385–389.

76. Van Wyk, J., Hoste, H., Kaplan, R.M., and Besier, R.B. (2006) Targeted selective treatment for worm management – how do we sell rational programs to farmers? *Vet. Parasitol.*, **139**, 336–346.

77. Knox, M.R., Besier, R.B., Le Jambre, L.F., Kaplan, R.M., Torres-Acosta, J.F., Miller, J., and Sutherland, I. (2012) Novel approaches for the control of helminth parasites of livestock VI: summary of discussions and conclusions. *Vet. Parasitol.*, **186**, 143–149.

78. Woods, D.J., Lauret, C., and Geary, T.G. (2007) Anthelmintic discovery and development in the animal health industry. *Expert Opin. Drug Discov.*, **2**, S25–S33.

79. Woods, D.J. and Knauer, C.S. (2010) Discovery of veterinary antiparasitic agents in the 21st century: a view from industry. *Int. J. Parasitol.*, **40**, 1177–1181.

80. http://www.ema.europa.eu/docs/en_GB/document_library/Scientific_guideline/2015/09/WC500193826.pdf (accessed 21 July 2016).

81. http://www.merck.com/about/featured-stories/mectizan.html (accessed 12 July 2016).

82. http://www.dndi.org/2014/media-centre/press-releases/pr-bayer-filaria (accessed 12 July 2016).

83. Sangster, N.C. (2001) Managing parasiticide resistance. *Vet. Parasitol.*, **98**, 89–109.

84. Molento, M.B. (2009) Parasite control in the age of drug resistance and changing agricultural practices. *Vet. Parasitol.*, **163**, 229–234.

85. Ahrens, D.C. (2006) Rede de Propriedades Familiares Agroecológicas (Agroecological Family Properties Network). Technical Bulletin, Vol. 68, IAPAR, Agron Inst of Paraná, 79 pp.

86. Dobson, R.J., Hosking, B.C., Besier, R.B., Love, S., Larsen, J.W.A., Rolfe, P.F., and Bailey, J.N. (2011) Minimising the development of anthelmintic resistance, and optimising the use of the novel anthelmintic monepantel, for the sustainable control of nematode parasites in Australian sheep grazing systems. *Aust. Vet. J.*, **89**, 160–166.

87. Roush, R.T. (1989) Designing resistance management programs: how can you choose? *Pestic. Sci.*, **26**, 423–441.

88. Hoste, H., Martinez-Ortiz-De-Montellano, C., Manolaraki, F., Brunet, S., Ojeda-Robertos, N., Fourquaux, I., Torres-Acosta, J.F.J., and Sandoval-Castro, C.A. (2012) Direct and indirect effects of bioactive tannin-rich tropical and temperate legumes against nematode infections. *Vet. Parasitol.*, **186**, 18–27.

89. Gordon, K.H. and Waterhouse, P.M. (2007) RNAi for insect-proof plants. *Nat. Biotechnol.*, **25**, 1231–1232.

90. http://www.motherjones.com/tom-philpott/2015/08/coming-farm-field-near-you-gene-silencing-pesticides-RNA-RNAi (accessed 14 July 2016).

91. Wright, M.G., Hoffmann, M.P., Kuhar, T.P., Gardner, J., and Pitcher, S.A. (2005) Evaluating risks of biological control introductions: a probabilistic risk-assessment approach. *Biol. Control*, **35**, 228–247.

92. Larsen, M. (2006) Biological control of nematode parasites in sheep. *J. Anim. Sci.*, **84**, E133–E139.

93. Abolins, S. and Wall, R. (2010) Control of livestock ectoparasites with entomopathogenic fungi: a review. *Acta Vet. Scand.*, **52**, S23.

94. da Silva, M.E., de Araujo, J.V., Braga, F.R., Borges, L.A., Soares, F.E.E., dos Santos Lima, W., and Guimaraes, M.P. (2013) Mycelial mass production of fungi *Duddingtonia flagrans* and *Monacrosporium thaumasium* under different culture conditions. *BMC Res. Notes*, **6**, 340.

2
Vaccination Against Ticks

Theo P.M. Schetters[*]

Abstract

Classically, tick infestation is controlled using acaricides that are administered as pour-ons, by dipping, or by spray race. Because of the continuous pressure of tick infestation and the decrease of active compound over time after administration, selection of relatively resistant ticks is a given. As a result of this, in all geographic areas resistant ticks have emerged, some of which are even multidrug resistant. As a consequence, there is also an upsurge in the prevalence of tick-transmitted diseases, which negatively impact the productivity of livestock. Vaccination against tick infestation would resolve these problems and does not select for drug-resistant tick strains.

Tick Evolution and Life Cycle

Ticks are arthropods that belong to the order Parasitiformes, which is one of the two orders of Acari. Parasitiform fossils of ticks have been found in amber specimens from the early Eocene (35–40 MYA), and one sample was dated 90–94 MYA [1]. The ticks in these specimens could be identified at the species level, indicating that speciation had already occurred. Ancestral ticks most likely developed from scavengers (feeding on dead organisms) to predators (feeding on lymph from arthropods). The latter is corroborated by the fact that cannibalism has been observed in soft ticks [2] and hard ticks (Trentelman and Schetters, unpublished observations). Later during evolution, ticks evolved to feed on blood from vertebrate hosts, usually only during a short portion of their life cycle, with the Ixodid ticks, for example, still spending 94–97% of their life off-host [3]. With the adaptation to feed on vertebrate hosts, ticks acquired the ability to transmit pathogens among vertebrate hosts, such as *Borrelia burgdorferi* (Lyme disease), tick-borne encephalitis virus, and *Babesia* parasites [4].

Ticks can be divided into hard ticks (Ixodidae) and soft ticks (Argasidae). Hard ticks have a keratinous plate on the back (scutum), which is absent in soft ticks.

[*]Corresponding author.

Ectoparasites: Drug Discovery Against Moving Targets, First Edition.
Edited by Charles Q. Meng and Ann E. Sluder

Ticks undergo different developmental stages, the first of which are larvae that hatch from fertilized eggs. Larvae usually feed on small vertebrates, after which they drop off the host. They moult to slightly larger nymphs that climb up in the vegetation to encounter a new host. Nymphs feed on a variety of hosts, large and small, depending on the tick species and habitat (see subsequent text). Adult ticks feed on larger vertebrate animals. Depending on the number of hosts that are visited during the life cycle, ticks are classified as three-host, two-host, and one-host ticks; the latter do not drop off at the larval and/or nymph stage but develop to fertile adults on a single host. After mating, the fully engorged females drop off and deposit the eggs in the vegetation [5].

The prevalence of ticks in a specific geographic area is dependent on a variety of climatic and ecological factors, the most important of which are humidity, environmental temperature, and the presence of suitable host species. When humidity and temperature are relatively low, ticks may become inactive and enter diapause (hibernation). Ticks become active again in spring time, which in the tropics is the start of the rainy season. Depending on such conditions and the particular tick species, several generations of ticks can be completed in 1 year; other species take more than 1 year to complete a full generation. Notably, one-host ticks that complete multiple generations in a single year can lead to individual hosts that harbor thousands of ticks, negatively affecting general health, growth, and production. Such biological differences influence the strategy and success of vaccination to control tick infestation.

Commercially Available Vaccines

With the discovery that one could immunize vertebrate hosts against feeding by hematophagous arthropods using a rough preparation of dead arthropods, the basis for vaccine development was created [6]. There is one commercially available vaccine on the market that is based on this principle (launched in 2005). The vaccine contains partially purified proteins from *Rhipicephalus microplus* larvae that are formulated in an oil-based adjuvant (TickVac®, Technochimica, http://www.soydelcampo.com/vademecum_veterinario/productos.php?id=6568; 21 September 2016). For initial vaccination, calves are immunized three times; at day 0, day 20, and day 60. To maintain the maximum level of immunity, revaccination every 6 months is recommended [7]. The efficacy varied between 10% and 79% reduction in fully engorged female ticks (average 44.3%). In addition, vaccination reduced the viability and number of eggs produced by female ticks. Although the protective mechanisms have not been elucidated, antibodies against a variety of tick antigens are presumed to inhibit tick development and impair viability when taken up with the blood meal from vaccinated animals.

Many of the antigens in such preparations are normally not injected into the host during feeding, and represent so-called concealed antigens [8]. Extensive purification studies using semi-engorged adult female ticks have led to the identification of an antigen from the surface of the midgut of *R. microplus* ticks (called Bm86) that was shown to induce protective immunity [9]. The biological function of this molecule is thus far unknown. The gene encoding for this antigen was identified and used to produce recombinant Bm86 protein in *Escherichia coli* [10]. The antigen, when formulated in a water-in-oil adjuvant, induced antibodies in the serum of vaccinated calves. These antibodies were shown to impair the integrity of the epithelial cells of the midgut, thereby affecting tick viability and leading to the death of most of the affected ticks [11, 12]. The vaccine is effective against all stages of *R. microplus*, including larvae (Trentelman *et al.*, 2016, submitted), resulting in recovery of fewer adult ticks from vaccinated compared to unvaccinated animals. Some of the larvae develop to adults that have a red appearance because of blood that leaks from the midgut into the hemocoel. This vaccine (TickGARD®) has been commercialized in Australia [13]. Results from the field showed that some animals did not respond upon vaccination. This problem was partially resolved by adding saponin to the formulation as an additional adjuvant (TickGARDPlus®; [14]). Importantly, vaccination did not protect cattle against infestation with the laboratory *R. microplus* A strain from Argentina [15]. Detailed analysis of the Bm86 gene indicated that there is some polymorphism, which could explain the variability in efficacy when used in different geographic areas [16]. The Australian vaccine was discontinued after reorganizations in the animal health industry.

A similar vaccine based on the same molecule has been developed in Cuba (Gavac®; [17]). This vaccine contains recombinant Bm86 protein that is produced in a *Pichia pastoris* expression system. An evaluation of these two vaccines after 10 years of use in the field revealed that efficacy was on average 50–60% reduction in engorged female ticks after the initial vaccination schedule [18]. The vaccine not only reduced the number of adult females feeding to repletion but also reduced the egg mass and viability of the progeny of females that survived on vaccinated animals. Extended use of the vaccine in the same herd further improved efficacy to approximately 75%, enabling an impressive reduction in the use of acaricides [19].

In order to circumvent problems with antigenic diversity among *R. microplus* strains, a number of synthetic peptides were produced, which contain three regions of the Bm86 protein that are conserved in strains from Colombia, Argentina, Venezuela, Uruguay, and Brazil. One of these peptides, SBm7462, induced protective immunity in calves, which was reflected in a reduction of fully engorged female ticks up to 61%. A relatively high amount of antigen (2 mg) was used with saponin as adjuvant [20].

Because commercial vaccines did not completely prevent tick infestation, unlike acaricides when these were first introduced, vaccination has not been widely adopted [18]. These results have stimulated research to develop vaccines with an improved efficacy profile.

Rational Tick Vaccine Development

In order to understand the basic principles of vaccine development against tick infestation, an understanding of the biology of ticks and their adaptation to different hosts and feeding behaviors is an important foundation. The advent of molecular tools for genetically modifying the expression of specific proteins (e.g., gene knockout and inhibition of the translation of proteins from mRNA using RNA-interference techniques) led to the discovery of a number of proteins that are vital to tick feeding and development. Ongoing research aims at evaluating the protective effect of these proteins when used to vaccinate animals against tick infestation. Some of these protective responses affect basic tick biology, whereas other responses influence processes at the tick–host interface (Figure 2.1). From the vast number of tick antigens that have been investigated, those that induced (partial) immunity against tick infestation are discussed here.

Tick Biology

The tick proteins that are targeted to kill ticks directly are normally not exposed to the immune system of the host (concealed antigens). An advantage is that during evolution the ticks had no opportunity to develop immune-evasive mechanisms. The disadvantage is that a natural infestation will not boost the immune responses that were induced against these concealed antigens. In order to maintain effective immunity, repeated booster immunizations are therefore required.

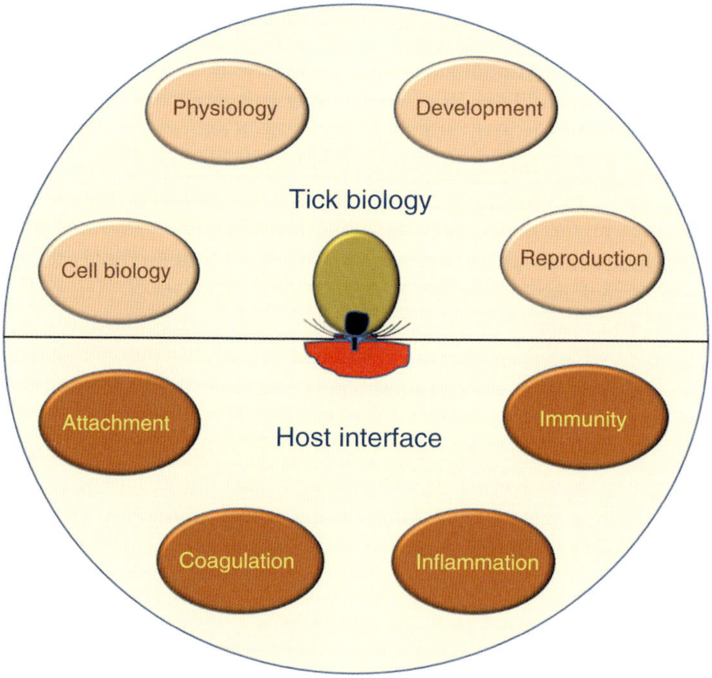

Figure 2.1 Schematic representation of the systems that are being targeted to develop vaccines against tick infestation.

Cell Biology

Subolesin/Akirin One of the first required proteins discovered was subolesin (originally called 4D8 antigen) in *Ixodes scapularis* ticks [21, 22]. This approximately 25 kDa protein appeared to be present in a wide range of tick species and was later discovered to be the homologue of akirin of other arthropods and vertebrates. Subolesin was subsequently shown to play a crucial role in gene transcription regulation. Vaccination of a variety of hosts (including cattle and deer) protected hosts against tick infestation (47% reduction in the number of fully engorged adult female ticks [23]). The mechanism of protection is unclear, since subolesin is an intracellular molecule that is not easily accessible for antibodies, although there is some evidence that antibodies can enter tick cells [24, 25].

Vaccination of rabbits with the full length recombinant subolesin of *Ornithodoros moubata* induced marginal protection [26]. An in-depth analysis showed that immune responses were directed against linear B-cell epitopes that were not considered to be related to the biologically active sites of subolesin. Therefore, four synthetic peptides derived from the area that comprises the putative biologically active site of subolesin were selected and evaluated for vaccine potential. Vaccination of rabbits with one of these synthetic peptides (Om1) conjugated to keyhole limpet hemocyanin (KLH), induced 70% and 83% protection against *Ornithodoros erraticus* and *O. moubata*, respectively [27].

Ribosomal P0 Antigen Rodríguez-Mallon *et al.* have described a cytoplasmic protein, P0 antigen, in Ixodid ticks that is involved in the assembly of ribosomes. The P0 protein is also involved in other metabolic processes, and was found to be associated with cell surface proteins of some Protozoa (*Plasmodium, Toxoplasma*) and other eukaryotic cells, a localization of which the function is unknown [28]. Importantly, when cattle were vaccinated with a 20 amino acid peptide derived from P0 that was conjugated to KLH, reduced numbers of engorged female *R. microplus* ticks were recovered, reduced engorged weight was observed, and ticks that survived produced less viable progeny. Overall protection against infestation reached 96%, which was slightly higher than that earlier obtained against *Rhipicephalus sanguineus* using the same antigen (90% overall protection; [29]).

Physiology

Ferritins Ticks have developed mechanisms to metabolize toxic iron-containing metabolites that are present in their blood meals, in which ferritins 1 and 2 play a major role. Ferritin 1 shows some homology to mammalian heavy-chain ferritins and facilitates intracellular uptake of nonheme iron. The translocation of nonheme iron from the tick gut to peripheral tissues is mediated by ferritin 2, orthologues of which are absent in vertebrates [30]. A recombinant form of the latter protein was used to vaccinate rabbits against *Ixodes ricinus*. Upon infestation, vaccinated rabbits exhibited 43% fewer engorged female ticks, reduced engorged tick weight, and reduced tick oviposition and egg fertility, resulting in overall protection of 98%. When cattle were vaccinated with recombinant ferritin 2 from *R. microplus*, the reduction in engorged female *R. microplus* ticks was 30%

(64% overall protection). The same animals, when infested with *Rhipicephalus annulatus*, showed 42% fewer engorged female *R. annulatus* ticks, (overall protection 72%; [31]). Using a similar approach for *Haemaphysalis longicornis*, partial protection was found in vaccinated rabbits as reflected in reduced engorged weights, oviposition, and hatching rate of the infesting ticks. However, no effect on tick numbers was observed [32].

Aquaporins A female Ixodid tick can imbibe 200–300 times its own body weight of blood. The blood meal is concentrated in the gut, and excess water is returned to the host in saliva. Aquaporins, also called water channels, play a major role in this process [33]. In *R. microplus*, three aquaporins have been described, and one of these (RmAQP1) has been tested as a vaccine candidate. RmAQP1 is mainly expressed in the synganglia of male and female ticks [34]. When cattle were vaccinated with a recombinant RmAQP1 antigen that was produced in *P. pastoris*, high levels of protection were obtained (average of 70% reduction in engorged female *R. microplus* ticks). Moreover, the level of protection was higher than that obtained with a Bm86-based vaccine preparation (49% reduction in engorged female ticks; [34]).

Development There is one report that specifically mentions an effect of vaccination on moulting. The antigen used was the *Rhipicephalus appendiculatus* homologue of the Bm86 midgut protein of *R. microplus*. Cattle were vaccinated with recombinant Ra86 that was produced in *P. pastoris*. Results showed that the moulting success of nymphs to the adult stage was significantly impaired [35]. Whether or not the inhibitory effect is specifically on the moulting process, or the result of reduced nymph viability that prevented further development, remains to be determined.

Reproduction

Mating Adult female ticks only feed to repletion when they have mated with male ticks. Male ticks pass on an engorgement factor (EF; voraxin) to females, which was first identified in *Amblyomma hebraeum* ticks [36]. Voraxin comprises two polypeptides termed AHEFα and AHEFβ. Vaccination of rabbits with recombinant forms of these polypeptides reduced the number of *A. hebraeum* female ticks that completed engorgement by 75%. Similarly, when rabbits were vaccinated with recombinant voraxin α of *R. appendiculatus*, the weight of engorged ticks was reduced by 40%. Importantly, the deduced amino acid sequence of voraxin α appeared to be highly conserved among *A. hebraeum*, *R. appendiculatus*, and *Dermacentor variabilis*, which holds promise for development of an effective broad-spectrum tick vaccine [37].

Egg Production Fully engorged female ticks produce many eggs that are deposited in the environment. For the production of these eggs, a number of molecules have been found to be crucial. In *R. microplus*, three enzymes contributing to egg production have been identified, comprising two aspartic peptidases (Boophilus yolk pro-cathepsin (BYC) and heme-binding aspartic peptidase), and a cysteine endopeptidase (vitellin-degrading cysteine endopeptidase (VTDCE); [38–40]). Cattle vaccinated with a vaccine containing only one of the different enzymes (recombinant proteins expressed in *E. coli*) exhibited limited protection [41–44]. However, when cattle in the field were vaccinated four times (with 3-week intervals) with a combined vaccine comprising *R. microplus* BYC and VTDCE antigens mixed with *H. longicornis* GST, a 35–62% reduction in engorged tick counts was found. The onset of immunity was evident as early as 2 weeks after the second vaccination, and lasted for more than 2 months after the final vaccination. This was also reflected in improved weight gain in vaccinated animals over the 127-day period of the study (vaccine group 39%, control group 25%; [45]).

Fewer vaccine studies have been done with soft ticks [46]. In one study, rabbits were vaccinated with purified tick egg yolk protein (vitellin) from *O. moubata*. After infestation with *O. moubata* nymphs and adults, no effect was observed on the number of ticks that engorged. However, the percentage of ticks that laid eggs was reduced by 60% and the time required to further develop eggs was increased from 40 to 49 days. It was hypothesized that specific antibodies crossed the midgut to the hemolymph where they bound to vitellogenin, thereby preventing uptake by oocytes [47].

Host Interface

During evolution ticks have developed a number of mechanisms to successfully attach to the host, to prevent coagulation to enable blood feeding, and to circumvent rejection by host immune responses. A number of tick molecules that play a crucial role in these processes have been identified and evaluated for use in vaccines against tick infestations. The advantage of these molecules is that a natural infection could potentially boost vaccine-induced immunity, because they are released in the host.

Attachment When adult ticks feed on a host they produce a cement cone underneath the dermis in order to anchor themselves to the host. Prevention of the formation of this cone would affect tick attachment and increase chances that ticks would drop off before feeding to repletion. A number of studies have shown the feasibility of such an approach. In *H. longicornis*, two cement cone proteins that protected vaccinated rabbits against tick infestation have been identified. The collagen-like P29 protein induced 40–56% mortality in *H. longicornis* larvae and nymphs, whereas vaccination against HL34 reduced survival of all developmental stages of the tick, including adults by 30% [48, 49]. In *R. appendiculatus*, a 15 kDa tick cement protein was discovered in saliva (64P). The native 64P tick protein is injected into the host during feeding. Vaccination of animals with a truncated form of this protein (called 64TRP) fused to glutathione S-transferase (GST) affected both feeding and the integrity of the tick midgut, which led to death [50, 51].

Coagulation Appearance of the first blood-sucking arthropods is estimated at 155 MYA, while the presumed origin of ticks is dated 120–90 MYA. The latter is based on the fact that the basal hard tick lineages (Ixodidae) have been identified only in Australia, suggesting that they evolved after the separation of Australia from Gondwanaland [52]. Although this might indicate that ticks evolved from such ancestral blood-sucking arthropods, available evidence suggests that adaptation of ticks to blood feeding happened independently [53]. It has been suggested that ancestral ticks originally fed on dead material, including arthropods (insects, etc.). In order to feed on such prey, ticks required enzymes to degrade the arthropod matrix, such as hyaluronidase, to aid digestion. In addition, as arthropods have a clotting mechanism in the hemolymph to minimize effects of external damage, only ticks that had developed mechanisms to prevent hemolymph clotting were able to feed successfully on live arthropods, because clotting would prevent uptake of hemolymph efficiently [54]. Indeed, a number of proteases have been described in ticks that prevent coagulation of hemolymph from arthropod targets. It has been hypothesized that during tick evolution, enzymes that interfere with the vertebrate coagulation system have derived from these ancestral proteases (Figure 2.2) [53]. These proteases are produced by the salivary glands, and injected into the host during feeding [55].

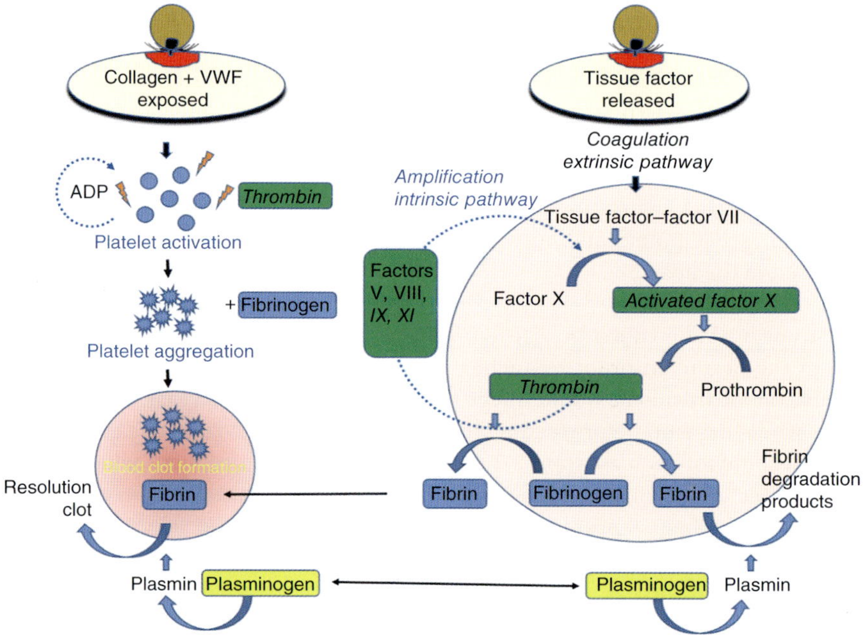

Figure 2.2 Simplified schematic of the blood coagulation processes that are modulated by tick proteins. Compounds in green fields and printed in italics are targets of tick serine protease inhibitors. Compounds in blue fields are substrates of tick proteases. Compounds in yellow are activated by tick proteins. ADP, adenosine diphosphate; VWF, von Willebrand factor.

Platelet activation is the most immediate event preventing blood loss when blood capillaries are damaged. Platelets adhere to the site of injury when they come into contact with collagen. Activation is mediated through a number of compounds, such as thrombin that also promotes coagulation. As a result of activation, platelets express certain integrins that are able to bind fibrinogen, leading to platelet aggregation and the formation of a plug (Figure 2.2). Activated and aggregated platelets release mediators that trigger the inflammatory response, thus creating a hostile environment for the tick [53]. During evolution, ticks have evolved mechanisms that interfere with this process at different levels [56]. Prevention of coagulation serves two purposes for the tick: maintaining blood flow to the attachment site and prevention of coagulation in the tick midgut to allow uptake and digestion of the blood meal. The tick molecules that affect these processes and that have met with some success when used to vaccinate hosts against tick infestation are described here.

Prevention of Blood Coagulation at the Attachment Site Tick metalloproteases have been shown to affect platelet aggregation and blood coagulation by fibrinolytic and gelatinase activity, thus diminishing blood coagulation and allowing blood feeding [57, 58]. Using a newly discovered metalloprotease of *H. longicornis* (HLMP1), rabbits vaccinated with recombinantly produced protein were partially protected against *H. longicornis* infestation. This was reflected in mortality of around 15% in nymphs and adults. Importantly, rabbits that had repeatedly been infested produced antibodies against this protease, indicating that in natural infection the native molecule is injected into the host and is immunogenic [59]. In *R. microplus* ticks, a metalloprotease that is expressed in the salivary gland was discovered. The protein was not detectable in ticks that had completed engorgement, suggesting that it is required during the process of feeding [60]. Although the precise biological activity of this protease has not been elucidated, it is tempting to speculate that it affects blood coagulation in a way similar to that of the protease of *H. longicornis* described earlier. Vaccination of cattle with a recombinant metalloprotease of *R. microplus* (BrBm-MP4) decreased the number of fully engorged female ticks by 43% and also affected size and viability of the egg mass, providing an overall protection of 60% against tick infestation [61].

Decrem *et al.* discovered a salivary gland protein in *I. ricinus* that accelerated fibrinolysis *in vitro*. The protein appeared to be a metalloprotease (Metis1), and antisera from repeatedly infested animals recognized the recombinant Metis1 protein, indicating that native Metis1 is injected at the injection site during feeding [62]. When rabbits were vaccinated with rMetis1, they were partially protected against tick infestation as reflected by partial reduction in female tick feeding, engorged weight, and subsequent oviposition. Vaccination did not affect attachment, nor nymphs or male ticks.

Prevot *et al.* identified a salivary gland serine protease inhibitor in *I. ricinus* (IRIS) that is a specific elastase inhibitor interfering with the intrinsic coagulation pathway and fibrinolysis, thereby disrupting platelet adhesion [63]. Animals that have been repeatedly infested produce antibodies against IRIS, which corroborates that the native protein is injected at the attachment site. When recombinant IRIS was used to vaccinate rabbits, partial immunity was induced against infestation by nymphs and adults as evidenced by increased feeding time, reduced engorgement, and approximately 30% mortality of ticks [64]. IRIS also affects the development of an acquired immune response (see subsequent text).

The salivary glands of *H. longicornis* produce longistatin, a protease that hydrolyzes fibrinogen and activates plasminogen [65]. This leads to shortage of fibrinogen for platelet aggregation and clot formation at the attachment site. In addition, activation of plasminogen leads to increased levels of plasmin, a molecule that degrades existing fibrin (clots; Figure 2.2). Immunization of mice with longistatin affected engorgement of adult female ticks (54% reduction) and post-engorgement body weight (11% reduction). In addition, moulting of nymphs was reduced by approximately 34%. The resulting overall protection was 73% [65].

The saliva of *Amblyomma americanum* ticks contains a serine protease inhibitor (AAS19), a highly conserved protein characterized by its functional domain being 100% conserved across tick species [66]. The protein inhibits trypsin-like proteases, including five of the eight serine protease factors in the blood clotting cascade (Figure 2.2). Vaccination of rabbits with recombinant AAS19 induced antibodies that reacted with native AAS19 protein. Upon subsequent infestation, significantly smaller blood meals were observed, especially after a second infestation. Importantly, 60% of ticks that engorged on immunized rabbits in the second post-vaccination infestation failed to lay eggs [66].

Two salivary gland proteins were discovered in the soft tick *O. moubata* that affected the hemostatic response of the host: an apyrase (OmAPY; hydrolyses adenosine diphosphate that activates platelets) and a disaggregin (OmMOU), both of which inhibit platelet aggregation (Figure 2.2). Vaccination of rabbits with recombinant forms of these proteins induced overall protection of 27% and 43%, respectively, mainly because of an effect on fertility (26% and 32%; [67]).

Prevention of Blood Coagulation in the Tick Midgut Ticks have also developed mechanisms that prevent blood coagulation in the midgut to facilitate blood digestion and uptake. Clearly, salivary components that are injected into the host to prevent coagulation are also imbibed with the blood meal. However, there are also tick molecules that are located in the midgut (not detectable in salivary glands and/or saliva) that interfere with blood coagulation.

A recombinant serine protease inhibitor from *H. longicornis*, HLS2, inhibited thrombin activity resulting in prolonged coagulation times. Rabbits vaccinated with HLS2 were partially protected against *H. longicornis* infestation (43–45% mortality in nymphs and adults). Interestingly, this protein was not expressed in salivary glands but instead in the hemolymph compartment [68].

Ixophilin is a thrombin inhibitor from *I. scapularis* that is expressed in the tick midgut. Immunization of mice against recombinant ixophilin prolonged feeding time and reduced engorgement weights of female ticks [69]. Homologous proteins have been described in *R. microplus* (boophilin; [70, 71]) and *H. longicornis* (hemalin; [72]).

R. appendiculatus serine protease inhibitors RAS 1–4 are expressed in the midgut, and two of these (RAS 3 and RAS 4) are predicted excretory proteins. RAS 1 and RAS 2 are equally expressed in midgut and salivary gland tissue, but appear to be concealed antigens [73, 74]. These results suggest that the RAS proteins play a role in blood digestion and uptake. Vaccination of cattle with a combination of RAS-1 and RAS-2 reduced the number of engorged nymphs by 60% and increased mortality of adult males and females by 43% and 28%, respectively [74].

Host Inflammatory and Immune Responses

Ticks damage host tissues when they force their mouth parts into the host skin to feed, which triggers the host's innate and adaptive immune systems. The inflammatory response promotes repair of the tissue, and involves the kallikrein, complement, and coagulation systems. In addition, cellular inflammatory responses are triggered, leading to attraction of granulocytes, macrophages, and lymphocytes to the site of injury [75]. Some of these responses induce itching and pain through release of bradykinin and subsequent prostaglandin production [76], which may increase grooming behavior at the site where the tick has attached, leading to subsequent removal of ticks. Detailed studies of *R. microplus* have shown that hosts differ in such primary (innate) local anti-tick responses that are induced upon tick bite, which could explain why some hosts are more sensitive to tick infestation than others [77, 78].

Subjects that have encountered tick bites repeatedly can develop some form of acquired immunity against subsequent bites, ranging from simple rejection of the parasite with little or no damage, to interference with feeding, increased feeding time, reduced engorged weight and egg mass, to death of the parasite [79]. Antibodies appear to play an important role in natural immunity. In a number of tick–host models (*I. ricinus*–rabbits, *D. variabilis*–rabbits, *R. microplus*–cattle), partial immunity could be transferred to naive subjects with serum from previously exposed animals [80–82]. Full expression of immunity, however, involves additional cellular responses, including hypersensitive responses to ticks. Basophil infiltration and degranulation followed by infiltration of eosinophils are characteristics of attachment sites where ticks are being rejected from various immune hosts [78].

Ticks modulate these innate and adaptive immune responses in order to prevent rejection and to complete feeding. Salivary gland extracts (SGEs) contain molecules that can potentially affect the humoral inflammatory defence systems (complement, coagulation, and kallikrein) and the recruitment and adhesion of leucocytes to endothelium, which are activated by injury inflicted by tick bite [83]. Indeed, Ribeiro and colleagues have shown that saliva from *Ixodes dammini* has anti-complement, anti-hemostatic, anti-prostaglandin, anti-interleukin-2, and other anti-inflammatory activities [84].

The composition of these actives in saliva is variable; whereas saliva of fully engorged female ticks does not have significant amounts of these enzymes, partially engorged female ticks do [85, 86]. Importantly, sera from bovines that had been repeatedly infested with *R. microplus* alleviated the inhibitory effects of tick saliva on coagulation, anti-thrombin activity, and modulation of endothelial cell activation [87]. This further shows that antibodies play an important role in the development of natural immunity, and stimulated research to develop vaccines that interfered with these defence mechanisms of the tick.

Inflammation Bradykinin is an important mediator in inflammation, playing a role in hemostasis and the occurrence of pain. Enzymes that hydrolyse any peptide bond in bradykinin (kininases such as angiotensin-converting enzyme (ACE)) abolish its activity. Proteins with kininase activity have been discovered in salivary glands of different tick species, and most of these are metalloproteases [61]. In *R. microplus*, a metalloprotease Bm91, which shares biochemical and enzymatic properties with ACE, has been evaluated as vaccine for cattle [88]. Vaccination of cattle with Bm91 did not affect the number of adult female ticks, but the ability to engorge and subsequently lay eggs was reduced [89]. When cattle were vaccinated with the combination of Bm91 and Bm86, there was an additive protective effect in impairing engorgement [90].

Histamine is released in response to tissue damage. It is mainly secreted by mast cells and basophils and increases the permeability of post-capillary blood vessels, thus allowing wound repair factors to pass into the damaged tissue [76]. The biological effect of histamine on tick feeding is bimodal: injection of histamine underneath the attachment site of *R. microplus* ticks interfered with feeding and led to detachment. This sensitivity to histamine declined when larval attachment had stabilized, and repeated injections had no effect on the weight of larvae after 3 days on the host [91]. In order to stay attached, Ixode ticks encode several histamine binding proteins to counteract the effect of histamine early in feeding. In contrast, later during the rapid phase of tick feeding, a tick histamine releasing factor (tHRF) is upregulated, facilitating blood flow for rapid engorgement. This protein has vaccine potential, as evidenced by the fact that transfer of antibodies against tHRF to mice drastically decreased feeding of *I. scapularis* ticks [92].

Immunity Pigs that had been vaccinated with SGE of the soft tick *O. moubata* recognized a protein of 44 kDa (Om44) that was not recognized by sera from repeatedly infected pigs. Functional characterization showed that the protein is a phospholipase A2 (renamed OmPLA2; [51]). The protein, an antagonist ligand of host P-selectin, prevents adhesion of platelets and leukocytes to vessel walls, and as such inhibits host hemostatic and inflammatory responses of the host. Subsequent vaccination studies with the recombinant OmPLA2 protein showed that feeding of adult *O. moubata* ticks was inhibited by almost 50% and fecundity by 44% [93].

The saliva of *O. moubata* contains a peptidase inhibitor (OmC2) that affects the function of antigen-presenting mouse dendritic cells by reducing the production of the pro-inflammatory cytokine tumor necrosis factor α and interleukin-12, and proliferation of antigen-specific $CD4^+$ T cells [94]. This suggests that suppression of this responsiveness facilitates tick survival. Indeed, blocking of

OmC2 by immunization significantly decreased the number of viable *O. moubata* in a mouse model (24% reduction of nymphs that moulted to adults; [95]).

The salivary gland serine protease inhibitor (IRIS) of *I. ricinus* that interferes with blood coagulation (described earlier) was also found to modulate T-lymphocyte and macrophage responsiveness. IRIS induced preferentially Th2-type immune responses and inhibited the production of pro-inflammatory cytokines [94]. The protective effects observed in animals that were vaccinated with IRIS could also be (in part) related to the abrogation of the modulation of the immune response by IRIS [64].

Anguita *et al.* showed that a protein in the saliva of Ixodes ticks (Salp15) inhibits interleukin-2 production upon T-cell receptor engagement [96] and suppresses the activity of dendritic cells [97]. Specific antibodies against Salp15 neutralized the immunosuppressive effects of Salp15 [98].

Future Developments

With the further development of molecular biological techniques it is envisaged that new vaccine candidate antigens will be discovered [99]. In order to increase the protective activity of tick vaccines, combining tick antigens that have shown partial protection into multivalent vaccine formulations will be pursued [50]. This will require closer collaboration of research groups, and initiatives to jointly develop tick vaccines for humans (ANTIDotE) and animals (CATVAC) have been taken [100, 101]. Effective vaccine formulations will be one of the tools to control tick infestations in animal husbandry, as such vaccines will complement the strategic use of acaricides and management practices.

References

1. Klompen, H. and Grimaldi, D. (2001) First Mesozoic record of a parasitiform mite: a larval argasid tick in Cretaceous amber (Acari: Ixodida: Argasidae). *Ann. Entomol. Soc. Am.*, **94**, 10–15.
2. Balashov, Y.S. (1972) Bloodsucking ticks (Ixodidae) – vectors of disease of man and animals. *Misc. Publ. Entomol. Soc. Am.*, **8**, 161–376.
3. Needham, G.R. and Teel, P.D. (1991) Off-host physiological ecology of Ixodid ticks. *Ann. Rev. Entomol.*, **36**, 659–681.
4. Jongejan, F. and Uilenberg, G. (2004) The global importance of ticks. *Parasitology*, **129**, S3–S14.
5. Mehlhorn, H. (ed.) (2001) *Encyclopedic Reference of Parasitology*, Springer-Verlag, Berlin/Heidelberg/New York, pp. 608–638.
6. Allen, J.R. and Humphreys, S.J. (1979) Immunization of guinea pigs and cattle against ticks. *Nature*, **280**, 491–493.
7. Betancourt, A., Patiñot, F., Torres, O., and Eugenio, B. (2004) Prueba de estado para evaluar la efectividad de Tickvac^{MK} contra la garrapata *Boophilus microplus*. ACOVEZ, Informe Especial, pp. 18–25.
8. Opdebeeck, J.P., Wong, J.Y.M., Jackson, L.A., and Dobson, C. (1988) Hereford cattle immunized and protected against *Boophilus microplus* with soluble and membrane-associated antigens from the midgut of ticks. *Parasite Immunol.*, **10**, 405–410.
9. Willadsen, P., McKenna, R.V., and Riding, G.A. (1988) Isolation from the cattle tick, *Boophilus microplus*, of antigenic material capable of eliciting a protective immunological response in the bovine host. *Int. J. Parasitol.*, **18**, 183–189.

10. Rand, K.N., Moore, T., Sriskantha, A., Spring, K. *et al.* (1989) Cloning and expression of a protective antigen from the cattle tick *Boophilus microplus*. *Proc. Natl. Acad. Sci. U. S. A.*, **86**, 9657–9961.
11. Agbede, R.I.S. and Kemp, D.H. (1986) Immunization of cattle against *Boophilus microplus* using extracts derived from adult female ticks; histopathology of ticks feeding on vaccinated cattle. *Int. J. Parasitol.*, **16**, 35–41.
12. Kemp, D.H., Pearson, R.D., Gough, J.M., and Willadsen, P. (1989) Vaccination against *Boophilus microplus*: localisation of antigens on tick gut cells and their interaction with the host immune system. *Exp. Appl. Acarol.*, **7**, 43–58.
13. Willadsen, P., Bird, P., Cobon, G.S., and Hungerford, J. (1995) Commercialisation of a recombinant vaccine against *Boophilus microplus*. *Parasitology*, **110**, S43–S50.
14. Jonsson, N., Matschoss, A.L., Pepper, P. *et al.* (2000) Evaluation of TickGARDPlus®, a novel vaccine against *Boophilus microplus* in lactating Holstein-Friesian cows. *Vet. Parasitol.*, **88**, 275–285.
15. de la Fuente, J. and Kocan, K.M. (2003) Advances in the identification and characterization of protective antigens for recombinant vaccines against tick infestations. *Expert Rev. Vaccines*, **2**, 583–593.
16. Garcia-Garcia, J.C., Montero, C., Redondo, M., Vargas, M. *et al* (2000) Control of ticks resistant to immunization with Bm86 in cattle vaccinated with the recombinant Bm95 isolated from the cattle tick, *Boophilus microplus*. *Vaccine*, **18**, 2275–2287.
17. de la Fuente, J., Rodríguez, M., Montero, C., Redondo, M. *et al.* (1999) Vaccination against ticks (*Boophilus* spp.): the experience with the Bm86-based vaccine GAVAC™. *Genet. Anal. Biomol. Eng.*, **15**, 143–148.
18. de la Fuente, J., Almazán, C., Canales, M., Pérez de la Lastra, J.M. *et al.* (2007) A ten-year review of commercial vaccine performance for control of tick infestations on cattle. *Anim. Health Res. Rev.*, **8**, 23–28.
19. Suarez, M., Rubi, J., Pérez, D., Cordova, V. *et al.* (2016) High impact and effectiveness of Gavac™ vaccine in the national program for control of bovine ticks *Rhipicephalus microplus* in Venezuela. *Livest. Sci.*, **187**, 48–52.
20. Patarroyo, J.H., Portela, R.W., De Castro, R.O., Couto Pimentel, J. *et al.* (2002) Immunization of cattle with synthetic peptides derived from the *Boophilus microplus* gut protein (Bm86). *Vet. Immunol. Immunopathol.*, **88**, 163–172.
21. Almazán, C., Kocan, K.M., Bergman, D.K., Garcia-Garcia, J.C. *et al.* (2003) Identification of protective antigens for the control of *Ixodes scapularis* infestations using cDNA expression library immunization. *Vaccine*, **21**, 1492–1501.
22. Almazán, C., Kocan, K.M., Blouin, E.F., and de la Fuente, J. (2005) Vaccination with recombinant tick antigens for the control of *Ixodes scapularis* adult infestations. *Vaccine*, **23**, 5294–5298.
23. de la Fuente, J., Moreno-Cid, J.A., Galindo, R.C., Almazan, C. *et al.* (2013) Subolesin/Akirin vaccines for the control of arthropod vectors and vector borne pathogens. *Transbound. Emerg. Dis.*, **60**, 172–178.
24. Tracey-Patte, P.D., Kemp, D.H., and Johnston, L.A.Y. (1987) *Boophilus microplus*: passage of immunoglobulin and albumin across the gut of cattle ticks feeding on normal or vaccinated cattle. *Res. Vet. Sci.*, **43**, 287–290.
25. Jasinskas, A. and Barbour, A.G. (2005) The Fc fragment mediates the uptake of immunoglobulin G from the midgut to hemolymph in the Ixodid tick *Amblyomma americanum* (Acari: Ixodidae). *J. Med. Entomol.*, **42**, 359–366.
26. Manzano-Román, R., Díaz-Martín, V., Oleaga, A., Siles-Lucas, M. *et al.* (2012) Subolesin/akirin orthologs from *Ornithodoros* spp. soft ticks: cloning, RNAi gene silencing and protective effect of the recombinant proteins. *Vet. Parasitol.*, **185**, 248–259.
27. Manzano-Román, R., Díaz-Martín, V., Oleaga, A., and Pérez-Sánchez, R. (2015) Identification of protective linear B-cell epitopes on the subolesin/akirin

orthologues of *Ornithodoros* spp. soft ticks. *Vaccine*, **33**, 1046–1055.

28. Rodríguez-Mallon, A., Encinosa, P.E., Méndez-Pérez, L., Bello, Y. *et al.* (2015) High efficacy of a 20 amino acid peptide of the acidic ribosomal protein P0 against the cattle tick, *Rhipicephalus microplus*. *Ticks Tick Borne Dis.*, **6**, 530–537.

29. Rodríguez-Mallon, A., Fernández, E., Encinosa, P.E., Bello, Y. *et al.* (2012) A novel antigen shown high vaccine efficacy against the dog tick *Rhipicephalus sanguineus*. *Vaccine*, **30**, 1782–1789.

30. Hajdusek, O., Sojka, D., Kopáček, P., Buresova, V. *et al.* (2009) Knockdown of proteins involved in iron metabolism limits tick reproduction and development. *Proc. Natl. Acad. Sci. U. S. A.*, **106**, 1033–1038.

31. Hajdusek, O., Almazan, C., Loosova, G., Villar, M. *et al.* (2010) Characterization of ferritin 2 for the control of tick infestations. *Vaccine*, **28**, 2993–2998.

32. Galay, R.L., Umemiya-Shirafuji, R., Bacolod, E.T., Maeda, H. *et al.* (2014) Evaluation and comparison of the potential of two ferritins as anti-tick vaccines against *Haemaphysalis longicornis*. *Parasit. Vectors*, **7**, 482.

33. Bowman, A.S. and Sauer, J.R. (2004) Tick salivary glands: function, physiology and future. *Parasitology*, **129**, S67–S81.

34. Guerrero, F.D., Andreotti, R., Bendele, K.G. *et al.* (2014) *Rhipicephalus (Boophilus) microplus* aquaporin as an effective vaccine antigen to protect against cattle tick infestations. *Parasit. Vectors*, **7**, 475.

35. Olds, C., Mwaura, S., Crowder, D., Odongo, D. *et al.* (2012) Immunization of cattle with Ra86 impedes *Rhipicephalus appendiculatus* nymphal-to-adult moulting. *Ticks Tick Borne Dis.*, **3**, 170–178.

36. Weiss, B.L. and Kaufman, W.R. (2004) Two feeding-induced proteins from the male gonad trigger engorgement of the female tick *Amblyomma hebraeum*. *Proc. Natl. Acad. Sci. U. S. A.*, **101**, 5874–5879.

37. Yamada, S., Konnai, S., Imamura, S., Ito, T. *et al.* (2009) Cloning and characterization of *Rhipicephalus appendiculatus* voraxinalpha and its effect as anti-tick vaccine. *Vaccine*, **27**, 5989–5997.

38. Logullo, C., Vaz, I., Sorgine, M.H., Paiva-Silva, G.O. *et al.* (1998) Isolation of an aspartic proteinase precursor from the egg of the hard tick, *Boophilus microplus*. *Parasitology*, **116**, 525–532.

39. Sorgine, M.H., Logullo, C., Zingali, R.B., Paiva-Silva, G.O. *et al.* (2000) A heme-binding aspartic proteinase from the eggs of hard tick *Boophilus microplus*. *J. Biol. Chem.*, **275**, 28659–28665.

40. Seixas, A., Dos Santos, P.C., Velloso, F.F., Da Silva Vaz, I. Jr. *et al.* (2003) A *Boophilus microplus* vitellin-degrading cysteine endopeptidase. *Parasitology*, **126**, 155–163.

41. Da Silva Vaz, I. Jr., Logullo, C., Sorgine, M., Velloso, F.F. *et al.* (1998) Immunization of bovines with an aspartic proteinase precursor isolated from *Boophilus microplus* eggs. *Vet. Immunol. Immunopathol.*, **66**, 331–341.

42. Leal, A.T., Seixas, A., Pohl, P.C., Ferreira, C.A.S. *et al.* (2006) Vaccination of bovines with recombinant Boophilus yolk pro-cathepsin. *Vet. Immunol. Immunopathol.*, **114**, 341–345.

43. Seixas, A., Oliveira, P., Termignoni, C., Logullo, C. *et al.* (2012) *Rhipicephalus (Boophilus) microplus* embryo proteins as target for tick vaccine. *Vet. Immunol. Immunopathol.*, **148**, 149–156.

44. Seixas, A., Leal, A.T., Nascimento-Silva, M.C.L., Masuda, A. *et al.* (2008) Vaccine potential of a tick vitellin-degrading enzyme (VTDCE). *Vet. Immunol. Immunopathol.*, **124**, 332–340.

45. Parizi, L.F., Reck, J. Jr., Oldiges, D.P., Guizzo, M.G. *et al.* (2012) Multi-antigenic vaccine against the cattle tick *Rhipicephalus (Boophilus) microplus*: a field evaluation. *Vaccine*, **30**, 6912–6917.

46. Díaz-Martin, V., Manzano-Román, R., Obolo-Mvoulouga, P., Oleaga, A. *et al.* (2015) Development of vaccines against Ornithodoros soft ticks: an update. *Ticks Tick Borne Dis.*, **6**, 211–220.

47. Chinzei, Y. and Minoura, H. (1988) Reduced oviposition in *Ornithodoros moubata* (Acari: Argasidae) fed on tick-sensitized and vitellin-immunized rabbits. *J. Entomol.*, **25**, 26–31.

48. Mulenga, A., Sugimoto, C., Sako, Y., Ohashi, K. *et al.* (1999) Molecular characterization of a *Haemaphysalis longicornis* tick salivary gland-associated 29-kilodalton protein and its effect as a vaccine against tick infestation in rabbits. *Infect. Immun.*, **67**, 1652–1658.
49. Tsuda, A., Mulenga, A., Sugimoto, C., Nakajima, M. *et al.* (2001) cDNA cloning, characterization and vaccine effect analysis of *Haemaphysalis longicornis* tick saliva proteins. *Vaccine*, **19**, 4287–4296.
50. Trimnell, A.R., Hails, R.S., and Nuttall, P.A. (2002) Dual action ectoparasite vaccine targeting 'exposed' and 'concealed' antigens. *Vaccine*, **20**, 3560–3568.
51. Trimnell, A.R., Davies, G.M., Lissina, O., Hails, R.S. *et al.* (2005) A cross-reactive tick cement antigen is a candidate broad-spectrum tick vaccine. *Vaccine*, **23**, 4329–4341.
52. Klompen, J.S.H., Black, W.C., Keirans, J.E., and Oliver, J.H. Jr. (1996) Evolution of ticks. *Ann. Rev. Entomol.*, **41**, 141–161.
53. Mans, B.J. and Neitz, A.W.H. (2004) Adaptation of ticks to a blood-feeding environment: evolution from a functional perspective. *Insect Biochem. Mol. Biol.*, **34**, 1–17.
54. Mulenga, A., Sugino, M., Nakajima, M., Sugimoto, C. *et al.* (2001) Tick-encoded serine protease inhibitors (serpins): potential target antigens for tick vaccine development. *J. Vet. Med. Sci.*, **63**, 1063–1069.
55. Francischetti, I.M.B., Sa-Nunes, A., Mans, B.J., Santos, I.M. *et al.* (2009) The role of saliva in tick feeding. *Front. Biosci.*, **14**, 2051–2088.
56. Law, J.H., Ribeiro, J.M.C., and Wells, M.A. (1992) Biochemical insights derived from insect diversity. *Annu. Rev. Biochem.*, **64**, 87–111.
57. Francischetti, I.M.B., Mather, T.N., and Ribeiro, J.M.C. (2003) Cloning of a salivary gland metalloprotease and characterization of gelatinase and fibrin(ogen)lytic activities in the saliva of the Lyme disease tick vector *Ixodes scapularis*. *Biochem. Biophys. Res. Commun.*, **305**, 869–875.
58. Chmelar, J., Calvo, E., Pedra, J.H.F., Francischetti, I.M. *et al.* (2012) Tick salivary secretion as a source of antihemostatics. *J. Proteome*, **75**, 3842–3854.
59. Imamura, S., da Silva Vaz Jr, I., Konnai, S., Yamada, S. *et al.* (2009) Effect of vaccination with a recombinant metalloprotease from *Haemaphysalis longicornis*. *Exp. Appl. Acarol.*, **48**, 345–358.
60. Ali, A., Tirloni, L., Isezaki, M., Seixas, A. *et al.* (2014) Reprolysin metalloproteases from Ixodes persulcatus, *Rhipicephalus sanguineus*, and *Rhipicephalus microplus* ticks. *Exp. Appl. Acarol.*, **63**, 559–578.
61. Ali, A., Parizi, L.F., Guizzo, M.G., Tirloni, L. *et al.* (2015) Immunoprotective potential of a *Rhipicephalus* (*Boophilus*) *microplus* metalloprotease. *Vet. Parasitol.*, **207**, 107–114.
62. Decrem, Y., Mariller, M., Lahaye, K., Blasioli, V. *et al.* (2008) The impact of gene knock-down and vaccination against salivary metalloproteases on blood feeding and egg laying by *Ixodes ricinus*. *Int. J. Parasitol.*, **38**, 549–560.
63. Prevot, P.P., Adam, B., Boudjeltia, K.Z., Brossard, M. *et al.* (2006) Anti-hemostatic effects of a serpin from the saliva of the tick *Ixodes ricinus*. *J. Biol. Chem.*, **281**, 26361–26369.
64. Prevot, P.P., Couvreur, B., Denis, V., Brossard, M. *et al.* (2007) Protective immunity against *Ixodes ricinus* induced by a salivary serpin. *Vaccine*, **25**, 3284–3293.
65. Anisuzzaman Islam, M.K., Alim, M.A., Miyoshi, T., Hatta, T. *et al.* (2012) Longistatin is an unconventional serine protease and induces protective immunity against tick infestation. *Mol. Biochem. Parasitol.*, **182**, 45–53.
66. Kim, T.K., Radulovic, Z., and Mulenga, A. (2016) Target validation of highly conserved *Amblyomma americanum* tick saliva serine protease inhibitor 19. *Ticks Tick Borne Dis.*, **7**, 405–414.
67. Díaz Martín, V. (2014) Análisis de la saliva de *Ornithodoros moubata* y producción de antígenos recombinants para el desarrollo de tests serológicos y vacunas antigarrepata. Doctoral thesis.

University of Salamanca, Spain, http://hdl.handle.net/10261/101385 921 September 2016.
68. Imamura, S., da Silva Vaz, I. Jr., Sugino, M., Ohashi, K. *et al.* (2015) A serine protease inhibitor (serpin) from *Haemaphysalis longicornis* as an anti-tick vaccine. *Vaccine*, **26**, 1301–1311.
69. Narasimhan, S., Perze, O., Mootien, S., DePonte, K. *et al.* (2013) Characterization of Ixophilin, a thrombin inhibitor from the gut of *Ixodes scapularis*. *PLoS One*, **8**, e68012.
70. Macedo-Ribeiro, S., Almeida, C., Calisto, B.M., Friedrich, T. *et al.* (2008) Isolation, cloning and structural characterisation of boophilin, a multifunctional Kunitz-type proteinase inhibitor from the cattle tick. *PLoS One*, **3** (2), e1624.
71. Sanches-Soares, T., Watanabe, R.M.O., Tanaka-Azevedo, M.M., Soares Torquato, R.J. *et al.* (2012) Expression and functional characterization of boophilin, a thrombin inhibitor from *Rhipicephalus (Boophilus) microplus* midgut. *Vet. Parasitol.*, **187**, 521–528.
72. Liao, M., Zhou, J., Gong, H., Boldbaatar, D. *et al.* (2009) Hemalin, thrombin inhibitor isolated from a midgut cDNA library from the hard tick *Haemaphysalis longicornis*. *J. Insect Physiol.*, **55**, 164–173.
73. Mulenga, A., Misao, O., and Sugimoto, C. (2003) Three serine proteinases from mid guts of the hard tick *Rhipicephalus appendiculatus*; cDNA cloning and preliminary characterization. *Exp. Appl. Acarol.*, **29**, 151–164.
74. Imamura, S., Namangala, B., Tajima, T., Tembo, M.E. *et al.* (2006) Two serine protease inhibitors (serpins) that induce a bovine protective immune response against *Rhipicephalus appendiculatus* ticks. *Vaccine*, **24**, 2230–2237.
75. Tatchell, R.J. and Moorhouse, D.E. (1968) The feeding process of the cattle tick *Boophilus microplus* (Canestrini). II. The sequence of host-tissue changes. *Parasitology*, **58**, 441–459.
76. Evans, S.W. and Whicher, J.T. (1992) An overview of the inflammatory response, in *Biochemistry of Inflammation* (eds S.W. Evans and J.T. Whicher), Kluwer Academic Publishers, Lancaster, pp. 1–16.
77. Carvalho, W.A., Bechara, G.H., Moré, D.D., Ferreira, B.B. *et al.* (2008) *Rhipicephalus (Boophilus) microplus*: distinct acute phase proteins vary during infestations according to the genetic composition of the bovine hosts, *Bos taurus* and *Bos indicus*. *Exp. Parasitol.*, **118**, 587–591.
78. Carvalho, W.A., Franzin, A.M., Abatepaulo, A.R.R., de Oliveira, C.J.F. *et al.* (2010) Modulation of cutaneous inflammation induced by ticks in contrasting phenotypes of infestation in bovines. *Vet. Parasitol.*, **167**, 260–273.
79. Akhtar, M., Muhammad, F., Lohdi, L.A., Hussain, I. *et al.* (2011) Immunity against ticks – a review. *Pak. Vet. J.*, **31**, 9–16.
80. Trager, W. (1939) Further observations on acquired immunity to the tick *Dermacentor variabilis* (Say). *Parasitology*, **25**, 137–139.
81. Brassard, M. (1977) Rabbits infested with the adults of *Ixodes ricinus*: passive transfer of resistance with immune serum. *Bull. Soc. Pathol. Exot.*, **70**, 289–294.
82. Roberts, J.A. and Kerr, J.D. (1976) Boophilus: passive transfer of resistance in cattle. *J. Parasitol.*, **62**, 485–489.
83. Kazimírová, M. and Štibrániová, I. (2013) Tick salivary gland compounds: their role in modulation of host defences and pathogen transmission. *Front. Cell. Infect. Microbiol.*, **3**, 1–19.
84. Ribeiro, J.M.C., Makoul, G.T., Levine, J., Robinson, D.R. *et al.* (1985) Antihemostatic, anti-inflammatory and immunosuppressive properties of the saliva of a tick, *Ixodes dammini*. *J. Exp. Med.*, **161**, 332–344.
85. Tirloni, L., Reck, J., Terra, R.M.S., Martins, J.R. *et al.* (2014) Proteomic analysis of cattle tick *Rhipicephalus (Boophilus) microplus* saliva: a comparison between partially and fully engorged females. *PLoS One*, **9**, e94831.
86. Kim, T.K., Tirloni, L., Pinto, A.F.M., Moresco, J. *et al.* (2016) *Ixodes scapularis* tick saliva proteins sequentially secreted every 24 h during blood feeding. *PLoS Negl. Trop. Dis.*, **10** (1), e0004323.

87. Reck, J. Jr., Berger, M., Marks, F.S., Zingali, R.B. *et al.* (2009) Pharmacological action of tick saliva upon haemostasis and the neutralization ability of sera from repeatedly infested hosts. *Parasitology*, **136** (11), 1339–1349.

88. Jarmey, J.M., Riding, G.A., Pearson, R.D., McKenna, R.V. *et al.* (1995) Carboxypeptidases from *Boophilus microplus*: a concealed antigen with similarity to angiotensinogen converting enzyme. *Insect Biochem.*, **25**, 969–974.

89. Riding, G.A., Jarmey, J., McKenna, R.V., Pearson, R. *et al.* (1994) A protective 'concealed' antigen from *Boophilus microplus*: purification, localisation and possible function. *J. Immunol.*, **153**, 5158–5166.

90. Willadsen, P., Smith, D., Cobon, G., and McKenna, R.V. (1996) Comparative vaccination of cattle against *Boophilus microplus* with recombinant antigen Bm86 alone or in combination with Bm91. *Parasite Immunol.*, **18**, 241–246.

91. Kemp, D.H. and Bourne, A. (1980) *Boophilus microplus*: the effect of histamine on the attachment of cattle-tick larvae – studies *in vivo* and *in vitro*. *Parasitology*, **80**, 487–496.

92. Dai, J., Narasimhan, S., Zhang, L., Liu, L. *et al.* (2010) Tick histamine release factor is critical for *Ixodes scapularis* engorgement and transmission of the lyme disease agent. *PLoS Pathog.*, **6** (11), e1001205.

93. Garcia-Varas, S., Manzano-Román, R., Fernández-Soto, P., Encinas-Grandes, A. *et al.* (2010) Purification and characterization of a P-selectin-binding molecule from the salivary glands of *Ornithodoros moubata* that indices protective anti-tick immune responses in pigs. *Int. J. Parasitol.*, **40**, 313–326.

94. Leboulle, G., Crippa, M., Decrem, Y., Mejri, N. *et al.* (2002) Characterization of a novel salivary immunosuppressive protein from *Ixodes ricinus* ticks. *J. Biol. Chem.*, **277**, 10083–10089.

95. Salát, J., Paesen, G.C., Řezáčová, P., Kotsyfakis, M. *et al.* (2010) Crystal structure and functional characterization of an immunomodulatory salivary cystatin from the soft tick *Ornithodoros moubata*. *Biochem. J.*, **429**, 103–112.

96. Anguita, J., Ramamoorthi, N., Hovius, J.W., Das, S. *et al.* (2002) Salp15, an *Ixodes scapularis* salivary protein, inhibits CD4(+) T cell activation. *Immunity*, **16**, 849–859.

97. Hovius, J.W., de Jong, M.A., den Dunnen, J., Litjens, M. *et al.* (2008) Salp15 binding to DC-SIGN inhibits cytokine expression by impairing both nucleosome remodelling and mRNA stabilization. *PLoS Pathog.*, **4** (2), e31.

98. Dai, J., Wang, P., Adusumilli, S., Booth, C.J. *et al.* (2009) Antibodies against a tick protein, Salp15, protect mice from the Lyme disease agent. *Cell Host Microbe*, **19**, 482–492.

99. de la Fuente, J., Kopáček, P., Lew-Tabor, A., and Maritz-Olivier, C. (2016) Strategies for new and improved vaccines against ticks and tick-borne diseases. *Parasite Immunol.*, **38** (12), 754–769. doi: 10.1111/pim.12339

100. Sprong, H., Trentelman, J., Seeman, I., Grubhoffer, L. *et al.* (2014) ANTIDotE: anti-tick vaccines to prevent tick-borne diseases in Europe. *Parasit. Vectors*, **7**, 77.

101. Schetters, T., Bishop, R., Crampton, R., Kopáček, P. *et al.* (2016) Cattle tick vaccine researchers join forces in CATVAC. *Parasit. Vectors*, **9**, 105.

3
Blocking Transmission of Vector-borne Diseases

*Sandra Schorderet-Weber, Sandra Noack, Paul M. Selzer, and Ronald Kaminsky**

Abstract

Vector-borne diseases are responsible for significant health problems in humans, as well as in companion and farm animals. Killing the vectors with ectoparasitic drugs before they have the opportunity to pass on their pathogens could be the ideal way to prevent vector-borne diseases. Blocking of transmission might work when transmission is delayed during a blood meal, as often happens in ticks. The recently described systemic isoxazolines have been shown to successfully prevent disease transmission under conditions of delayed pathogen transfer. However, if the pathogen is transmitted immediately at bite as it is the case with most insects, blocking transmission becomes only possible if ectoparasiticides prevent the vector from landing on or, at least, from biting the host. Chemical entities exhibiting repellent activity in addition to fast killing, like pyrethroids, could prevent pathogen transmission even in cases of immediate transfer. Successful blocking depends on effective action in the context of the extremely diverse life cycles of vectors and vector-borne pathogens of medical and veterinary importance which are summarized in this review. This complexity leads to important parameters to consider for ectoparasiticide research and when considering the ideal drug profile for preventing disease transmission.

Introduction

Blood-feeding ectoparasites are responsible for severe aggravation through their constant attempts to get blood from their hosts. Besides causing discomfort, allergic reactions, skin damage, and pain, many ectoparasites are also vectors of life-threatening or debilitating diseases caused by the transmission of a wide variety of pathogens, that is, viruses, bacteria, protozoans, and worms, adding to their economic and emotional impact on human and animal health [1]. Therefore,

*Corresponding author.
Note: This article has been prepublished in the International Journal for Parasitology: Drugs and Drug Resistance 7 (2017) 90–109.

Ectoparasites: Drug Discovery Against Moving Targets, First Edition.
Edited by Charles Q. Meng and Ann E. Sluder

requirements for new ectoparasitic drugs should include not only the control of ectoparasites for a certain period of time but also address their ability to block the transmission of the various vector-borne pathogens by a rapid onset of action. In this scope, "speed of kill" has become an important commercial differentiator for recent marketed products [2–6] and many studies have been designed for testing the ability of those products to block transmission of some important pathogens of cats like *Bartonella henselae* [7], and of dogs like *Dipylidium caninum* [8], *Leishmania infantum* [9], *Ehrlichia canis* [10], *Borrelia burgdorferi*, *Anaplasma phagocytophilum* [11], and *Babesia canis* [12, 13]. These studies all report a complete prevention of pathogen transmission by fast elimination of the vector. These promising results confirm that a rapid onset of action should be an essential component of a novel drug profile. However, due to the diversity and specificity of vector–parasite interactions, the blocking characteristics of those ectoparasiticides may not be sufficient to control other major pathogens transmitted by vectors to humans and companion and farm animals. The arthropod can be either a mechanical vector, that is, a simple carrier for dispersion, or a biological vector, within which the pathogen undergoes asexual and/or sexual multiplication before being transferred to a mammalian host. In the latter situation, pathogens need time to undergo development inside the vector and reach their infective stage. This depends to a major part on environmental conditions like temperature and humidity, and on the ability of the vector to survive long enough to harbor the matured infectious stage to be transmitted at next bite. Blocking pathogen transmission during that period has been tried with success as seen with *Ixodes scapularis* and *B. burgdorferi* [14]. Treating only the mammalian host with an efficient drug is a simpler option, but ensuring that a new drug is able to reliably block pathogen transfer remains very challenging. Nevertheless, there is a window of time for an ectoparasitic drug to prevent disease transmission (Figure 3.1). This time period differs in length for each pathogen and vector, and can last from mere seconds to weeks.

Here we catalog a fair number of ectoparasite vectors and the respective transmitted pathogens of medical and veterinary importance. In addition, we complement that list with published information on the pathogen transmission time.

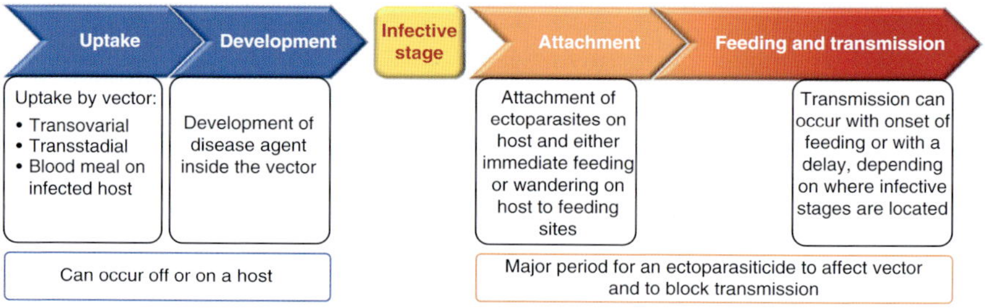

Figure 3.1 Generic sketch of transmission of diseases by ectoparasites (vectors). Blocking of transmission can, in principle, occur at every stage, but most drugs aim to interfere during the "Attachment" phase and/or "Feeding and Transmission" phase.

Based on these results, we propose several characteristics for an effective ectoparasitic drug profile.

Vector-borne Transmitted Pathogens

Major pathogens of medical and veterinary importance are listed in Tables 3.1–3.3. A short description of their development in the vector and timing of transmission is given when available. For each category of mammalian hosts, vectors are listed according to their importance in disease transmission. Many pathogens are zoonotic, with companion or farm animals becoming reservoirs in close contact with human populations, thus highlighting the practicality of employing common strategies for both human and animals to control pathogen transmission. The tables demonstrate the diversity of the vector–pathogen interactions. In most cases, the pathogen will undergo a multiplication, a change in morphology, and maturation in the vector. Very often, the infectious pathogen is waiting in the vector's salivary glands and will be passed on to the host together with the saliva immediately at bite. On the other hand, there are a few organisms, like *Rickettsia* sp. [44], *Anaplasma* sp. [45–47], *Borrelia* sp. [47, 51], or *Babesia* sp. [52, 53], that need an activation step for migration into the salivary glands, multiplication within the salivary glands, or a maturation phase all triggered by the onset of the blood meal. Interestingly, these pathogens all mature in ticks, which are slow blood-feeding arthropods and typically need days of host attachment to fully engorge.

Transmission Time Is Considerably Different Between Insects and Ticks

When considering ectoparasites in relation to the pathogens transmitted and the time needed to transfer the pathogens after biting the host, a clear difference between insects and ticks is noticeable (Tables 3.1–3.3). Many, if not all, holometabolic insects like mosquitoes [18], tsetse flies [26], fleas [32], or sand flies [21], which undergo complete metamorphosis, almost always transfer the respective pathogens immediately at bite. By contrast, some ticks can require host attachment time periods of several hours, extending up to days in some instances before transmission of pathogens occurs. As hard ticks (*Ixodidae*), sometimes also referred to as hardbacked ticks, feed only once before molting to the next stage, ingested pathogens will have to survive the molting process and be transferred transstadially (i.e., *Babesia* sp. [52]; *Ehrlichia* sp., [48, 63]). It may be difficult for the pathogen to develop, migrate, or mature while the physical and metabolic changes take place during the vector's molting process. The pathogen will also have to survive for an extended period in the tick vector that might not find the next host immediately and could stay unfed for weeks or months. Those microorganisms may then need a reactivation from some kind of dormant condition to resume their development. Temperature change due to the tick attachment to a warm-blooded animal [44], or fresh blood entering the tick may be the signal for the pathogen to multiply [45], migrate to the salivary glands [51],

Table 3.1 Human vector-transmitted pathogens.

Vector	Pathogen class		Disease	Pathogen development phases and timing in vector	Pathogen-to-host transfer timing at vector bite	References
Mosquitoes (Culicidae) *Aedes* sp. *Culex* sp. *Anopheles* sp. *Ochlerotatus* sp.	Arboviruses	Togaviridae	Chikungunya, Ross River	Replication occurs first in midgut cells, followed by dissemination to other organs including salivary glands with additional multiplication cycles. Development success in vector is dependent on temperature, vector competence, and viral dose at infection. An EIP (extrinsic incubation period) is defined for each vectorvirus combination. For dengue, an average of 7 to 12 days reported, but could be as early as 4 days	Immediate transmission at next blood meal once viruses have infected salivary glands	[15–17]
		Flaviridae	Zika, Yellow fever, Dengue, West Nile, etc.			
		Bunyaviridae	La Cross, Rift Valley			
	Protozoans	*Plasmodium* sp.	Malaria	Only sexual stages (gametocytes) survive in vector. Fertilization results in an ookinete that moves out of the midgut lumen and settles in the midgut outer epithelium.	Immediate transmission at next blood meal once sporozoites are present in the salivary ducts	[18, 19]

			Transforms into an oocyst. Asexual multiplication occurs (sporogony). Cyst opens and sporozoites migrate to salivary glands through hemocoel. They are imbedded into a parasitophorus vacuole until they are released into the salivary ducts. About 2 weeks are needed from ingestion of gametocytes to migration of sporozoites to salivary glands. Timing dependent on parasite-mosquito species combination		
Nematodes (Filariae)	*Wuchereria bancrofti* *Brugia malayi*	Lymphatic filariasis	Ingested microfilariae (mf) cross the vector midgut wall to enter the thoracic muscles. Subsequent molting to L1, L2, and to the L3 infective stage. L3 migrate back into the hemocoel, then to the head and mouth parts. No active injection by vector. L3 penetrate the host skin at biting site. Development from mf to L3 takes at least 10–11 days	Immediate transmission at next blood meal once L3 have reached the mosquito mouthparts	[20]

(*continued overleaf*)

Table 3.1 (*Continued*)

Vector	Pathogen class		Disease	Pathogen development phases and timing in vector	Pathogen-to-host transfer timing at vector bite	References
Sand flies (Phlebotominae) *Lutzomyia* sp. *Phlebotomus* sp.	Protozoans	*Leishmania* sp.[a)] Subgenera *Leishmania* sp. *Vannia* (New World only)	Cutaneous and visceral leishmaniasis	Amastigotes (intracellular in host macrophages) taken up mainly from skin at insect bite. Changes within the vectors internal environment trigger the transformation into motile procyclic promastigotes that multiply in blood meal. After a few days, differentiation into highly motile elongated nectomonad promastigotes *Leishmania*: Nectomonad promastigotes migrate to the anterior part of the midgut and break out of the peritrophic membrane. They move to the cardia and transform into leptomonad promastigotes, which further multiply and produce a promastigote secretory gel. Some attach and transform into haptomonad promastigotes. Some differentiate into infective metacyclic promastigotes	Immediate transmission at next bite once the gel containing the pathogens is blocking the foregut of the vector, in a way that the vector has to expel the gel into the host to be able to feed	[21, 22]

				Vannia: Similar to *Leishmania*, except for a concentration and replication step in the hindgut. Attachment as haptomonad promastigotes. Migration to foregut and establishment in the cardia as leptomonad promastigotes		
				Both subgenera: The gel containing the infective metacyclic forms obstructs the anterior midgut, forcing regurgitation at next bite prior to feeding, releasing the pathogen into the host		
				Almost 1–2 weeks are needed between ingestion of amastigotes and regurgitation of the infective metacyclic promastigotes		
Black flies (Simuliidae) *Simulium* sp.	Nematodes (Filariae)	*Onchocerca volvulus*	River blindness	Ingested mf cross the vector midgut wall to enter the thoracic muscles. Subsequent molting to L1, L2, and to the L3 infective stage. L3 migrate back to the hemocoel, then to the head and mouth parts. No active injection by vector. L3 penetrate the host skin at biting site	Immediate transmission at next blood meal once L3 have reached the vector mouthparts	[23, 24]

(*continued overleaf*)

Table 3.1 (Continued)

Vector	Pathogen class		Disease	Pathogen development phases and timing in vector	Pathogen-to-host transfer timing at vector bite	References
				Development in vector takes 6–11 days, depending on temperature and vector species. The simulids need to feed 1–2 times before the infective L3 are fully developed		
Biting midges (*Ceratopogonidae*) *Culicoides* sp.	Nematodes (Filariae)	*Mansonella perstans*	Mansonellosis	Life cycle similar to *O. volvulus*. Development in vector takes 7–9 days	Immediate transmission at next blood meal once L3 have reached the vector mouthparts	[25]
Tsetse flies (*Glossinidae*) *Glossina* sp.	Protozoans	*Trypanosoma gambiense*, *T. rhodesesiense*	African trypanosomiasis	Ingestion by vector of bloodstream trypanosomes. Transformation into procyclic trypomastigotes and intense multiplication in midgut from day 3 after feeding. From day 6, migration starts from hindgut to foregut, pharynx, and finally to salivary glands. Metacyclic trypomastygotes are the infective stage, detectable in salivary glands from day 12 after feeding and can be injected at next blood meal. Flies infective for the rest of their lives	Immediate transmission at next blood meal once infective metacyclic forms are developed in salivary glands	[26, 27]

Tabanids (Tabanidae) *Chrisops* sp.	Nematodes (Filariae)	*Loa loa*	African eye worm	Ingested mf exsheath in midgut and migrate predominantly to abdominal fat bodies. Subsequent molting to L1, L2, and then to the infective L3. L3 migrate back into the hemocoel, then to the head and mouthparts. Development time is temperature dependent, requiring 7–10 days	Immediate transmission at next blood meal once L3 have reached the vector mouthparts	[28]
True bugs (Hemiptera, Reduviidae) *Rhodnius* sp. *Triatoma* sp.	Protozoans	*Trypanosoma cruzi*[a)]	Chagas disease	Bloodstream trypomastigotes ingested by vector. Change to spheromastigotes and then to epimastigotes. Active multiplication in hindgut. Transformation into infective metacyclic forms, released with feces or Malpighian secretions. Infection via rubbing feces over skin lesions, contact with mucosae (mouth, nose, eye), or ingestion of the whole bug. Development timing is temperature and vector species dependent. At least 15–30 days are needed to detect infective metacyclic forms in the hindgut. Timing is shorter in immature instars (6–15 days)	Immediate transmission once infective forms are present in feces	[29–31]

(*continued overleaf*)

Table 3.1 (Continued)

Vector	Pathogen class		Disease	Pathogen development phases and timing in vector	Pathogen-to-host transfer timing at vector bite	References
Fleas (Siphonaptera, Pulicidae) *Xenopsylla* and *Pulex* sp. (plague) *Ctenocephalides* sp.	Bacteria	*Yersinia pestis*	Plague	After ingestion, bacteria multiply in the midgut until the total blockage of the flea proventricule is achieved. Host infection occurs via regurgitation by the flea or via direct contact and aerosol during an epidemic. In the flea vector, 4–16 days are needed to complete proventricule blockage. Depending on temperature and flea species, timing to reach proventricule blockage can be much longer	Immediate transmission at next bite once proventricule blockage is achieved	[32]
		Rickettsia felis	Cat flea typhus	Transmitted by *C. felis*. Ingestion by feeding on an infected host. Multiplication in midgut cells and dissemination in the flea tissues, including ovaries and salivary glands. Migration to salivary glands takes 7–14 days but transmission has been reported to occur as soon as 12h after infection	Between 12–24h for transmission via cofeeding. Timing for infection in host not measured per se. Could be immediate at next bite once salivary glands are invaded	[33–35]

			feeding (surely within 24h) via cofeeding with infected fleas. This early phase transmission seems to be mechanical. Transovarial transmission also occurs in the flea vector. Mosquitoes (*Anopheles gambiense*) now also suspected to be vector		
	Bartonella henselae[a)]	Cat scratch disease	Transmitted by *C. felis*. Pathogen ingested via an infected blood meal. Acquisition starts 3h after feeding begins. Replication occurs in gut cells. Bacteria survive during entire flea life span. Detected in feces 24h after first feeding starts. Survival in flea feces estimated to be at least 3 days. Host infection through exposure with flea feces, ingestion of infected fleas or flea feces, scratching or biting of a flea contaminated carrier animal	Immediate transmission via exposure to contaminated feces. In unfed fleas starting to feed, 24h delay before infected feces are released	[36]

(*continued overleaf*)

Table 3.1 (Continued)

Vector	Pathogen class		Disease	Pathogen development phases and timing in vector	Pathogen-to-host transfer timing at vector bite	References
Lice (Phthiraptera, Pediculidae) *Pediculus* sp.	Bacteria	*Rickettsia prowazekii*	Epidemic typhus	The pathogen develops in gut cells and is released in the gut lumen when the cells break. The insect feces are infectious. Host infection occurs when the skin damaged by scratching comes into contact with infected feces A minimum of 5 days required between feeding on infected blood and first release in feces. Narrow infection time window as infected lice die prematurely due to gut cell burst and perforation	Immediate transmission if infected feces in contact with wounds due to scratching	[37]
		Bartonella quintana	Trench fever	The pathogen multiplies in gut lumen and in epithelial gut cells, then shed in feces. Infection via skin damaged by scratching, contact with eyes mucosa, or if wounds are in contact with contaminated bedding or clothes. Five to 8 days needed between feeding on an infected host and detection of the pathogen in lice feces	Immediate transmission if infected feces in contact with wounds due to scratching	[38, 39]

Hard ticks (Ixodidae) *Amblyomma* sp. *Dermacentor* sp. *Ixodes* sp. *Ripicephalus* sp. *Hyalomma* sp. *Haemaphysalis* sp.	Arboviruses	Bunyaviridae[a)]	Crimean-Congo hemorrhagic fever (Old World)	Transmitted by *Hyalomma* sp. Virus persistence in the vector through transstadial and transovarial transmission. Venereal transmission from male to female ticks also occurs. Intrastadial virus transfer via cofeeding demonstrated. Development time in vector not measured. Contamination also reported via direct contact with an infected host, raw meat or milk ingestion, aerosol, etc.	Immediate transmission at next blood meal once viruses have reached and multiplied in salivary glands	[40, 41]
			Heartland virus (United States)	*Amblyomma americanum* seems to be main vector. Virus persistence in the vector through transstadial and transovarial transmission. Intrastadial virus transfer via cofeeding demonstrated. Development in vector not described in detail. Virus detected in midgut epithelial cells after infection feeding	Transmission timing not reported, but likely to be similar to other Bunyaviridae	[42]
		Flaviridae	Tickborne encephalitis (TBE)	Transmitted by *Ixodes* sp. Virus persistence in the vector through transstadial and transovarial transmission. Virus amplification via cofeeding on infected reservoirs. Virus multiplication	Transmission occurs presumably as soon as feeding starts, as salivary glands are invaded prior to feeding	[40, 43]

(continued overleaf)

Table 3.1 (Continued)

Vector	Pathogen class		Disease	Pathogen development phases and timing in vector	Pathogen-to-host transfer timing at vector bite	References
				takes place first in gut cell, then other tissue cells are invaded with further replication. Salivary glands are invaded relatively late, probably after molting as this organ undergoes resorption and regeneration during molting. Virus detected in salivary glands before the next blood meal starts. Transmission cases through infected milk and derivate are also reported		
	Bacteria	*Rickettsia conorii* *R. rickettsii*	Fièvre boutonneuse Rocky Mountain spotted fever	*R. conorii* transmitted by many tick genera. *R. rickettsii* transmitted by *Dermacentor* sp. The pathogen can multiply in almost all organs of the vector. If present in ovaries, transovarial transmission can occur. Persistence in the vector also through transstadial transmission. The pathogen is avirulent in ticks that have not fed for a long time period.	At least 10h of tick feeding are needed before the pathogen becomes infective again and can be successfully transmitted	[44]

		Reactivation can be triggered by the temperature increase that typically occurs during blood feeding on a warmblooded vertebrate		
Anaplasma phagocytophylum[a)]	Human granulocytic anaplasmosis	Transmitted by *Ixodes* sp. Persistence in the vector through transstadial transmission, but not transovarial. Acquisition by the vector within 24h of blood feeding. Multiplication in vector during and after acquisition feeding, and triggered again by next blood meal	Transmission does not take place before 36–48h post tick feeding. In the laboratory, transmission has occasionally been shown to occur within 24h of attachment	[45–47]
Ehrlichia ewingii[a)]	Human granulocytic ehrlichiosis	Transmitted by *Amblyomma americanum*. Life cycle in ticks not described. Closely related to *E. chaffeensis*. Development time in vector not measured. In one transmission study, pathogens were detected in host only 11–28 days after the beginning of exposure to adult ticks having acquired infection at the nymph stage. The speed of detection is depending on the size of the inoculum	Timing not measured	[48, 49]

(*continued overleaf*)

Table 3.1 (Continued)

Vector	Pathogen class		Disease	Pathogen development phases and timing in vector	Pathogen-to-host transfer timing at vector bite	References
		Ehrlichia chaffeensis[a)]	Human monocytic ehrlichiosis	Transmitted by *A. americanum* nymphs and adults. Persistence in the vector through transstadial transmission, not transovarial. Development time in vector not measured. In one transmission study, pathogens were detected in host only 7–12 days after the beginning of exposure to adult ticks having acquired infection at the nymph stage	Timing not measured	[48, 50]
		Borrelia burgdorferi sensu lato complex[a)]	Lyme disease	Transmitted by *Ixodes* sp. Persistence in the vector through transstadial transmission, not transovarial. After ingestion of infected blood, the spirochetes multiply in the ticks midgut by binary fission. They survive the vector molting. They migrate from the tick midgut to the salivary glands within 24h of the start of the transmission feeding	Most transmission occurs between 48h and 72h after tick attachment. But some studies report infection as early as 16h post attachment	[47, 51]

Protozoans	*Babesia divergens*[a)]	Human babesiosis (*European Union*)	*Ixodes ricinus* is the main vector. Persistence in the vector through transovarial and transstadial transmission. Only adult ticks seem to be able to acquire infection, but all stages can transmit. The sexual part of the life cycle occurs in the vector. Ingested gametocytes fuse to give rise to immobile zygotes that transform into mobile kinetes. They enter the hemolymph, disseminate into various tissues including muscles, epidermis, Malpighian tubules, and ovaries in adults. They undergo an additional asexual multiplication step and further dissemination as secondary kinetes. In salivary glands, kinetes continue to multiply asexually. The maturation into infective haploid sporozoites happens only after transmission feeding starts. In nymph ticks, sporozoites were detected in salivary glands from the third day of feeding	Like other *Babesia*, transmission is delayed to the second half of the tick blood meal. Transmission reported from day 3 of feeding	[52–54]

(*continued overleaf*)

Table 3.1 (Continued)

Vector	Pathogen class		Disease	Pathogen development phases and timing in vector	Pathogen-to-host transfer timing at vector bite	References
		Babesia microti[a)]	Human babesiosis (United States)	Transmitted by *Ixodes* sp. Persistence in the vector through transstadial transmission over one stage only, not transovarial. Nymphs are the main transmission stage. The sexual part of the life cycle occurs in the vector. The ingested gametocytes fuse into zygotes that move to the gut epithelial cells, then to the hemolymph as ookinetes and reach the salivary glands. They establish and become a multinucleate sporoblast after asexual multiplication. Maturation of infective sporozoites starts only after the tick host begins to feed again	Transmission success increases with feeding time. Minimal infectious dose of sporozoites detected 48h after transmission feeding starts. Maximum numbers of sporozoites found in salivary glands 60h after feeding start	[52, 55, 56]
Hard and soft ticks (Ixodidae and Argasidae)	Bacteria	*Coxiella burnetti*[a)]	Q fever	Transmitted by many tick genera. Persistence in the vector through transovarial and transstadial transmission. Multiplication in midgut cells.	Timing not known in feeding ticks	[57]

				The bacteria are released in tick feces when the tick begins to feed again. Transmission via an arthropod vector is very rare; occurs mostly through aerosol or from parturient fluids released by infected vertebrate hosts. The pathogen persists in the environment for weeks and can be spread by the wind		
Soft ticks (Argasidae) *Ornithodoros* sp.	Bacteria	*Borrelia duttoni (Old World); B. hermsii, B, turicatae, B. parkeri (New World)*	tick-borne relapsing fever	*Ornithodoros moubata* (*B. duttoni*): After ingestion with the blood meal, the pathogens enter the hemolymph and invade numerous tissues including synganglion, salivary glands, ovaries, and coxal organs. Transmission via saliva in nymph ticks, and mainly via coxal fluid contamination of tick	Transmission can occur within minutes, and has even been shown happening as quickly as 30s after tick bite	[58, 59]

(*continued overleaf*)

Table 3.1 (Continued)

Vector	Pathogen class		Disease	Pathogen development phases and timing in vector	Pathogen-to-host transfer timing at vector bite	References
				bite in adult ticks. Persistence in the vector through transovarial and transstadial transmission *O. hermsi* (*B. hermsii*): Transmission via saliva in all tick instars. Persistence in vector mainly through transstadial transmission, very rare transovarial transmission reported		
Lice (Phthiraptera, Pediculidae) *Pediculus humanis*		*Borrelia recurrentis*	Lice-borne relapsing fever	Ingestion by feeding on an infected host. From the midgut, spirochetes infect the body cavity and multiply without invading other tissues. No transovarial transmission. Transmission occurs when lice are crushed during scratching and spirochete-infected hemolymph is released onto the host skin	Immediate	[60]
Chigger mites (Trombiculidae) *Leptotrombidium* sp.	Bacteria	*Orientia tsutsugamushi*	Oriental scrub typhus	Chiggers feed cnly once on infected host. Only larvae are parasitic. Chiggers do not feed on blood. They inject digestive	Timing not known	[61]

				fluids to digest the hosts tissues and feed on serum exudates. Details on the pathogen life cycle in the vector not described. The pathogen is likely to be inoculated into the extracellular exudates during feeding. Persistence in the vector mainly through transovarial transmission. Transstadial and cofeeding transmission have also been shown		
Tabanids, mosquitoes, fleas, hard ticks	Bacteria	*Francisella tularensis*[a]	Tularemia	Main ways of transmission via tick bites and direct contact with a contaminated animal, mainly rabbits and hares, but occurs also via insect bites, ingestion of contaminated food, or aerosol *Ticks*: *Dermacentor variabilis* is the main vector. Persistence in the vector through transstadial transmission, although infected nymph ticks suffer high mortality due to the pathogen. Transovarial transmission also reported	*Ticks*: Can occur within 1 day after an adult tick infected as nymph begins to feed	[62]

a) Zoonotic diseases.

Table 3.2 Companion animal vector-transmitted pathogens.

Vector	Pathogen class		Disease	Pathogen development phases and timing in vector	Pathogen-to-host transfer timing at vector bite	References
Hard ticks (Ixodidae) *Rhipicephalus* sp. *Dermacentor* sp. *Haemaphysalis* sp. *Amblyomma* sp. *Ixodes* sp.	Bacteria	*Ehrlichia ewingii*[a)]	Canine granulocytic ehrlichiosis	Transmitted by *Amblyomma americanum*. Life cycle in ticks is not described. Closely related to *E. chaffeensis*. Development time in vector not measured. In one transmission study, the pathogens were detected in host only 11–28 days after the beginning of exposure to adult ticks having acquired infection at the nymph stage. The speed of detection is dependent on the size of the inoculum	Timing not measured	[48, 49]
		Ehrlichia chaffeensis[a)]	Canine monocytic ehrlichiosis	Transmitted by *A. americanum* nymphs and adults. Persistence in vector through transstadial transmission, not transovarial.	Timing not measured	[48, 50]

		Development time in vector not measured. In one transmission study, pathogen detected in host only 7–12 days after beginning of exposure to adult ticks having acquired infection at the nymph stage		
Ehrlichia canis	Canine ehrlichiosis	*Rhipicephalus sanguineus* is the main vector. *D. variabilis* also reported to be vector. Development in ticks not investigated in detail. Pathogen likely to multiply within gut cells, hemocytes, and salivary glands. Persistence in the vector through transstadial transmission. Intrastadial infection reported (infection among ticks of same stage cofeeding). Importance of male ticks in the epidemiology of the disease as they can move from host to host and could transmit the pathogen more efficiently	Host infection can occur as soon as 3h post tick attachment	[63, 64]
Anaplasma platys	Canine cyclic thrombocytopenia	*R. sanguineus* is the main vector. Rickettsia-like organism. Life cycle in ticks not described	Timing not measured, could be within 2 days, likely hours	[65]

(*continued overleaf*)

Table 3.2 (Continued)

Vector	Pathogen class		Disease	Pathogen development phases and timing in vector	Pathogen-to-host transfer timing at vector bite	References
		Borrelia burgdorferi sensu lato complex[a)]	Lyme disease	Transmitted by *Ixodes* sp. Persistence in the vector through transstadial transmission, not transovarial. After ingestion of infected blood, spirochetes multiply in the ticks midgut by binary fission. They survive the vector molting. They migrate from the midgut to the salivary glands within 24h of the start of the transmission feeding	Most transmission occurs between 48h and 72h after tick attachment. But some studies report infection as early as 16h post attachment	[47, 51]
	Protozoans	*Babesia canis (European Union), B. vogeli (United States)*	Canine babesiosis	*R. sanguineus* (United States, European Union), *D. reticulatus* (European Union), and *H. leachi* (Africa) are reported to be vectors. Persistence in the vector through transovarial and transstadial transmission. The sexual part of the life cycle occurs in the vector. Ingested	At least 48h of feeding are needed before transmission occurs. But if male ticks have already been feeding once, transmission could be immediate on the next host visited	[6668]

		gametocytes fuse into zygotes, transform into mobile kinetes that enter the hemolymph and disseminate into various tissues including muscles, epidermis and Malpighian tubules. They undergo a further division cycle. Secondary ookinetes invade almost all tick tissues, including ovaries in adults and salivary glands. In salivary glands, asexual multiplication by binary fission occurs. The maturation into infective sporozoites happens only after transmission feeding starts. Sporozoites formation in salivary glands takes 2–3 days		
Babesia microti[a)]	Canine babesiosis	Transmitted by *Ixodes* sp. Persistence in the vector through transstadial transmission over one stage only, not transovarial. Nymphs are the main transmission stage. The sexual part of the life cycle occurs in the vector. The ingested gametocytes fuse into zygotes that move to the	Transmission success increases with feeding time. Minimal infectious dose of sporozoites reported 48h after transmission feeding start. Maximum numbers of sporozoites detected 60h after feeding start	[52, 55, 56]

(*continued overleaf*)

Table 3.2 (Continued)

Vector	Pathogen class	Disease	Pathogen development phases and timing in vector	Pathogen-to-host transfer timing at vector bite	References
			gut epithelial cells, then to the hemolymph as ookinetes and reach the salivary glands. They establish and become a multinucleate sporoblast after asexual multiplication. Maturation of infective sporozoites starts only after the tick host begins to feed again		
	Babesia vulpes (= *Theileria annae*)		Tick vector not known but likely to be *Ixodes* sp. and *R. sanguineus*. Life cycle in vector not described but likely to be similar to *B. microti* as genetically closely related to it	Timing not measured, but likely to be delayed like other *Babesia* sp.	[69]
	Hepatozoon canis	Canine hepatozoonosis	*R. sanguineus* is the main vector. Infection by ingestion of gamonts from an infected dog. Persistence in the vector through transstadial transmission, not transovarial. Mature oocysts	Immediate transmission	[70]

			containing infective sporozoites located in the hemocoel. Dogs get infected through oral ingestion of a tick containing oocysts. Oocysts break inside the dogs digestive system, releasing the infective sporozoites. Transmission success is more dependent on temperature more than blood meal size dependent. Success is higher if the tick has been feeding for some days, being heated by the ingested blood, but transmission also works when unfed ticks are ingested		
	Cytauxzoon felis	Cat theileriosis	*A. americanum* is the main vector. Persistence in the vector through transstadial transmission. Details of the life cycle in vector not reported	Transmission occurs within 48h of feeding	[71]

(*continued overleaf*)

Table 3.2 (Continued)

Vector	Pathogen class		Disease	Pathogen development phases and timing in vector	Pathogen-to-host transfer timing at vector bite	References
Mosquitoes (Culicidae) *Aedes* sp. *Ochlerotatus* sp. *Anopheles* sp. *Culex* sp.	Nematodes (Filariae)	*Dirofilaria immitis*	Heartworm	Ingested microfilariae (mf) cross the vector midgut wall to enter the Malpighian tubules. Subsequent molting to L1, L2, and to the L3 infective stage. L3 migrate back to the hemocoel, then to the head and mouth parts. No active injection by vector. L3 penetrate the host skin at biting site. Development from mf to L3 lasts about 15–17 days (*Aedes aegypti*), temperature and mosquito species dependent	Immediate transmission at next blood meal once L3 have reached the mosquito mouthparts	[72, 73]
Sand flies (Phlebotominae) *Lutzomyia* *Phlebotomus*	Protozoans	*Leishmania* sp.[a)]	Cutaneous and visceral leishmaniasis	Amastigotes (intracellular in host macrophages) taken up mainly from skin at insect bite. Changes within the vectors internal environment trigger the transformation into motile procyclic	Immediate transmission at next bite once the gel containing the pathogens is blocking the foregut of the vector	[21, 22]

promastigotes that multiply in blood meal. After a few days, differentiation into highly motile elongated nectomonad promastigotes. They migrate to the anterior part of the midgut and break out of the peritrophic membrane. They move to the cardia and transform into leptomonad promastigotes, which further multiply and produce a promastigote secretory gel. Some attach and transform into haptomonad promastigotes. Some differentiate into the infective metacyclic promastigotes. The gel containing the infective metacyclic forms obstructs the anterior midgut, forcing regurgitation at next bite prior to feeding, releasing the pathogen into the host

At least 1–2 weeks are needed between ingestion of amastigotes and regurgitation of the infective metacyclic promastigotes

(*continued overleaf*)

Table 3.2 (Continued)

Vector	Pathogen class		Disease	Pathogen development phases and timing in vector	Pathogen-to-host transfer timing at vector bite	References
True bugs (Hemiptera, Reduviidae) *Rhodnius* sp.	Protozoans	*Trypanosoma cruzi*[a)]	Chagas disease	Bloodstream trypomastigotes ingested by vector. Change to spheromastigotes and then to epimastigotes. Active multiplication in hindgut. Transformation into infective metacyclic forms, released with feces or Malpighian secretions. Infection via rubbing feces over skin lesions, contact with mucosae (mouth, nose, eye), or ingestion of the whole bug. Development timing is temperature and vector species dependent. At least 15–30 days are needed to detect infective metacyclic forms in the hindgut. Timing is shorter in immature instars (6–15 days)	Immediate transmission once infective forms are present in feces	[29–31]

Fleas (Siphonaptera, Pulicidae) *Ctenocephalides* sp.	Bacteria	*Bartonella henselae*[a)]	Cat scratch disease	Transmitted by *C. felis*. Pathogen ingested via an infected blood meal. Acquisition starts 3h after feeding begins. Replication occurs in gut cells. Bacteria survive during entire flea life span. Detected in feces 24h after first feeding starts. Survival in flea feces estimated to be at least 3 days. Host infection through exposure with flea feces, ingestion of infected fleas or flea feces, scratching or biting of a flea contaminated carrier animal	Immediate transmission via exposure to contaminated feces. In unfed fleas starting to feed, 24h delay before infected feces are released	[36]

(*continued overleaf*)

Table 3.2 (Continued)

Vector	Pathogen class		Disease	Pathogen development phases and timing in vector	Pathogen-to-host transfer timing at vector bite	References
Fleas (Siphonaptera, Pulicidae) *Ctenocephalides* sp. Lice (Phthiraptera, Trichodectidae) *Trichodectes* sp.	Cestodes	*Dipylidium caninum*	Dog tapeworm	Vector gets infected at larval stage through cestode egg ingestion. Development in fleas is temperature dependent. With temperature lower than 30°C, the infective metacestode is not ready when the adult fleas emerge. The flea will need to survive and stay on a host a few days to allow completion of the development of the metacestode, triggered by the higher temperature of the host. Blood meal has no effect on development. Dog infection through ingestion of the parasitized flea	Immediate transmission once the infective larva is mature	[74]

a Zoonotic diseases.

Table 3.3 Farm animal vector-transmitted pathogens.

Vector	Pathogen class		Disease	Pathogen development phases and timing in vector	Pathogen-to-host transfer timing at vector bite	References
Hard ticks (Ixodidae) *Rhipicephalus* sp. *Dermacentor* sp. *Amblyomma* sp. *Ixodes* sp. *Hyalomma* sp.	Viruses	Bunyaviridae[a)]	Crimean-Congo hemorrhagic fever	Transmitted by *Hyalomma* sp. Virus persistence in the vector through transstadial and transovarial transmission. Venereal transmission from male to female ticks also occurs. Transmission via cofeeding demonstrated. Development time in vector not measured. Contamination can also occur via direct contact with an infected host, raw meat or milk ingestion, aerosol, etc.	Immediate transmission at next blood meal once viruses have reached and multiplied in the salivary glands	[40, 41]
		Flaviviridae[a)]	Louping ill	Transmitted by *Ixodes ricinus*. Virus persistence in the vector through transstadial transmission, no evidence of transovarial transmission. Closely related to tickbone encephalitis (TBE) virus. Transmission between vectors via cofeeding demonstrated. Development time in vector not measured	Immediate transmission at next feeding once viruses have reached and multiplied in salivary glands	[75]

(*continued overleaf*)

Table 3.3 (Continued)

Vector	Pathogen class		Disease	Pathogen development phases and timing in vector	Pathogen-to-host transfer timing at vector bite	References
	Bacteria	*Anaplasma phagocytophylum*[a)]	Granulocytic anaplasmosis	Transmitted by *Ixodes* sp. Persistence in the vector through transstadial transmission, but not transovarial. Acquisition by the vector within 24h of blood feeding. Multiplication in vector during and after acquisition feeding, and triggered again by next blood meal	Transmission does not take place before 36–48h tick feeding, but was shown in the laboratory to occasionally occur within 24h of attachment already	[45–47]
		Ehrlichia (= *Cowdria*) *ruminantium*	Heartwater	Transmitted by *Amblyomma* sp. Persistence in the vector through transstadial transmission. Transstadial transmission can happen over one or more stages depending on tick species. No interstadial transmission reported and transovarial transmission not sure. Bacteria ingested with the blood meal and enter the gut cells into which they multiply by binary fission in inclusion bodies. Migration to other organs like hemocytes, Malpighian tubules, and salivary glands. Bacteria colonies detected in salivary glands only after transmission feeding start	Transmission reported to occur from the second day of feeding in nymphs, and from the fourth day in adult ticks	[76–78]

	Borrelia theileri	Bovine borreliosis	Transmitted by *Rhipicephalus microplus*. Persistence in the vector through transstadial and transovarial transmission. Spirochetes ingested or present in ticks, multiply in hemocytes and hemolymph, with a tropism for ovaries and central ganglion. Also, transitory presence of the bacteria can be detected in other organs. Salivary glands may be invaded prior to or during the next blood meal. Larvae do not transmit infection as spirochetes are too scarce to sufficiently invade salivary glands. The pathogen multiplication is proportional to the level of tick nutrition. Highest in adult females.	Timing not measured.	[79]
Protozoans	*Babesia divergens*[a)]	Bovine babesiosis (European Union)	*Ixodes ricinus* is the main vector. Persistence in the vector through transovarial and transstadial transmission. Only adult ticks seem to be able to acquire infection, but all stages can transmit. The sexual part of the life cycle occurs in the vector. Ingested gametocytes fuse to give rise to immobile zygotes that transform into mobile kinetes, They	Like other *Babesia*, transmission is delayed to the second half of the tick blood meal. Transmission reported from day 3 of feeding	[52–54]

(*continued overleaf*)

Table 3.3 (Continued)

Vector	Pathogen class	Disease	Pathogen development phases and timing in vector	Pathogen-to-host transfer timing at vector bite	References
			enter the hemolymph, disseminate into various tissues including muscles, epidermis, Malpighian tubules, and ovaries in adults. They undergo an additional asexual multiplication step and further dissemination as secondary ookinetes. In salivary glands, kinetes continue to multiply asexually. Maturation into infective haploid sporozoites happens only after transmission feeding starts. In nymph ticks, sporozoites were detected in salivary glands from the third day of feeding		
	Babesia bovis *B. bigemina*	Bovine babesiosis (Rest of the World)	Transmitted by *Rhipicephalus (Boophilus)* sp. that are one host ticks. Persistence in the vector through transovarial and transstadial transmission. The sexual part of the life cycle occurs in the vector. Ingested gametocytes develop into gametes (ray bodies) that fuse to form diploid zygotes and	In tick larvae, transmission delayed due to maturation of sporozoites taking place after tick feeding starts. Nine days needed, in *B. bigemina*, 2–3 days in *B. bovis*.	[80, 81]

			enter the gut cells. Asexual multiplication and generation of kinetes that will migrate to invade other tissues, including salivary glands and tick oocytes. A secondary multiplication step occurs in these tissues. In salivary glands, kinetes transform into multinucleate stages that break up to form infective sporozoites. Maturation into infective sporozoites will start only when the tick is feeding again *B. bigemina*: 9 days required for the development of infective sporozoites. Transmission possible only at nymph and adult stage *B. bovis*: infective stages detected 2–3 days after feeding starts. Larvae can also transmit infection		
	Theileria parva	East Coast fever	*Rhipicephalus appendiculatus* is the main vector for *T. parva*. *T. annulata*, and *T. lestoquardi* transmitted by *Hyalomma* sp. Persistence in the vector through transstadial transmission only. The sexual part of the life cycle	*T. annulata*: 2 days feeding required for infective sporozoites to be ready for transmission. Can take up to 6–9 days	[82, 83]

(*continued overleaf*)

Table 3.3 (Continued)

Vector	Pathogen class	Disease	Pathogen development phases and timing in vector	Pathogen-to-host transfer timing at vector bite	References
	Theileria annulata	Tropical theileriosis (cattle)	occurs in vector. Ingested merozoites undergo gamogony. The resulting gametocytes fuse into zygotes that enter the gut cells. Maturation into ookinetes without obvious multiplication. Ookinetes move out of the gut cells and migrate to the salivary glands cells through the hemolymph. Transformation into multinucleate sporonts. Ticks are molting at this stage. Development stopped until the next blood meal begins, triggering a secondary multiplication and differentiation into infective sporozoites	in ticks having starved for months. No information on the other species	
	Theileria lestoquardi	Old World tropical theileriosis (sheep)			

Hard ticks (Ixodidae) Tabanids, stable flies, mosquitoes	Bacteria	*Anaplasma marginale*	Cattle anaplasmosis, erythrocytic anaplasmosis	*Ticks*: transmitted by many tick genera. Persistence in the vector through interstadial and transstadial transmission. No transovarial transmission. Importance of male ticks (i.e., *Dermacentor* sp.) that remain persistently infected and feed on different hosts. Acquisition by the vector within 24h of blood feeding. Multiplication in gut cells, then in other tissues, including salivary glands. Multiplication triggered again by next blood meal *Insects*: mechanical transmission via contact with infected mouthparts. Transmission also via contaminated fomites (i.e., needles)	*Ticks*: transmission does not take place before 36–48h of tick feeding, but was shown in the laboratory to occasionally occur after 24h	[84, 85]
Hard and soft ticks (Ixodidae and Argasidae)		*Coxiella burnetti*[a)]	Q fever	Transmitted by many tick genera. Persistence in the vector through transovarial and transstadial transmission. Multiplication in midgut cells. The bacteria are released in tick feces when the tick begins to feed again. Transmission via an arthropod vector is very rare, occurs mostly through aerosol or from parturient fluids released by infected vertebrate hosts. The pathogen persists in the environment for weeks, and can be spread by the wind	Timing not known in feeding ticks	[57]

(*continued overleaf*)

Table 3.3 (Continued)

Vector	Pathogen class		Disease	Pathogen development phases and timing in vector	Pathogen-to-host transfer timing at vector bite	References
Tsetse flies (Glossinidae) *Glossina* sp. Tabanids, stable flies (*T. vivax*)	Protozoans	*Trypanosoma brucei, T. congolense, T. vivax*	Nagana (cattle) Sheep trypanosomiasis (*T. vivax*)	General life cycle in vector similar to human trypanosomes (Table 3.1) *T. brucei*: 12 days to 2 weeks needed to complete the cycle in flies. Metacyclic forms found in salivary glands. Flies remain infective for life afterwards *T. congolense*: do not infect salivary glands. Epimastigotes attach to the labrum and hypopharynx of the proboscis, proliferate and mature into infective metacyclic forms *T. congolense*: infective metacyclic forms detected in saliva 21 days after infection feeding. Flies remain infective straight at bite afterwards *T. vivax*: after ingestion, only elongated forms survive in foregut and migrate to the cibarium. They transform into epimastigotes that migrate to the proboscis, attach and multiply. Free premetacyclic forms are generated by asymmetric fission of the epimastigotes. Host infection	Immediate transmission at next blood meal once infective metacyclic forms have matured	[27, 86–88]

				process not clear: either premetacyclic forms invade the hypopharynx and differentiate into infective metacyclic forms and/or metacyclic forms remaining in the alimentary canal are expelled via regurgitation before blood meal ingestion. At least 3 days are needed from infection feeding to detect infective forms in the hypopharynx. In addition to transmission by tsetse flies, mechanical transmission by other biting flies like stable flies and tabanids also occurs. In tabanids, the pathogen can survive in the crop or midgut up to 5–7h, and be regurgitated during early stages of feeding		
Tabanids (Tabanidae) *Tabanus* sp., *Chrysops* sp. Stable flies (Muscidae) *Stomoxys* sp.	Protozoans	*Trypanosoma evansi*	Surra	Mechanical transmission of the pathogen during restart of feeding on a different host after feeding interruption on an infected host. Infected blood remains in mouth parts and is reinjected with saliva into the next host. Blood meal regurgitation also shown. Trypanosomes survive up to 30min on mouthparts. Survival in midgut can be hours, up to 48h in *Stomoxys* flies allowing pathogen regurgitation in a delayed transmission mode	Immediate transmission at insect bite	[87]

(*continued overleaf*)

Table 3.3 (Continued)

Vector	Pathogen class		Disease	Pathogen development phases and timing in vector	Pathogen-to-host transfer timing at vector bite	References
Black flies (Simuliidae) *Simulium* sp. Biting midges (Ceratopogonidae) *Culicoides* sp.	Nematodes (Filariae)	*Onchocerca lienalis, O. ochengi, O. dukei, (O. gutturosa) O. gibsoni, O. gutturosa*	Bovine onchocercosis	Ingested microfilariae (mf) cross the vector midgut wall to enter the thoracic muscles. Subsequent molting to L1, L2, and to the L3 infective stage. L3 migrate back into the hemocoel, then to the head and mouth parts. No active injection by vector. L3 penetrate the host skin at biting site Development time to infective L3 is temperature dependent: *O. ochengi* in about 6 days, *O. dukei* in 6–9 days, *O. gibsoni* in about 6 days at 30°C, *O. gutturosa* in 13–15 days at 23°C, but up to 19 days have been reported. *O. lienalis* similar to *O. gutturosa*	Immediate transmission at next blood meal once L3 have reached the insect mouthparts	[89, 90]
Biting midges (Ceratopogonidae) *Culicoides* sp.	Viruses	Reoviridae	Bluetongue	To be transmitted, the virus ingested need to enter the midgut cells, replicate in them, escape into the hemocoel, and finally invade and replicate into the salivary glands. Development is temperature, virus serotype, and vector species dependent. The extrinsic incubation period (EIP) varies from 4 to more than 20 days. Vector remains infective for life	Immediate transmission once salivary glands are infected	[91, 92]

Face flies (Muscidae) *Musca automnalis*	Bacteria	*Moraxella bovis*	Bovine keratoconjunctivitis	Transmission via direct contact, through feces or regurgitation of the bacteria by the vector. Regurgitation seems to play a major role. Bacteria accumulate in the fly crop	Immediate transmission, with success depending on fly numbers feeding at same time	[93]
Tabanids, mosquitoes, fleas, hard ticks	Bacteria	*Francisella tularensis*[a]	Tularemia	Main ways of transmission via tick bites and direct contact with a contaminated animal, mainly rabbits and hares, but occurs also via insect bites, ingestion of contaminated food or aerosol *Ticks*: *Dermacentor variabilis* is the main vector. Persistence in the vector through transstadial transmission, although infected nymph ticks suffer high mortality due to the pathogen. Transovarial transmission also reported	*Ticks*: can occur within 1 day of an adult tick infected as nymph begins to feed	[62]

a) Zoonotic diseases.

or finish its maturation and be ready for transmission [52]. This last step might take hours or days, and gives opportunities to block the transfer. In this context, fast-acting ectoparasiticides could be effective at preventing disease transmission in ticks [8–13, 71].

In soft ticks (*Argasidae*), also referred to as softbacked ticks, pathogens face conditions similar to those in hard ticks (i.e., survival through molting, long periods of fasting, transstadial transmission) but also have to adapt to additional constraints. Soft ticks like *Ornithodoros* are fast blood feeders that need only minutes to fully engorge. Adults feed many times, and females lay eggs in small batches after each blood meal. They develop through more than one nymphal stage, increasing the number of opportunities for transmitting pathogens during their life span [58]. Fast feeding implies that pathogens cannot go through an activation step during the blood meal like that previously discussed for hard ticks, but rather have to be ready in the salivary glands to be transferred as soon as feeding starts. As an example, *B. duttoni* infecting soft ticks is transmitted from within 30 s to a few minutes after feeding starts [59], whereas *B. burgdorferi* is only transmitted by hard ticks after 24–48 h on average [47, 58]. Thus, a drug with an onset of action within a few hours might be sufficient for blocking transmission by hard ticks, but not for preventing transmission by soft ticks. In the latter case, preventing the vector from accessing the host with a repellent could be a more effective solution.

Some major pathogens of hemimetabolic insects like true bugs (*Reduviidae*) or lice (*Phthiraptera*) can develop in either immature stages or adults. *Trypanosoma cruzi* [29], *Rickettsia prowazekii* [37], or *Bartonella quintana* [38] are transmitted to their host via infected feces rubbed on wounded skin. Killing the vector before it gets time to produce infected feces could be possible using a drug with very fast onset of action. In the case of lice, such a drug could also have a massive impact on lice populations that do not move easily from one host to another, and therefore reduce the inflammation and scratching that are the real cause of infection. In *Reduviidae*, blocking transmission via killing the insect before releasing infected feces may also work. However, as *Reduviidae* are fast feeders, release of the feces could occur within the first minutes of a blood meal. It remains to be demonstrated whether preventing access to a host and subsequent biting with a repellent drug can effectively block *T. cruzi* transmission.

Pathogens of most holometabolic insects develop and multiply in an adult individual that has a life expectancy on the order of days or weeks. Their development can start immediately after ingestion and needs to reach the infective stage within the life span of the insect vector. In these cases, the pathogen strategy appears to be different and infectious stages are transmitted often within seconds to the mammalian host. Blocking transmission is therefore more challenging, and avoiding insect bite via a repellent drug could be the best option.

Drug Profile for Blocking Pathogen Transmission

The principal feature of an ectoparasitic drug aiming to block transmission should certainly be a very fast onset of action. This requirement is generally understood by the animal health industry, and most products marketed recently have been tested and compared for the speed of their onset of action [2–6]. Recent compounds deriving from the fairly new chemical class of isoxazolines ([94], Chapter 12) exhibit their ectoparasitic action against both, insects and acari of veterinary importance, within hours, and certainly reduce the risk of disease transmission of hard tick pathogens that are not immediately passed on to the host such as *Babesia* sp. [12, 13] or *Borrelia* sp. [11]. Such a beneficial effect was shown especially for canine borreliosis [11, 94]. Based on those results, one could hypothesize that isoxazolines may also be able to prevent human borreliosis (Lyme disease). However, to date many unknowns remain, including the pharmacokinetic behavior and safety of the drug in humans. Although effective at eliminating some tick infestations and consequently blocking pathogen transmission, systemic ectoparasiticides may be more limited in controlling those pathogens that are transmitted within a few hours or immediately after the vector's bite. For example, in a comparative study on the ability to block *E. canis* transmission from *Rhipicephalus sanguineus* ticks by orally administered isoxazolines compared against topically applied products containing synthetic pyrethroids, the tested systemic isoxazoline ectoparasiticides gave insufficient protection of dogs from pathogen transfer [95]. Despite being considered "old drugs," synthetic pyrethroids (i.e., permethrin, deltamethrin, flumethrin) exhibit features that would, in principle, be close to an ideal drug profile. In addition to having a fast onset of action on many insects and tick species, some pyrethroids are also irritant or repulsive for a variety of ectoparasites [96]. It appears that a combination of repellency and parasiticidal activity could be the best way to prevent pathogen transmission, independently from the transfer time at bite. Synthetic pyrethroids have been shown to efficiently block transmission of *Leishmania* sp. in dogs by repelling and killing sandflies [9, 97]. They are also widely used for impregnating bed nets and clothing to prevent insect bites and disease transmission to humans [98, 99]. They have been added to some recently marketed products for companion animals, to act as repellents and/or speed up the onset of action [4, 5]. However, widespread resistance in many vectors (including mosquitoes, lice, true bugs, and ticks) and safety issues [100, 101] disqualify them for longer term use and motivate the search for novel drugs displaying an equivalent profile with improved safety. Designing and developing new and safe ectoparasiticide drugs able to effectively block fast transmitted vector-borne pathogens is still on a wish list and remains extremely challenging. In our opinion, such novel ectoparasitic drug for animal health should combine features of fast killing, long persistency, and repellency to both acari and insects. Additional constraints may be encountered if any new ectoparasiticide should be considered for human use. Beyond identifying a relevant application, it is not clear if humans would accept a persistent drug exposure to achieve a long-lasting protection period. In principle, repellency combined with long-term persistence is very difficult to achieve in a single compound, constituting a challenge as big as achieving a very rapid onset of action. In addition, a drug with only repellent activity would have the disadvantage of having no impact on vector populations. The fast killing and long-lasting persistence already achieved with the isoxazolines would allow prevention of important tick-borne diseases. Additional repellency or deterrent activity would be efficacious at preventing insect-borne pathogens that are transmitted rapidly upon biting. Combining all of these activities would be the ideal profile for an ectoparasiticide. Achieving that goal might not be possible with a single chemical entity but may be possible with a combination of molecules, bearing in mind the challenges of maintaining a good safety profile for the host and for the environment. Hurdles remain extremely high, however, and other complementary measures targeting the pathogen itself via specific drugs or vaccines should definitely be investigated in parallel.

Conclusion

Meeting the requirements for new ectoparasiticides, including prevention of transmission of pathogens, is challenging if possible at all. Transmission in companion animals of some major tick-borne pathogens can be now controlled with compounds of the isoxazoline class because of their fast onset of action. Extending the use of this class of molecules to humans and farm animals may help control some tick-borne zoonotic diseases. For other pathogens, mainly those transferred to the host by insects immediately at bite and by soft ticks, the speed of kill by isoxazolines is insufficient to effectively prevent pathogen transmission. Most insect vectors have little time for feeding before being chased away or being killed by the host, and therefore, in most cases, blood feeding and associated pathogen transmission begins immediately upon landing. In this situation, drugs having repellent or deterrent activity that hinders the vector from biting or landing on the host would be more successful at preventing disease transmission. Solutions could, therefore, be different depending on the vector, the associated pathogens and the speed of transmission. In an ideal situation, a drug or a combination of chemical entities should prevent the vector from access, or at least from biting the host. If the vector eventually succeeds in reaching the host, killing by the drug should happen very rapidly. Repellent efficacy combined with parasiticidal activity seems to be the ideal drug profile for successfully preventing vector-borne diseases in humans, pets, and livestock. This easy statement unfortunately hides major difficulties, especially if the repellent effect has to be long-lasting for weeks or months. Due to those substantial difficulties, the search for new vaccines or drugs targeting the pathogen should not be left aside. Novel alternative approaches, for example, ones based on regulators of the immune system like the Toll pathway of the vector [102] should also continue to be explored.

Acknowledgments

The authors thank Dr Léonore Lovis for her support in the literature search, and Dr Ann E. Sluder, Dr Charles Meng, and Dr Stephen Peterson for their critical assessment of the manuscript.

References

1. Mehlhorn, H. (2008) Encyclopedia of Parasitology, 3rd edn, Springer-Verlag, Berlin/Heidelberg.
2. Halos, L., Lebon, W., Chalvet-Monfray, K., Larsen, D. *et al.* (2014) Immediate efficacy and persistent speed of kill of a novel oral formulation of afoxolaner (NexGard™) against induced infestations with *Ixodes ricinus* ticks. *Parasit. Vectors*, **7**, 452–455.
3. Wengenmayer, C., Williams, H., Zschiesche, E., Moritz, A. *et al.* (2014) The speed of kill of fluralaner (Bravecto™) against *Ixodes ricinus* ticks on dogs. *Parasit. Vectors*, **7**, 525–529.
4. Beugnet, F., Halos, L., Liebenberg, J., and Fourie, J. (2016) Assessment of the prophylactic speed of kill of Frontline Tri-Act(®) against ticks (*Ixodes ricinus* and *Rhipicephalus sanguineus*) on dogs. *Parasite*, **23**, 2–6.
5. Blair, J., Fourie, J.J., Varloud, M., and Horak, I.G. (2016) Efficacy and speed of kill of a topically applied formulation of dinotefuran-permethrin-pyriproxyfen against weekly tick infestations with *Rhipicephalus sanguineus* (*sensu lato*) on dogs. *Parasit. Vectors*, **9**, 283–289.
6. Six, R.H., Geurden, T., Carter, L., Everett, W.R. *et al.* (2016) Evaluation of the speed of kill of sarolaner (Simparica™) against induced infestations of three species of ticks (*Amblyomma maculatum, Ixodes scapularis, Ixodes ricinus*) on dogs. *Vet. Parasitol.*, **222**, 37–42.
7. Bradbury, C.A. and Lappin, M.R. (2010) Evaluation of topical application of 10% imidacloprid-1% moxidectin to prevent *Bartonella henselae* transmission from cat fleas. *J. Am. Vet. Med. Assoc.*, **236**, 869–873.
8. Fourie, J.J., Crafford, D., Horak, I.G., and Stanneck, D. (2013) Prophylactic treatment of flea-infested dogs with an imidacloprid/flumethrin collar (Seresto®, Bayer) to preempt infection with *Dipylidium caninum*. *Parasitol. Res.*, **112** (Suppl 1), 33–46.
9. Brianti, E., Gaglio, G., Napoli, E., Falsone, L. *et al.* (2014) Efficacy of a slow-release imidacloprid (10%)/flumethrin (4.5%) collar for the prevention of canine leishmaniosis. *Parasit. Vectors*, **7**, 327–336.
10. Jongejan, F., de Vos, C., Fourie, J.J., and Beugnet, F. (2015) A novel combination of fipronil and permethrin (Frontline Tri-Act®/Frontect®) reduces risk of transmission of *Babesia canis* by *Dermacentor reticulatus* and of *Ehrlichia canis* by *Rhipicephalus sanguineus* ticks to dogs. *Parasit. Vectors*, **8**, 602–611.
11. Honsberger, N.A., Six, R.H., Heinz, T.J., Weber, A. *et al.* (2016) Efficacy of sarolaner in the prevention of *Borrelia burgdorferi* and *Anaplasma phagocytophilum* transmission from infected *Ixodes scapularis* to dogs. *Vet. Parasitol.*, **222**, 67–72.
12. Beugnet, F., Halos, L., Larsen, D., Labuschagné, M. *et al.* (2014) The ability of an oral formulation of afoxolaner to block the transmission of *Babesia canis* by *Dermacentor reticulatus* ticks to dogs. *Parasit. Vectors*, **7**, 283–289.
13. Taenzler, J., Liebenberg, J., Roepke, R.K., and Heckeroth, A.R. (2016) Prevention of transmission of *Babesia canis* by *Dermacentor reticulatus* ticks to dogs after topical administration of fluralaner spot-on solution. *Parasit. Vectors*, **9**, 234–236.
14. Eisen, L. and Dolan, M.C. (2016) Evidence for personal protective measures to reduce human contact with blacklegged ticks and for environmentally based control methods to suppress host-seeking blacklegged ticks and reduce infection with Lyme disease spirochetes in tick vectors and rodent reservoirs. *J. Med. Entomol.*, **43**, 1063–1092. 10.1093/jme/tjw103..
15. Watts, D.M., Burke, D.S., Harrison, B.A., Whitmire, R.E. *et al.* (1987) Effect of temperature on the vector efficiency of *Aedes aegypti* for dengue 2 virus. *Am. J. Trop. Med. Hyg.*, **36**, 143–152.
16. Salazar, M.A., Richardson, J.H., Sanchez-Vargas, I., Olson, K.E. *et al.* (2007) Dengue virus type 2: replication and tropisms in orally infected *Aedes aegypti* mosquitoes. *BMC Microbiol.*, **7**, 9.

17. Kilpatrick, A.M., Meola, M.A., Moudy, R.M., and Kramer, L.D. (2008) Temperature, viral genetics, and the transmission of West Nile Virus by *Culex pipiens* mosquitoes. *PLoS Pathog.*, **4**, e1000092. doi: 10.1371/journal.ppat.1000092
18. Vlachou, D., Schlegelmilch, T., Runn, E., Mendes, A. *et al.* (2006) The developmental migration of *Plasmodium* in mosquitoes. *Curr. Opin. Genet. Dev.*, **16**, 384–391.
19. Meis, J.F., Wismans, P.G., Jap, P.H., Lensen, A.H. *et al.* (1992) A scanning electron microscopic study of the sporogonic development of *Plasmodium falciparum* in *Anopheles stephensi. Acta Trop.*, **50**, 227–236.
20. Paily, K.P., Hoti, S.L., and Das, P.K. (2009) A review of the complexity of biology of lymphatic filarial parasites. *J. Parasit. Dis.*, **33**, 3–12.
21. Bates, P.A. (2007) Transmission of *Leishmania* metacyclic promastigotes by phlebotomine sand flies. *Int. J. Parasitol.*, **37**, 1097–1106.
22. Killick-Kendrick, R. (1990) The life-cycle of *Leishmania* in the sandfly with special reference to the form infective to the vertebrate host. *Ann. Parasitol. Hum. Comp.*, **65** (Suppl 1), 37–42.
23. World Health Organization (1976) Epidemiology of onchocerciasis. Report of a WHO-Expert Committee. *World Health Organ. Tech. Rep. Ser.*, **597**, 1–94.
24. Takaoka, H., Ochoa, J.O., Juarez, E.L., and Hansen, K.M. (1982) Effects of temperature on development of *Onchocerca volvulus* in *Simulium ochraceum*, and longevity of the simuliid vector. *J. Parasitol.*, **68**, 478–483.
25. Simonsen, P.E., Onapa, A.W., and Asio, S.M. (2011) *Mansonella perstans* filariasis in Africa. *Acta Trop.*, **120** (Supplement 1), S109–S120.
26. Van Den Abbeele, J., Claes, Y., van Bockstaele, D., Le Ray, D. *et al.* (1999) *Trypanosoma brucei* spp. development in the tsetse fly: characterization of the post-mesocyclic stages in the foregut and proboscis. *Parasitology*, **118**, 469–478.
27. Gibson, W. and Baileys, M. (2003) The development of *Trypanosoma brucei* within the tsetse fly midgut using green fluorescent trypanosomes. *Kinetoplastid Biol. Dis.*, **2**, 1–13.
28. Orihel, T.C. and Lowrie, R.C. Jr. (1975) *Loa loa*: development to the infective stage in an American deerfly, *Chrysops atlanticus. Am. J. Trop. Med. Hyg.*, **24**, 610–615.
29. Krinsky, W.L. (2008) True bugs (Hemiptera), in Medical and Veterinary Entomology, 2nd e edn (eds G.R. Mullen and L.A. Durden), Academic Press, pp. 83–99.
30. Perlowagora-Szumlewicz, A. and de Carvalho Moreira, C.J. (1994) *In vivo* differentiation of *Trypanosoma cruzi* –1. Experimental evidence of the influence of vector species on metacyclogenesis. *Mem. Inst. Oswaldo Cruz*, **89**, 603–618.
31. Bern, C., Kjos, S., Yabsley, M.J., and Montgomery, S.P. (2011) *Trypanosoma cruzi* and Chagas' Disease in the United States. *Clin. Microbiol. Rev.*, **24**, 655–681.
32. Gage, K.L. and Kosoy, M.Y. (2005) Natural history of plague: perspective for more than a century of research. *Annu. Rev. Entomol.*, **50**, 505–528.
33. Thepparit, C., Hirunkanokpun, S., Popov, V.L., Foil, L.D. *et al.* (2013) Dissemination of bloodmeal acquired *Rickettsia felis* in cat fleas, *Ctenocephalides felis. Parasit. Vectors*, **6**, 149–155.
34. Brown, L.D., Banajee, K.H., Foil, L.D., and Macaluso, K.R. (2016) Transmission mechanisms of an emerging insect-borne rickettsial pathogen. *Parasit. Vectors*, **9**, 237–248.
35. Angelakis, E., Mediannikov, O., Parola, P., and Raoult, D. (2016) *Rickettsia felis*: the complex journey of an emergent human pathogen. *Trends Parasitol.*, **32**, 554–564.
36. Bouhsira, E., Franc, M., Boulouis, H.J., Jacquiet, P. *et al.* (2013) Assessment of persistence of *Bartonella henselae* in *Ctenocephalides felis. Appl. Environ. Microbiol.*, **79**, 7439–7444.
37. Houhamdi, L., Fournier, P.-E., Fang, R., Lepidi, H. *et al.* (2002) An experimental model of human body louse infection with *Rickettsia prowazekii. J. Infect. Dis.*, **186**, 1639–1646.
38. Byam, W. and Lloyd, L.L. (1920) Trench fever: its epidemiology and endemiology. *Proc. R. Soc. Med.*, **13**, 1–27.

39. Fournier, P.-E., Minnick, M.F., Lepidi, H., Salvo, E. *et al.* (2001) Experimental model of human body louse infection using green fluorescent protein-expressing *Bartonella quintana. Infect. Immun.*, **69**, 1876–1879.

40. Nuttall, P.A., Jones, L.D., Labuda, M., and Kaufman, W.R. (1994) Adaptations of arbovirus to ticks. *J. Med. Entomol.*, **31**, 1–9.

41. Charrel, R.N., Attoui, H., Butenko, A.M., Clegg, J.C. *et al.* (2004) Tick-borne virus diseases of human interest in Europe. *Clin. Microbiol. Infect.*, **10**, 1040–1055.

42. Godsey, M.S., Savage, H.M., Burkhalter, K.L., Bosco-Lauth, A.M. *et al.* (2016) Transmission of Heartland virus (Bunyaviridae: *Phlebovirus*) by experimentally infected *Amblyomma americanum* (Acari: Ixodidae). *J. Med. Entomol.*, **53**, 1226–1233.

43. Karbowiak, G. and Biernat, B. (2016) The role of particular tick developmental stages in the circulation of tick-borne pathogens affecting humans in Central Europe. 2. Tick-borne encephalitis virus. *Ann. Parasitol.*, **62**, 3–9.

44. Hayes, S.F. and Burgdorfer, W. (1982) Reactivation of *Rickettsia rickettsii* in *Dermacentor andersoni* ticks: an ultrastructural analysis. *Infect. Immun.*, **37**, 779–785.

45. Hodzic, E., Fish, D., Maretzki, D.Z., De Silva, A.M. *et al.* (1998) Acquisition and transmission of the agent of human granulocytic ehrlichiosis by *Ixodes scapularis* ticks. *J. Clin. Microbiol.*, **36**, 3574–3578.

46. Katavolos, P., Armstrong, P.M., Dawson, J.E., and Telford, S.R. 3rd (1998) Duration of tick attachment required for transmission of granulocytic ehrlichiosis. *J. Infect. Dis.*, **177**, 1422–1425.

47. des Vignes, F., Piesman, J., Hoffernan, R., Schulze, T.L. *et al.* (2001) Effect of tick removal on transmission of *Borrelia burgdorferi* and *Ehrlichia phagocytophila* by *Ixodes scapularis* nymphs. *J. Infect. Dis.*, **183**, 773–778.

48. Paddock, C.D. and Childs, J.E. (2003) *Ehrlichia chaffeensis*: a prototypical emerging pathogen. *Clin. Microbiol. Rev.*, **16**, 37–64.

49. Yabsley, M.J., Adams, D.S., O'Connor, T.P., Chandrashekar, R. *et al.* (2011) Experimental primary and secondary infections of domestic dogs with *Ehrlichia ewingii. Vet. Microbiol.*, **150**, 315–321.

50. Varela-Stokes, A.S. (2007) Transmission of *Ehrlichia chaffeensis* from Lone Star ticks (*Amblyomma americanum*) to white tailed deer (*Odocoileus virginianus*). *J. Wildl. Dis.*, **43**, 376–381.

51. Kahl, O., Janetzki-Mittmann, C., Gray, J.S., Jonas, R. *et al.* (1998) Risk of infection with *Borrelia burgdorferi sensu lato* for a host in relation to the duration of nymphal *Ixodes ricinus* feeding and the method of tick removal. *Zentralbl. Bakteriol.*, **287**, 41–52.

52. Homer, M.J., Aguilar-Delfin, I., Telford, S.R. 3rd, Krause, P.J. *et al.* (2000) Babesiosis. *Clin. Microbiol. Rev.*, **13**, 451–469.

53. Zintl, A., Mulcahy, G., Skerrett, H.E., Taylorm, S.M. *et al.* (2003) *Babesia divergens*, a bovine blood parasite of veterinary and zoonotic importance. *Clin. Microbiol. Rev.*, **16**, 622–636.

54. Donnelly, J. and Pierce, M.A. (1975) Experiments on the transmission of *Babesia divergens* to cattle by the tick *Ixodes ricinus. Int. J. Parasitol.*, **5**, 363–367.

55. Piesman, J. and Spielman, A. (1982) *Babesia microti*: infectivity of parasites from ticks for hamsters and white-footed mice. *Exp. Parasitol.*, **53**, 242–248.

56. Gray, J., von Stedingk, L.V., Gürtelschmid, M., and Granström, M. (2002) Transmission studies of *Babesia microti* in *Ixodes ricinus* ticks and gerbils. *J. Clin. Microbiol.*, **40**, 1259–1263.

57. Maurin, M. and Raoult, D. (1999) Q fever. *Clin. Microbiol. Rev.*, **12**, 518–553.

58. Schwan, T.G. and Piesman, J. (2002) Vector interactions and molecular adaptations of Lyme disease and relapsing fever spirochetes associated with transmission by ticks. *Emerg. Infect. Dis.*, **8**, 115–121.

59. Dworkin, M.S., Schwan, T.G., Anderson, D.E. Jr., and Borchardt, S.M. (2008) Tick-Borne Relapsing Fever. *Infect. Dis. Clin. North Am.*, **22**, 449–468.

60. Piesman, J. and Schwan, T.G. (2010) Ecology of *Borreliae* and their arthropod vectors, in Borrelia: Molecular Biology, Host Interactions and Pathogenesis (eds D.S. Samuels and J.D. Radolf), Caister Academic Press, pp. 251–278.

61. Lerdthusnee, K., Khlaimanee, N., Monkanna, T., Sangjun, N. *et al.* (2002) Efficiency of *Leptotrombidium* chiggers (*Acari: Trombiculidae*) at transmitting *Orientia tsutsugamushi* to laboratory mice. *J. Med. Entomol.*, **39**, 521–525.

62. Reese, S.M., Petersen, J.M., Sheldon, S.W., Dolan, M.C. *et al.* (2011) Transmission efficiency of *Francisella tularensis* by adult American dog ticks (*Acari: Ixodidae*). *J. Med. Entomol.*, **48**, 884–890.

63. Stich, R.W., Schaefer, J.J., Bremer, W.G., Needham, G.R. *et al.* (2008) Host surveys, Ixodid tick biology and transmission scenarios as related to the tick-borne pathogen, *Ehrlichia canis*. *Vet. Parasitol.*, **158**, 256–273.

64. Fourie, J.J., Stanneck, D., Luus, H.G., Beugnet, F. *et al.* (2013) Transmission of *Ehrlichia canis* by *Rhipicephalus sanguineus* ticks feeding on dogs and on artificial membranes. *Vet. Parasitol.*, **197**, 595–603.

65. Dantas-Torres, F., Capelli, G., Giannelli, A., Ramos, R.A.N. *et al.* (2013) Efficacy of an Imidacloprid/flumethrin collar against fleas, ticks and tick-borne pathogens in dogs. *Parasit. Vectors*, **6**, 245–253.

66. Schein, E., Mehlhorn, H., and Voigt, W.P. (1979) Electron microscopical studies on the development of *Babesia canis* (Sporozoa) in the salivary glands of the vector tick *Dermacentor reticulatus*. *Acta Trop.*, **36**, 229–241.

67. Mehlhorn, H. and Schein, E. (1984) The piroplasms: life cycle and sexual stages, in Advances in Parasitology, vol. **23**, Academic Press, pp. 37–103.

68. Heile, C., Hoffmann-Köhler, P., Wiemann, A., and Schein, E. (2007) Ubertragungszeiten von durch Zecken übertragenen Erregern beim Hund: Borrelien, Anaplasmen/Ehrlichien und Babesien. *Prakt. Tierarzt*, **88**, 584–590.

69. Baneth, G., Florin-Christensen, M., Cardoso, L., and Schnittger, L. (2015) Reclassification of *Theileria annae* as *Babesia vulpes* sp. nov. *Parasit. Vectors*, **8**, 207–213.

70. Baneth, G., Samish, M., Alekseev, E., Aroch, I. *et al.* (2001) Transmission of *Hepatozoon canis* to dogs by naturally fed or percutaneously-injected *Ripicephalus sanguineus* ticks. *J. Parasitol.*, **87**, 606–611.

71. Reichard, M.V., Thomas, J.E., Arther, R.G., Hostetler, J.A. *et al.* (2013) Efficacy of an Imidacloprid 10%/Flumethrin 4.5% collar (Seresto, Bayer) for preventing the transmission of *Cytauxzoon felis* to domestic cats by *Amblyomma americanum*. *Parasitol. Res.*, **112**, S11–S20.

72. Taylor, A.E. (1960) The development of *Dirofilaria immitis* in the mosquito *Aedes aegypti*. *J. Helminthol.*, **34**, 27–38.

73. Montarsi, F., Ciocchetta, S., Devine, G., Ravagnan, S. *et al.* (2015) Development of *Dirofilaria immitis* within the mosquito *Aedes (Finlaya) koreicus*, a new invasive species for Europe. *Parasit. Vectors*, **8**, 177–185.

74. Pugh, R.E. (1987) Effects on the development of *Dipylidium caninum* and on the host reaction to this parasite in the adult flea (*Ctenocephalides felis felis*). *Parasitol. Res.*, **73**, 171–177.

75. Jeffries, C.L., Mansfield, K.L., Phipps, L.P., Wakeley, P.R. *et al.* (2014) Louping ill virus: an endemic tick-borne disease of Great Britain. *J. Gen. Virol.*, **95**, 1005–1014.

76. Kocan, K.M. and Bezuidenhout, J.D. (1987) Morphology and development of *Cowdria ruminantium* in *Amblyomma* ticks. *Onderstepoort J. Vet. Res.*, **54**, 177–182.

77. Bezuidenhout, J.D. (1987) Natural transmission of heartwater. *Onderstepoort J. Vet. Res.*, **54**, 349–351.

78. Allsopp, B.A. (2010) Natural history of *Ehrlichia ruminantium*. *Vet. Parasitol.*, **167**, 123–135.

79. Smith, R.D., Brener, J., Osorno, M., and Ristic, M. (1978) Pathobiology of *Borrelia theileri* in the tropical cattle tick, *Boophilus microplus*. *J. Invertebr. Pathol.*, **32**, 182–190.

80. Bock, R., Jackson, L., de Vos, A., and Jorgensen, W. (2004) Babesiosis of cattle. *Parasitology*, **129**, S247–S269.

81. Howell, J.M., Ueti, M.W., Palmer, G.H., Scoles, G.A. *et al.* (2007) Transovarial transmission efficiency of *Babesia bovis* tick stages acquired by *Rhipicephalus (Boophilus) microplus* during acute infection. *J. Clin. Microbiol.*, **45**, 426–431.

82. Schein, E., Buscher, G., and Friedhoff, K.T. (1975) Entwicklung von *Theileria annulata* (Dschunkowsky and Luhs, 1904) in *Hyalomma anatolicum excavatum* (Koch, 1844). II. Die Entwicklung in Hämolymphe und Speicheldrüsen. *Z. Parasitenkd.*, **56**, 287–303.

83. Nicholson, W.L., Sonenshine, D., Lane, R.S., and Uilenberg, G. (2008) Ticks (Ixodida), in Medical and Veterinary Entomology, 2nd edn (eds G.R. Mullen and L.A. Durden), Academic Press, pp. 493–542.

84. Kocan, K.M., de la Fuente, J., Blouin, E.F., and Garcia-Garcia, J.C. (2004) *Anaplasma marginale* (Rickettsiales: *Anaplasmaceae*): recent advances in defining host-pathogen adaptations of a tick-borne rickettsia. *Parasitology*, **129**, S285–S300.

85. Kocan, K.M., de la Fuente, J., Blouin, E.F., Coetzee, J.F. *et al.* (2010) The natural history of *Anaplasma marginale. Vet. Parasitol.*, **167**, 95–107.

86. Peacock, L., Cook, S., Ferris, V., Bailey, M. *et al.* (2012) The life cycle of *Trypanosoma (Nannomonas) congolense* in the tsetse fly. *Parasit. Vectors*, **5**, 109–121.

87. Desquesnes, M., Dargantes, A., Lai, D.-H., Lun, Z.-R. *et al.* (2013) *Trypanosoma evansi* and Surra: a review and perspectives on transmission, epidemiology and control, impact, and zoonotic aspects. *Biomed. Res. Int.*, **2013**, Article ID 321237, 20 pages. 10.1155/2013/321237.

88. Ooi, C.-P., Schuster, S., Cren-Travaillé, C., Bertiaux, E. *et al.* (2016) The cyclical development of *Tripanosoma vivax* in the tsetse fly involves an asymmetric division. *Front. Cell. Infect. Microbiol.*, **6**, 115. doi: 10.3389/fcimb.2016.00115

89. Eichler, D.A. (1973) Studies on *Onchocerca gutturosa* (Neumann, 1910) and its development in *Simulium ornatum* (Meigen, 1818). 3. Factors affecting the development of the parasite in its vector. *J. Helminthol.*, **47**, 73–88.

90. Kettle, D.S. (1995) Other helminths transmitted by insects (Chapter 32), in Medical and Veterinary Entomology, 2nd edn (ed. D.S. Kettle), CAB International.

91. Wittmann, E.J., Mellor, P.S., and Baylis, M. (2002) Effect of temperature on the transmission of orbiviruses by the biting midge, *Culicoides sonorensis. Med. Vet. Entomol.*, **16**, 147–156.

92. Wilson, A.J. and Mellor, P.S. (2009) Bluetongue in Europe: past, present and future. *Philos. Trans. R. Soc. B*, **364**, 2669–2681.

93. Glass, H.W. Jr. and Gerhardt, R.R. (1984) Transmission of *Moraxella bovis* by regurgitation from the crop of the face fly *(Diptera: Muscidae). J. Econ. Entomol.*, **77**, 399–401.

94. Weber, T. and Selzer, P.M. (2016) Isoxazolines: a novel chemotype highly effective on ectoparasites. *ChemMedChem*, **11**, 270–276.

95. Jongejan, F., Crafford, D., Erasmus, H., Fourie, J.J. *et al.* (2016) Comparative efficacy of oral administrated afoxolaner (NexGard™) and fluralaner (Bravecto™) with topically applied permethrin/imidacloprid (Advantix®) against transmission of *Ehrlichia canis* by infected *Rhipicephalus sanguineus* ticks to dogs. *Parasit. Vectors*, **9**, 348–361.

96. Mencke, N. (2006) Acaricidal and repellent properties of permethrin, its role in reducing transmission of vector-borne pathogens. *Parassitologia*, **48**, 139–140.

97. Ferroglio, E., Poggi, M., and Trisciuoglio, A. (2008) Evaluation of 65% permethrin spot-on and deltamethrin-impregnated collars for canine *Leishmania infantum* infection prevention. *Zoonoses Public Health*, **55**, 145–148.

98. Curtis, C.F., Jana-Kara, B., and Maxwell, C.A. (2003) Insecticide treated nets: impact on vector populations and relevance of initial intensity of transmission and pyrethroid resistance. *J. Vector Borne Dis.*, **40**, 1–8.

99. Banks, S.D., Murray, N., Wilder-Smith, A., and Logan, J.G. (2014) Insecticide-treated clothes for the control of vector-borne diseases: a review on effectiveness and safety. *Med. Vet. Entomol.*, **28** (Suppl. 1), 14–25.

100. Anadón, A., Martínez-Larrañaga, M.R., and Martínez, M.A. (2009) Use and abuse of pyrethrins and synthetic pyrethroids in veterinary medicine. *Vet. J.*, **182**, 7–20.

101. Peterson, R.K.D., Barber, L.M., and Schleier, J.J. III (2011) Net risk: a risk assessment of long-lasting insecticide bed nets used for malaria management. *Am. J. Trop. Med. Hyg.*, **84**, 951–956.

102. Garver, L.S., Dong, Y., and Dimopoulos, G. (2009) Caspar controls resistance to *Plasmodium falciparum* in diverse anopheline species. *PLoS Pathog.*, **5** (3), e1000335. doi: 10.1371/journal.ppat.1000335. Epub 2009 Mar 13.

4
The Threat and Reality of Drug Resistance in the Cattle Tick *Rhipicephalus (Boophilus) microplus*

Heinz Sager, *Léonore Lovis, Christian Epe, and Ronald Kaminsky*

Abstract

Use of acaricides is one of the key strategies to protect cattle herds on pasture against tick infestation with *Rhipicephalus (Boophilus) microplus*. Control of cattle ticks with chemical compounds started at the end of the nineteenth century with the use of arsenic. This acaricide class was followed by organochlorines, organophosphates, amidines, synthetic pyrethroids, phenylpyrazols, macrocyclic lactones, growth regulators, and natural products. Unfortunately, the introduction of new acaricide classes has been followed by reports of treatment failure and finally the identification of resistance. Responsible for resistance are genetic changes in a cattle tick population that cause modifications to the target site, increased metabolism or sequestration of the acaricide, or reduced ability of the acaricide to penetrate through the outer protective layers of the tick's body. The mechanisms of resistance have been widely examined for some classes like the synthetic pyrethroids and organophosphates, while for other acaricides many aspects still remain unclear.

Nowadays, resistance can be found to almost all acaricide classes in the endemic areas. As a consequence, alternative strategies have been proposed to reduce the dependency on chemical products. While the use of indicine cattle breeds – which are less susceptible to tick-infestation – needs to be balanced with reduced productivity, other strategies like tick vaccines, may simply not be available in some countries or regions. Finally, the use of acaricides will remain an important pillar for control of cattle ticks, which increases the pressure on their sustainable use.

The Cattle Tick *Rhipicephalus (Boophilus) microplus*

The cattle tick *Rhipicephalus* (*Boophilus*) *microplus* is widespread in tropical and subtropical countries and has a major economic impact on the cattle industry due to blood feeding as well as to transmission of pathogens. Due to its

*Corresponding author.

Ectoparasites: Drug Discovery Against Moving Targets, First Edition.
Edited by Charles Q. Meng and Ann E. Sluder

importance, different measures are taken to reduce the tick burden on the animals as well as to control the spread of the ticks. The currently available tools for tick control in cattle consist of chemical acaricides of which different application methods and various formulations exist. In addition, nonchemical strategies are based on cattle breeds which are less susceptible to tick infestation, vaccines against ticks and tick-borne diseases, as well as rotations between livestock and crops [1]. Among these different options, chemical control is by far the most important strategy to control *R. (B.) microplus* infestations. However, resistance of *R. (B.) microplus* against acaricides has been acknowledged as a threat for successful tick control.

As a one-host tick, all life cycle stages of *R. (B.) microplus* develop on the same host and only the engorged females will drop off to lay eggs. Large numbers of larvae are usually present on the pasture in late spring, while successive generations of larvae occur through the summer till autumn and early winter [2]. The duration of the life cycle is strongly dependent on the climatic conditions, but it can be assumed that an average of four generations per year is not unusual in endemic areas [3].

History of Acaricidal Compound Classes: Their Introduction on the Market and Emergence of Resistance

Control of cattle ticks with chemical compounds started at the end of the nineteenth century with the use of arsenic. This class of compounds was followed by the organochlorines (OCs), organophosphates (OPs), amidines (AMs), synthetic pyrethroids (SPs), phenylpyrazols (PYZs), macrocyclic lactones (MLs), growth regulators, and natural products. The introduction of new acaricide classes has been followed by the emergence of resistance (Figure 4.1). In the following section, the different acaricide classes are presented in their order of market introduction.

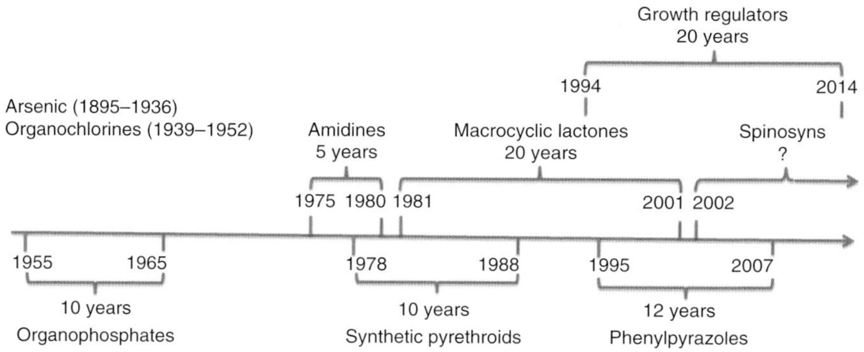

Figure 4.1 Chronological order of introduction of acaricides for use against cattle ticks (green date markers) and the first documentation of resistance against the respective class (red date markers).

Arsenicals

Arsenicals were the first compounds used for tick control. They were introduced in 1895 and the standard dose was 0.2% arsenic tetroxide [4]. In 1896, a local farmer in Queensland, Australia, initiated the use of an arsenic-based dipping vat for acaricidal treatments [5]. This successful arsenic dip was very rapidly adopted in other countries such as the United States, South Africa, and Cuba and dipping became a widespread practice [5]. Dipping had to be very frequent due to the very short residual effect (24 h) of arsenic [6]. After around 40 years of use, in 1936 the first case of *R. (B.) microplus* resistance to arsenic was reported [7].

Organochlorines

OC insecticides were the first synthetic organic insecticides (and acaricides) to be marketed in the late 1930s and 1940s for the control of ticks on cattle [8]. Dichlorodiphenyltrichloroethane (DDT) and benzene hexachloride (BHC) were the first of this group of chemicals to be used as acaricides [9]. A third important group of OCs is the cyclodienes, which include dieldrin and toxaphene [10]. OCs demonstrated a high efficacy, had a long residual activity and a large spectrum of action, and offered the advantages of being less toxic than arsenicals and being available at low costs. The first case of resistance to OCs was observed in Brazil in 1952 [11]. A decade later, in 1962, the use of all OCs was banned for tick control because of residues in meat, milk, and in the environment [4], resulting from their low biodegradability and their affinity for fat tissues.

Organophosphates and Carbamates

OPs began to be used as ectoparasiticides in the mid-1950s [12]. They first allowed control of ticks which had become OC resistant. Major compounds of this class were ethion, chlorpyrifos, chlorfenvinfos, and coumaphos. OPs were less stable and less persistent than OCs, although they may be very toxic to mammals [10, 13]. In contrast to OCs, they offered the advantages of being biodegradable, and therefore not accumulating in the environment, and of being much less lipophilic than OCs. Residual effect of OPs against tick reinfestation was 2 to 3 days [6]. OP resistance appeared first in Australia in the mid-1960s [14–16] and is nowadays widespread in the whole endemic area of *R. (B.) microplus*.

Carbamates comprise two main compounds used for tick control, which are carbaryl and propoxur. Carbaryl has low mammalian toxicity but may be carcinogenic and is often combined with other active ingredients [10]. Carbamates and OPs share the same mode of action, which is based on the inhibition of acetylcholine (AChE) and related esterases. As a consequence, cattle ticks resistant to carbamates will show cross-resistance to OPs, and vice versa [17].

Formamidines

AMs began to be used for tick control in the mid-1970s [18]. Today, amitraz is the most frequently used molecule of this class. It has a narrow spectrum of action but is very effective against ticks, with residual effectiveness against tick reinfestation for 9 days. It presents minimal toxicity to cattle and humans, is rapidly degraded in the environment, and has no meat withholding period [19].

Amitraz resistance appeared 4 to 10 years after its initial use in different parts of the world and was first identified in the early 1980s in Australia [18]. Since then, resistance has also been reported in Mexico, South America, South Africa, India, and New Caledonia. However, in 2007, Jonsson and Hope reported that amitraz was still one of the most popular acaricides for the control of cattle ticks in Australia, Latin America, and Southern Africa [19].

Synthetic Pyrethroids

SPs were introduced in the mid- to late 1970s and since then have been widely used [8]. SPs have very effective insecticidal and acaricidal activities while presenting a very low toxicity to mammals, and also are well biodegraded.

In the late 1980s, resistance was already observed in Australia [20] and Brazil [21, 22]. Nowadays, SP resistance is extremely common. Widespread resistance has been found throughout all the countries where resistance studies have been carried out and reported in the literature, that is, Mexico, Central and South America, South Africa, Australia, India, and New Caledonia. As for the OPs, SP resistance was also observed in the tick strain associated with an outbreak in Texas [23].

Macrocyclic Lactones

MLs were introduced to the market in 1981 [24]. They are divided into two categories [10]: (i) the avermectins, such as ivermectin, doramectin, abamectin, eprinomectin; and (ii) the milbemycins, including moxidectin and milbemycin oxime. MLs are active systemically against ticks and have the advantages of presenting a longer residual activity than SPs and of being active against a wide range of arthropods and nematodes [10]. Resistance to doramectin and ivermectin was first reported in Brazil in 2001 [25], followed by reports of ivermectin resistance in Mexico [26]. More recent data show the presence of ivermectin resistance in India and indicate emerging resistance against MLs in Uruguay, Argentina, and South Africa [27–29].

Phenylpyrazoles

Fipronil is the only phenylpyrazole compound to be used in livestock for the control of cattle ticks, with use starting in the mid-1990s [30]. Fipronil has a long residual activity [10] and its field efficacy persists up to five weeks [31]. Resistance to fipronil was reported for the first time in 2007 in Uruguay [32], and later in Brazil [33, 34].

Growth Regulators

Growth regulators are the most recent acaricide classes, of which fluazuron became the first representative available on the acaricidal market in 1994 in Australia. They have a completely different mode of action compared to all other previously introduced classes. Based on their mode of action, they are divided into (i) benzoylphenyl ureas, which are chitin synthesis inhibitors; (ii) triazine/pyrimidine derivatives, acting as chitin inhibitors; and (iii) juvenile hormone analogs [35]. Fluazuron is a benzoylphenyl urea providing long-term protection against *R. (B.) microplus* (6–12 weeks) [36]. In contrast, the other benzoylphenyl urea compounds have a relatively low efficacy against ticks. Fluazuron cannot be used for dairy cows because it is highly lipophilic and therefore accumulates in body fat tissues and milk. Recent data show the appearance of fluazuron resistance in Brazil and Colombia [37, 38].

Natural Products

Spinosad is a natural product containing a mixture of two fermentation metabolites produced by the actinomycete *Saccharopolyspora spinosa* [39, 40]. It has been shown to be effective against *R. (B.) microplus* larvae and nymphs [39] and is registered for tick control in some countries in Latin America, such as Brazil since 2002 [41]. Spinosad has the advantages of being rapidly biodegraded in soil, of exhibiting a low cross-resistance with other chemicals, and of presenting a reduced risk for workers [39].

Mechanisms of Resistance in Cattle Ticks

Resistance can arise through several mechanisms that are generally classified in three main categories: target site insensitivity, increased metabolic detoxification, and reduced cuticular penetration [42]. Penetration resistance is a reduced susceptibility of the target organism to penetration by an acaricide. Although it has been reported in *R. (B.) microplus* [43], this mechanism has not been much studied. In contrast, target-site and metabolic resistances are more common and have been widely studied for some classes of compounds, such as the SPs. Target-site resistance is due to single nucleotide substitution in the gene coding for the target molecule of an acaricide, resulting in an amino acid mutation which confers a lower susceptibility to the acaricidal compound [42]. Metabolic resistance occurs when an individual has an increased ability to detoxify or sequester an acaricide. Three main enzyme families are known to be involved in this type of resistance: the cytochrome P450s, esterases, and glutathione S-transferases. Chemicals known as synergists are often used to detect metabolic resistance. Piperonyl butoxide (PBO), triphenyl phosphate (TPP), and diethylmaleate (DEM) are the three most commonly used synergists and are considered to be specific inhibitors for cytochrome P450s, esterases, and glutathione S-transferases, respectively [42]. Synergist studies are usually carried out using the larval packet test (see Chapter 5) by comparing a compound in the presence and absence of a synergist [44–47]. An observed impact on the toxicity of a chemical compound in the presence of a synergist is an indicator for the involvement of the corresponding enzyme class in the degradation of the acaricidal compound [42]. Synergist studies are useful guides to determining if metabolic resistance mechanisms are present, but do not provide a definitive determination of the mechanism of metabolic resistance.

Studying resistance mechanisms of ticks to acaricides provides important information. First, the knowledge of the biochemical or molecular mechanisms of resistance allows the development of tools for rapid detection of resistance in the field that can be integrated into a resistance management program. Secondly, the knowledge of the resistance mechanism will allow the selection of the best alternative compounds to overcome the resistance. Furthermore, if the mechanism of resistance is known, the useful lifetime of a pesticide may be extended once resistance emerges by combination with an appropriate synergist. Finally, resistant ticks are valuable tools to study the mode of action of acaricides [48].

Resistance mechanisms to OP and amitraz have been studied, but not yet elucidated. For OP, more than one acetylcholinesterase may be targeted by OP and thus be potentially involved in resistance mechanisms [49, 50]. Several studies attributed *R. (B.) microplus* OP resistance to target-site insensitivity, although no mutation has been identified yet [42]. Some authors provided evidence of potential metabolic resistance [46, 51, 52]. For amitraz, target-site resistance has been suspected, although it has not been unequivocally demonstrated [42], and synergist studies have shown that metabolic resistance plays a role in some strains [53, 54].

Resistance mechanisms to SP are better understood: both target-site- and metabolism-based resistances have been identified in *R. (B.) microplus*. Generally, when both mechanisms coexist, target-site resistance is the more important one. Four single nucleotide substitutions have been shown to confer pyrethroid resistance: first, a mutation located in domain III of the *R. (B.) microplus* sodium channel and resulting in a phenylalanine to isoleucine amino acid substitution was identified in Mexican tick populations [55]. More recently, three additional mutations have been identified in other tick populations, leading to leucine to isoleucine, glycine to valine, and methionine to threonine amino acid substitutions, respectively, in domain II of the sodium channel [56–58]. These four mutations are associated with different resistance phenotypes and have different geographical distributions. The molecular aspects of metabolic pyrethroid resistance are not yet well defined in *R. (B.) microplus*. Overproduction of an esterase that hydrolyzed permethrin, designated CzEst9, was observed by Jamroz *et al.* and Pruett *et al.* [59, 60]. In addition, synergist studies with PBO have indicated that cytochrome P450s also play a role in pyrethroid resistance in some strains, although the exact molecular mechanisms have not yet been elucidated [45].

Fipronil acts on both the 4-aminobutyric acid (GABA)-gated chloride channel and the glutamate-gated chloride channel [61]. This activity on dual targets probably plays a role in delaying or preventing the buildup of high levels of resistance. However, one of these targets is shared with the cyclodiene class of pesticides and low levels of fipronil resistance can be associated with resistance to dieldrin in *Drosophila melanogaster* [62]. Thus far no mechanistic studies have been performed to confirm reports of fipronil resistance in *R. (B.) microplus* and no information on potential mechanisms of resistance is available.

The situation is very similar for MLs, where resistance has been described for *R. (B.) microplus* in the field, but studies of resistance mechanisms are at a very early stage. Despite the assumption that fipronil and the MLs both act on the glutamate-gated chloride channel and the GABA-gated chloride channel, thus far cross resistance has not been observed. Isolates of *R. (B.) microplus* that were either resistant to MLs or fipronil have proved sensitive to the other class [27, 63].

There are only few and very recent reports on resistance to the growth regulator fluazuron, which has the consequence that no information on potential resistance mechanisms is available up to now. There has been as yet no reported resistance of ticks to spinosad.

Resistance Management

The availability of appropriate tools to determine acaricide resistance in ticks is a prerequisite for giving recommendations to farmers on treating cattle successfully in the short term and on reducing the risk of further development of resistance in the long term. As shown by the history of the different acaricide classes, each market introduction (with the exception of spinosyns, up to now) has been followed by the development of resistance. Therefore, when a new class of compounds is launched, the question is not whether resistance will appear, but rather when resistance will appear. An additional problem is the persistence of resistance once established in a population. There is a general assumption that a reversion of acaricide resistance in ticks does not occur. As an exception, the AM class is the only acaricide group in which a resistance reversion has been reported [41, 64]. However, the level of resistance of a *R. (B.) microplus* strain that was not exposed to amitraz for 10 years did not change significantly [3]. This finding supports the hypothesis that reversion of acaricide resistance in ticks is very unlikely. In many organisms it has been described that resistance mechanisms impose fitness costs. In such a scenario, the resistant individuals would be selected and become dominant during exposure, but become less frequent over time in the absence of the respective chemical class. It is difficult to estimate how much this applies to cattle ticks. Temeyer and colleagues described in their article that ticks maintain a large and diverse assortment of AChE alleles available for rapid recombination and selection, which potentially reduces fitness costs associated with individual mutations [65]. This would mean that – at least for OPs – the prevalence of resistant ticks within a population will not change considerably in the absence of the respective chemical class.

Hence, strategies to delay the emergence and spread of resistance to a new or existing acaricide are crucial. One of the most important factors is the training of the farmers. There is a need for widespread understanding of how to correctly dose and use the available acaricides. In addition, there are several recommendations for slowing the emergence of resistance when using acaricides: (i) Reducing the frequency of treatments. This requirement, however, is often conflicting with the producers wishing a high level of tick control [1]. (ii) Limiting the number of ticks exposed to chemical treatments, for example, by using a threshold approach. In such an approach, treatments are applied only when a predetermined number of engorged females is reached on each animal [1]. (iii) Rotation of acaricides with different modes of action [66]. This must be done with great care to avoid the selection of multiresistant strains, however. Rotation should be done after follow-up with efficiency tests or at least every two years [67]. (iv) The use of combination products. The combination should ideally demonstrate an additive or even synergistic effect, although this can raise the risk of underdosing the individual compounds and consequently increasing the risk of developing resistance. It is therefore of utmost importance to respect the correct dose of each compound. (v) Monitoring of ticks with bioassays, biochemical or molecular tests to detect developing acaricide resistance at an early stage and to adapt treatment strategies accordingly.

Furthermore, when possible, chemical control of tick populations should not be employed as a stand-alone strategy, but should be integrated in a pest management program designed to decrease the frequency of treatment application and to delay the onset of resistance. Nonchemical methods which can be combined with chemical control are: (i) The use of tick-resistant cattle breeds. The variation in resistance to tick infestation is most marked between *Bos taurus* and *Bos indicus* cattle. The latter carry between five and ten times less ticks than the former when held under the same conditions. One explanation may be found in the strong T-cell-mediated response directed against larval stages that has been demonstrated in indicine cattle. Other parameters like the skin properties (epidermal growth factors, collagens and other matrix components such as lumican) may also contribute to variation in host resistance [68]. The inferiority of indicine cattle breeds in productivity may be partially overcome by cross breeding with taurine cattle [69]. (ii) The use of cattle tick vaccines. In Australia and in Cuba, a genetically engineered tick vaccine based on the antigen Bm86 was launched in the 1990s [70, 71]. The vaccine induced antibodies to bind to the Bm86 molecule on tick intestinal cells causing them to lyse, and thereby interfere with the blood-feeding activity of the tick. In order to keep a high level of protection, the vaccination had to be repeated every 10–12 weeks. The vaccine had no immediate effect on the numbers of ticks and reduced infestations only marginally. But due to the effect on the reproductive capacity of the females, the number of surviving ticks was considerably reduced in the following generations. A limiting factor was the large variation of responsiveness to the Bm86-antigen between animals. It was feared that nonresponders may contribute disproportionately to the reestablishment of tick populations. Efforts were made to improve the protection, for example, by combining with additional antigens like Bm95 [72, 73]. Different groups and companies are working on the development of new *R. (B.) microplus* vaccines. Where commercially available, the vaccine can be deployed to reduce the dependency on acaricides and to delay the appearance of acaricide-resistant ticks. Specifically, acaricides can be employed in combination with vaccination to reduce the frequency of chemical treatments. (iii) Rotation between crops and livestock. This involves removal of all livestock hosts from pastures for a period of time long enough to ensure death of most of free-living ticks [1]. (iv) Biological control. This last option is the least developed for cattle ticks. However, the use of the fungus *Metarhizium anisopliae*, already commercialised as an insecticide, seems to be successful against *R. (B.) microplus* in laboratory bioassays [1].

Whatever strategy is selected, the people directly involved, that is, in most cases the farmers, have to understand the benefit (and also the limits) of the strategies so that they are willing to implement the control measures. Studies have shown that the incorrect use of chemicals is widespread and accelerates the development of resistance [74]. Furthermore, recent surveys in Brazil among milk producers showed that producers had only limited knowledge of the tick life cycle and about the products they were using, and were not aware of the suboptimal practices that can promote the development of resistance [67, 75]. Therefore, to ensure a sustainable use of the acaricidal compounds that are currently available and still effective, and to extend their life span, educating farmers about ticks, tick control, tick resistance, and integrated management programs is crucial.

References

1. FAO (2004) Ticks: Acaricide Resistance: Diagnosis, Management and Prevention, pp. 25–77. Resistance Management and Integrated Parasite Control in Ruminants: Guidelines, ftp://ftp.fao.org/docrep/fao/010/ag014e/ag014e05.pdf (accessed 04 November 2016).
2. Walker, A.R., Bouattour, A., Camicas, J.L., Estrada-Pena, A. *et al.* (2003) Ticks of Domestic Animals in Africa: A Guide to Identification of Species. Bioscience Reports, Edinburgh, Scotland, UK.
3. Maciel, W.G., Lopes, W.D., Cruz, B.C., Gomes, L.V. *et al.* (2015) Ten years later: evaluation of the effectiveness of 12.5% amitraz against a field population of *Rhipicephalus* (*Boophilus*) *microplus* using field studies, artificial infestation (Stall tests) and adult immersion tests. *Vet. Parasitol.*, **214**, 233–241.
4. Waltisbuhl, D., Farnsworth, B., and Kemp, D. (2005) Pesticides used against ectoparasites (ticks, flies, lice) of cattle in Australia, in *Ectoparasiticide Use in Contemporary Australian Livestock Production* (ed. P.A. Holdsworth), Avcare Limited, Canberra, pp. 45–57.
5. Angus, B.M. (1996) The history of the cattle tick *Boophilus microplus* in Australia and achievements in its control. *Int. J. Parasitol.*, **26**, 1341–1355.
6. Mitchell, M. (1996) Acaricide resistance – back to basics. *Trop. Anim. Health Prod.*, **28**, 53S–58S.
7. Legg, J. (1947) Recent developments in tick and buffalo-fly control. *Aust. Vet. J.*, **23**, 181–185.
8. Graf, J.F., Gogolewski, R., Leach-Bing, N., Sabatini, G.A. *et al.* (2004) Tick control: an industry point of view. *Parasitology*, **129**, S427–S442.
9. George, J.E., Pound, J.M., and Davey, R.B. (2004) Chemical control of ticks on cattle and the resistance of these parasites to acaricides. *Parasitology*, **129**, S353–S366.
10. Taylor, M.A. (2001) Recent developments in ectoparasiticides. *Vet. J.*, **161**, 253–268.
11. Freire, J.J. (1953) Arseno e cloro resistência e emprego de tiofosfato de dietilparanitrofenila (Parathion) na luta anticarrapato *Boophilus microplus* (Canestrini 1887). *Bol. Dir. Prod. Anim.*, **9**, 3–31.
12. Andreotti, R. (2010) Diagnóstico de resistência de *Boophilus microplus* a piretróides, Embrapa Gado de Corte, Campo Grande, pp. 1–36.
13. Cremlyn, R. (1978) Pesticides in the environment, in *Pesticides: Preparation and Mode of Action*, John Wiley & Sons, Inc., pp. 210–221.

14. Shaw, R.D. and Malcolm, H.A. (1964) Resistance of *Boophilus microplus* to organophosphorus insecticides. *Vet. Rec.*, **76**, 210–211.
15. Shaw, R.D. (1966) Culture of an organophosphorus-resistant strain of *Boophilus microplus* (Can.) and an assessment of its resistance spectrum. *Bull. Entomol. Res.*, **56**, 389–405.
16. Roulston, W.J., Stone, B.F., Wilson, J.T., and White, L.I. (1968) Chemical control of an organophosphorus – and carbamate-resistant strain of *Boophilus microplus* (Can.) from Queensland. *Bull. Entomol. Res.*, **58**, 379–392.
17. Li, A.Y., Davey, L.R., and George, J.E. (2005) Carbaryl resistance in Mexican strains of the southern cattle tick (Acari: Ixodidae). *J. Econ. Entomol.*, **98** (2), 552–556.
18. Nolan, J. (1981) Current developments in resistance to amidine and pyrethroid tickicides in Australia, in *Proceedings of Tick Biology and Control Conference* (eds G.B. Whitehead and J.D. Gibson), Tick Research Unit, Rhodes University, Grahamstown, pp. 109–114.
19. Jonsson, N.N. and Hope, M. (2007) Progress in the epidemiology and diagnosis of amitraz resistance in the cattle tick *Boophilus microplus*. *Vet. Parasitol.*, **146**, 193–198.
20. Nolan, J., Wilson, J.T., Green, P.E., and Bird, P.E. (1989) Synthetic pyrethroid resistance in field samples in the cattle tick (*Boophilus microplus*). *Aust. Vet. J.*, **66**, 179–182.
21. Leite, R.C. (1988) *Boophilus microplus* (Canestrini 1887) susceptibilidade, usa atual e retrospectivo de carrapaticidas em propiedades das regiões fisiográficas da Baixada do Grande Rio e Rio de Janeiro. Uma abordagem epidemiológica. Tese de doutorado Universidade Federal Rural do Rio de Janeiro, Rio de Janeiro.
22. Laranja, R.J., Martins, J.R., Ceresér, V.H., Corrêo, B.L. *et al.* (1989) Identificação de uma estirpe de *Boophilus microplus* resistente a carrapaticidas piretroides, no Estado de Rio Grande de Sul. 83.
23. Miller, R.J., Davey, R.B., and George, J.E. (2007) First report of permethrin-resistant *Boophilus microplus* (Acari: Ixodidae) collected within the United States. *J. Med. Entomol.*, **44**, 308–315.
24. Geary, T.G. (2005) Ivermectin 20 years on: maturation of a wonder drug. *Trends Parasitol.*, **21**, 530–532.
25. Martins, J.R. and Furlong, J. (2001) Avermectin resistance of the cattle tick *Boophilus microplus* in Brazil. *Vet. Rec.*, **149**, 64.
26. Perez-Cogollo, L.C., Rodriguez-Vivas, R.I., Ramirez-Cruz, G.T., and Miller, R.J. (2010) First report of the cattle tick *Rhipicephalus microplus* resistant to ivermectin in Mexico. *Vet. Parasitol.*, **168**, 165–169.
27. Castro-Janer, E., Rifran, L., Gonzalez, P., Niell, C. *et al.* (2011) Determination of the susceptibility of *Rhipicephalus* (*Boophilus*) *microplus* (Acari: Ixodidae) to ivermectin and fipronil by Larval Immersion Test (LIT) in Uruguay. *Vet. Parasitol.*, **178**, 148–155.
28. Lovis, L., Reggi, J., Berggoetz, M., Betschart, B. *et al.* (2013) Determination of acaricide resistance in *Rhipicephalus* (*Boophilus*) *microplus* (Acari: Ixodidae) field populations of Argentina, South Africa, and Australia with the Larval Tarsal Test. *J. Med. Entomol.*, **50**, 326–335.
29. Singh, N.K., Singh, H., Jyoti, Prerna, M. *et al.* (2015) First report of ivermectin resistance in field populations of *Rhipicephalus* (*Boophilus*) *microplus* (Acari: Ixodidae) in Punjab districts of India. *Vet. Parasitol.*, **214**, 192–194.
30. Davey, R.B., Ahrens, E.H., George, J.E., Hunter, J.S. *et al.* (1998) Therapeutic and persistent efficacy of fipronil against *Boophilus microplus* (Acari: Ixodidae) on cattle. *Vet. Parasitol.*, **74**, 261–276.
31. Davey, R.B., George, J.E., Hunter, J.S.I., and Jeannin, P. (1999) Evaluation of a pour-on formulation of fipronil against *Boophilus annulatus* (Acari: Ixodidae) under natural South Texas field conditions. *Exp. Appl. Acarol.*, **23**, 351–364.
32. Cuore, U., Trelles, A., Sanchis, J., Gayo, V. *et al.* (2007) Primer diagnostico de resistencia al Fipronil en la garrapata comun del ganado *Boophilus microplus*. *Veterinaria*, **42**, 35–41.

33. Castro-Janer, E., Martins, J.R., Mendes, M.C., Namindome, A. *et al.* (2010) Diagnoses of fipronil resistance in Brazilian cattle ticks (*Rhipicephalus* (*Boophilus*) *microplus*) using *in vitro* larval bioassays. *Vet. Parasitol.*, **173**, 300–306.
34. Castro-Janer, E., Rifran, L., Gonzalez, P., Piaggio, J. *et al.* (2010) *Rhipicephalus* (*Boophilus*) *microplus* (Acari: Ixodidae) resistance to fipronil in Uruguay evaluated by *in vitro* bioassays. *Vet. Parasitol.*, **169**, 172–177.
35. Graf, J.F. (1993) The role of insect growth regulators in arthropod control. *Parasitol. Today*, **9**, 471–474.
36. Bull, M.S., Swindale, S., Overend, D., and Hess, E.A. (1996) Suppression of *Boophilus microplus* populations with fluazuron – an acarine growth regulator. *Aust. Vet. J.*, **74**, 468–470.
37. Puerta, J.M., Chaparro, J.J., Lopez-Arias, A., Arroyave, S.A. *et al.* (2015) Loss of *in vitro* efficacy of topical commercial acaricides on *Rhipicephalus microplus* (Ixodida: Ixodidae) from antioquian farms, Colombia. *J. Med. Entomol.*, **52** (6), 1309–1314.
38. Reck, J., Klafke, G.M., Webster, A., Dall'Agnol, B. *et al.* (2014) First report of fluazuron resistance in *Rhipicephalus microplus*: a field tick population resistant to six classes of acaricides. *Vet. Parasitol.*, **201** (1–2), 128–136.
39. Davey, R.B., George, J.E., and Snyder, D.E. (2001) Efficacy of a single whole-body spray treatment of spinosad, against *Boophilus microplus* (Acari: Ixodidae) on cattle. *Vet. Parasitol.*, **99**, 41–52.
40. Miller, R.J., White, W.H., Davey, R.B., George, J.E. *et al.* (2011) Efficacy of spinosad against acaricide-resistant and -susceptible *Rhipicephalus* (*Boophilus*) *microplus* and acaricide-susceptible *Amblyomma americanum* and *Dermacentor variabilis*. *J. Med. Entomol.*, **48**, 358–365.
41. Jonsson, N.N., Miller, R.J., Kemp, D.H., Knowles, A. *et al.* (2010) Rotation of treatments between spinosad and amitraz for the control of *Rhipicephalus* (*Boophilus*) *microplus* populations with amitraz resistance. *Vet. Parasitol.*, **169**, 157–164.
42. Guerrero, F.D., Lovis, L., and Martins, J.R. (2012) Acaricide resistance mechanisms in *Rhipicephalus microplus*. *Rev. Bras. Parasitol. Vet.*, **21**, 1–6.
43. Schnitzerling, H.J., Nolan, J., and Hughes, S. (1983) Toxicology and metabolism of some synthetic pyrethroids in larvae of the cattle tick *Boophilus microplus* (Can.). *Pestic. Sci.*, **14**, 64–72.
44. Crampton, A.L., Green, P., Baxter, G.D., and Barker, S.C. (1999) Monooxygenases play only a minor role in resistance to synthetic pyrethroids in the cattle tick, *Boophilus microplus*. *Exp. Appl. Acarol.*, **23**, 897–905.
45. Miller, R.J., Davey, R.B., and George, J.E. (1999) Characterization of pyrethroid resistance and susceptibility to coumaphos in Mexican *Boophilus microplus* (Acari: Ixodidae). *J. Med. Entomol.*, **36**, 533–538.
46. Li, A.Y., Davey, R.B., Miller, R.J., and George, J.E. (2003) Resistance to coumaphos and diazinon in *Boophilus microplus* (Acari: Ixodidae) and evidence for the involvement of an oxidative detoxification mechanism. *J. Med. Entomol.*, **40**, 482–490.
47. Li, A.Y., Davey, R.B., Miller, R.J., Guerrero, F.D. *et al.* (2008) Genetics and mechanisms of permethrin resistance in the Santa Luiza strain of *Boophilus microplus* (Acari: Ixodidae). *J. Med. Entomol.*, **45**, 427–438.
48. Scott, L.G. (1990) Investigating mechanisms of insecticide resistance: methods, strategies, and pitfalls, in *Pesticide Resistance in Arthropods* (eds R.T. Roush and B.E. Tabasnik), Chapman Hall, New York, pp. 39–57.
49. Baffi, M.A., de Souza, G.R., de Sousa, C.S., Ceron, C.R. *et al.* (2008) Esterase enzymes involved in pyrethroid and organophosphate resistance in a Brazilian population of *Rhipicephalus* (*Boophilus*) *microplus* (Acari, Ixodidae). *Mol. Biochem. Parasitol.*, **160**, 70–73.
50. Temeyer, K.B., Pruett, J.H., and Olafson, P.U. (2010) Baculovirus expression, biochemical characterization and organophosphate sensitivity of

rBmAChE1, rBmAChE2, and rBmAChE3 of *Rhipicephalus (Boophilus) microplus*. *Vet. Parasitol.*, **172**, 114–121.

51. Villarino, M.A., Waghela, S.D., and Wagner, G.G. (2003) Biochemical detection of esterases in the adult female integument of organophosphate-resistant *Boophilus microplus* (Acari: Ixodidae). *J. Med. Entomol.*, **40**, 52–57.
52. Saldivar, L., Guerrero, F.D., Miller, R.J., Bendele, K.G. *et al.* (2008) Microarray analysis of acaricide-inducible gene expression in the southern cattle tick, *Rhipicephalus (Boophilus) microplus*. *Insect Mol. Biol.*, **17**, 597–606.
53. Knowles, C.O. and Roulston, W.J. (1973) Toxicity to *Boophilus microplus* of formamidine acaricides and related compounds, and modification of toxicity by certain insecticide synergists. *J. Econ. Entomol.*, **66**, 1245–1251.
54. Li, A.Y., Davey, R.B., Miller, R.J., and George, J.E. (2004) Detection and characterization of amitraz resistance in the southern cattle tick, *Boophilus microplus* (Acari: Ixodidae). *J. Med. Entomol.*, **41**, 193–200.
55. He, H., Chen, A.C., Davey, R.B., Ivie, G.W. *et al.* (1999) Identification of a point mutation in the para-type sodium channel gene from a pyrethroid-resistant cattle tick. *Biochem. Biophys. Res. Commun.*, **261**, 558–561.
56. Morgan, J.A., Corley, S.W., Jackson, L.A., Lew-Tabor, A.E. *et al.* (2009) Identification of a mutation in the para sodium channel gene of the cattle tick *Rhipicephalus (Boophilus) microplus* associated with resistance to synthetic pyrethroid acaricides. *Int. J. Parasitol.*, **39**, 775–779.
57. Jonsson, N.N., Cutulle, C., Corley, S.W., and Seddon, J.M. (2010) Identification of a mutation in the para-sodium channel gene of the cattle tick *Rhipicephalus microplus* associated with resistance to flumethrin but not to cypermethrin. *Int. J. Parasitol.*, **40**, 1659–1664.
58. Stone, N.E., Olafson, P.U., Davey, R.B., Buckmeier, G. *et al.* (2014) Multiple mutations in the para-sodium channel gene are associated with pyrethroid resistance in *Rhipicephalus microplus* from the United States and Mexico. *Parasit. Vectors*, **7**, 456–469.
59. Jamroz, R.C., Guerrero, F.D., Pruett, J.H., Oehler, D.D. *et al.* (2000) Molecular and biochemical survey of acaricide resistance mechanisms in larvae from Mexican strains of the southern cattle tick, *Boophilus microplus*. *J. Insect Physiol.*, **46**, 685–695.
60. Pruett, J.H., Guerrero, F.D., and Hernandez, R. (2002) Isolation and identification of an esterase from a Mexican strain of *Boophilus microplus* (Acari: Ixodidae). *J. Econ. Entomol.*, **95**, 1001–1007.
61. Zhao, X., Yeh, J.Z., Salgado, V.L., and Narahashi, T. (2004) Fipronil is a potent open channel blocker of glutamate-activated chloride channels in cockroach neurons. *J. Pharmacol. Exp. Ther.*, **310** (1), 192–201.
62. Bloomquist, J.R. (1994) Cyclodiene resistance at the insect GABA receptor/chloride channel complex confers broad cross resistance to convulsants and experimental phenylpyrazole insecticides. *Arch. Insect Biochem. Physiol.*, **26** (1), 69–79.
63. Lopes, W.D., Cruz, B.C., Teixeira, W.F., Felippelli, G. *et al.* (2014) Efficacy of fipronil (1.0 mg/kg) against *Rhipicephalus (Boophilus) microplus* strains resistant to ivermectin (0.63 mg/kg). *Prev. Vet. Med.*, **115** (3–4), 88–93.
64. Foil, L.D., Coleman, P., Eisler, M., Fragoso-Sanchez, H., Garcia-Vasquez, Z., Guerrero, F.D., Jonsson, N.N., Langstaff, I.G., Li, A.Y., Machila, N., Miller, R.J., Morton, J., Pruett, J., and Torr, S. (2004) Factors that influence the prevalence of acaricide resistance and tick-borne diseases. *Vet. Parasitol.*, **125**, 163–181.
65. Temeyer, K.B., Olafson, P.U., and Pruett, J.H. (2012) Sequence polymorphism in acetylcholinesterase transcripts and genotyping survey of BmAchE1 in laboratory and Mexican strains of *Rhipicephalus (Boophilus) microplus*. *J. Med. Entomol.*, **49** (3), 555–562.

66. Thullner, F., Willadsen, P., and Kemp, D. (2007) Acaricide rotation strategy for managing resistance in the tick *Rhipicephalus* (*Boophilus*) *microplus* (Acarina: Ixodidae): laboratory experiment with a field strain from Costa Rica. *J. Med. Entomol.*, **44**, 817–821.

67. da Rocha, C.M., Leite, R.C., Bruhn, F.R., Guimaraes, A.M. *et al.* (2011) Perceptions of milk producers from Divinopolis, Minas Gerais, regarding *Rhipicephalus* (*Boophilus*) *microplus* control. *Rev. Bras. Parasitol. Vet.*, **20**, 295–302.

68. Jonsson, N.N., Piper, E.K., and Constantinoiu, C.C. (2014) Host resistance in cattle to infestation with the cattle tick *Rhipicephalus microplus*. *Parasite Immunol.*, **36**, 553–559.

69. Utech, K.B.W., Wharton, R.H., and Kerr, J.D. (1978) Resistance to *Boophilus microplus* (Canestrini) in different breeds of cattle. *Aust. J. Agric. Res.*, **29**, 885–895.

70. Willadsen, P., Bird, P., Cobon, G.S., and Hungerford, J. (1995) Commercialisation of a recombinant vaccine against *Boophilus microplus*. *Parasitology*, **110** (Suppl), S43–S50.

71. Canales, M., Enríquez, A., Ramos, E., Cabrera, D. *et al.* (1997) Large-scale production in *Pichia pastoris* of the recombinant vaccine Gavac against cattle tick. *Vaccine*, **15**, 414–422.

72. Pengally, C. (1999) Antiparasitics: Products and Markets. Animal Pharm Reports, PJB Publications, London.

73. García-García, J.C., Montero, C., Redondo, M., and Vargas, M. (2000) Control of ticks resistant to immunization with Bm86 in cattle vaccinated with the recombinant antigen Bm95 isolated from the cattle tick, *Boophilus microplus*. *Vaccine*, **18**, 2275–2287.

74. Amaral, M.A., Rocha, C.M., Faccini, J.L., Furlong, J. *et al.* (2011) Strategic control of cattle ticks: milk producers' perceptions. *Rev. Bras. Parasitol. Vet.*, **20**, 148–154.

75. da Rocha, C.M., Leite, R.C., Bruhn, F.R., Guimaraes, A.M. *et al.* (2011) Perceptions about the biology of *Rhipicephalus* (*Boophilus*) *microplus* among milk producers in Divinopolis, Minas Gerais. *Rev. Bras. Parasitol. Vet.*, **20**, 289–294.

5
Monitoring Drug Sensitivity in Cattle Ticks

Leonore Lovis, Christian Epe, Ronald Kaminsky, and Heinz Sager[*]

Abstract

Use of acaricides is one of the key strategies to protect cattle herds on pasture against tick infestation with *Rhipicephalus (Boophilus) microplus*. Unfortunately, the constant exposure to drugs has led to a selection of resistance to most of the chemical classes. Under these conditions it is even more important to know which drugs (or acaricide classes) can still be used and to what extent they provide protection. While testing drugs *in vivo* is costly and does not provide information on susceptibility to alternative acaricides, the *in vitro* tests seem more appropriate: they are less expensive and provide more data on the level of resistance against various chemical classes. Hence, resistance can be confirmed and differentiated using *in vitro* bioassays, as well as through the use of biochemical or molecular tests to detect metabolic- and target-site-based resistance, respectively. This chapter describes the currently available assays for determining the drug sensitivity status of tick populations, and discusses the advantages and limitations of the various assays.

Introduction

The cattle tick *Rhipicephalus* (*Boophilus*) *microplus* is widespread in tropical and subtropical countries and has a major economic impact on the cattle industry due to blood feeding as well as to transmission of pathogens. Because of its economic importance, different measures are taken to reduce the tick burden on the animals as well as to control the spread of the ticks. The currently available tools for tick control consist of chemical acaricides, of which different application methods and various formulations exist. In addition, nonchemical strategies consist of cattle breeds being less susceptible to tick infestation, vaccines against ticks and tick-borne diseases, as well as periodic rotations between livestock and crops [1].

[*]Corresponding author.

Ectoparasites: Drug Discovery Against Moving Targets, First Edition.
Edited by Charles Q. Meng and Ann E. Sluder

However, among these different options, chemical control is by far the most important and applied strategy to control *R. (B.) microplus* infestations.

Historically, the introduction of any new chemical acaricide class including the organochlorines (OCs), organophosphates (OPs), amidines (AMIs), synthetic pyrethroids (SPs), phenylpyrazoles, and macrocyclic lactones (MLs) has been followed by the emergence of resistance to those chemical classes (see Chapter 4). Acquisition of accurate information on the drug sensitivity of the involved tick populations is fundamental for recommending and applying optimal acaricide protection of cattle. Furthermore, once resistant tick populations have been identified, they may be used for the discovery of new acaricide classes with distinct modes of action to overcome the resistance to currently available products.

How to Identify Acaricide Resistance

Monitoring of cattle tick resistance to acaricides is essential on a local and global scale. It is important to recognize that tick control failure is not always due to resistance but can be caused by faulty equipment, inappropriate dosing, or use of forged products or expired chemicals. A confirmation of the observed lack of efficacy is therefore crucial in order to advise farmers on alternative chemical classes and other means of controlling the tick population. For the former, an evaluation of the susceptibility of the ticks to a spectrum of compounds is of utmost importance. Suspected resistance can be tested by retreating cattle (*in vivo*) with the same acaricide after ensuring that application procedures and doses are correct [2]. However, this procedure is costly and does not provide

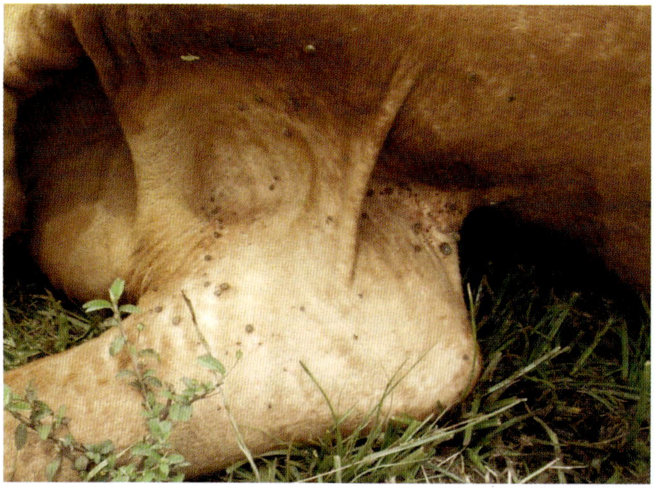

Figure 5.1 The cattle tick *R. (B.) microplus* has attachment sites where it may not be easily detected. In the present case, a heifer was captured on pasture and laid down for determination of the tick burden. Using such a technique before and after acaricide treatment to identify resistance is very resource intensive and time consuming.

information on susceptibility to alternative acaricides (Figure 5.1). Therefore, *in vitro* tests seem more appropriate, because they are less expensive and provide more information on the level of resistance against various acaricides. Hence, resistance can be confirmed and differentiated using *in vitro* bioassays, as well as through the use of biochemical or molecular tests to detect metabolic- and target-site-based resistance, respectively. The objective of this chapter is to describe the currently available assays for determining the drug sensitivity status of tick populations and to discuss the advantages and limitations of the various assays.

Bioassays

Bioassays are based on *in vitro* exposure of ticks (larvae or engorged females) to a single dose or to linear dose escalations of an acaricidal compound. The type of contact with the active ingredient (AI) and its duration differ among the tests. Use of bioassays in ectoparasiticide discovery is discussed in Chapter 8. In this chapter we review the particular application of bioassays in diagnosing acaricide resistance, for which an ideal bioassay should meet several requirements [1]: The diagnostic test should be sensitive enough to identify resistance early in its emergence and should cover the full range of chemical groups that are in use. The test should be simple and inexpensive, require a low number of engorged female ticks, and be time-effective. In addition, it should provide rapid and reliable results, and be suitable for standardization among laboratories in many countries. However, none of the currently available tests meet all these requirements.

In order to have unified standards, the Food and Agriculture Organization of the United Nations (FAO) adopted and recommended in 1975 the use of the larval packet test (LPT) bioassay [2], and provided a standardized protocol available since 1999. This test was recommended by the FAO for surveys and for definitive confirmation of a diagnosis of resistance despite some limitations due to its laborious nature. However, despite these efforts, in 2004 the FAO pointed out that "a lack of standardized techniques for diagnosing acaricide resistance appears to be the main difficulty in creating and maintaining a tick resistance monitoring system" [1]. A survey carried out by an FAO Working Group on Parasite Resistance (WGPR) revealed that the method most widely used to diagnose acaricide resistance in laboratories was not the LPT but an adult test called adult immersion test (AIT) [1]. As a consequence, the FAO decided to provide a protocol for the AIT and recommended this test for preliminary screening in addition to the LPT.

Two other *in vitro* tests which have proved to be useful for the determination of acaricide resistance are the larval immersion test (LIT) [3] and the larval tarsal test (LTT) [4]. Although neither test has been officially recommended by the FAO, both have proved their usefulness in various studies [5–8].

The key characteristics of the currently available tests are summarized here.

Adult Immersion Test

The AIT was first developed by Drummond [9]. The principle of the initial protocol was to expose engorged female ticks to a range of dilutions of an acaricide and to assess the effect of a treatment on fecundity and fertility, comparing treated and untreated ticks (Figure 5.2a). In the following years, several modifications were made to the test. The duration of exposure of the ticks to the acaricide varies between 30 s and 30 min and in many cases a single dose is used instead of a dilution series. The evaluation of oviposition and larval hatching is normally done after 4–5 weeks. In order to shorten the assay time, a recommended standardized assay focuses on oviposition 7 days after treatment [1]. However, various protocols of the AIT are currently in use, despite the efforts of the FAO for standardization.

Larval Packet Test

The LPT was first developed by Stone and Haydock [10]. In this test, tick larvae are exposed to chemically impregnated filter papers and their subsequent mortality is quantified after 24 h. The filter papers are folded and sealed with clips in order to form packets which harbor the larvae (Figure 5.2b). The handling of larvae to place them in the packets with a paintbrush is delicate due to their agility. In order to simplify the method, FAO recommended discriminating doses (DDs) of various compounds and even offered an acaricide resistance testing kit in the early years of 2000 [1]. Although this kit is no longer distributed and the application of the DD proposed by the FAO is not common practice, the standardized LPT protocol is widespread and, in contrast to the AIT, is carried out with only little variation between the different laboratories [1].

Larval Immersion Test

The LIT was first developed by Shaw [3] and later modified by Sabatini *et al.* [11]. In this test, tick larvae are immersed in acaricide dilutions and then incubated for 24 h before the assessment of mortality (Figure 5.2c). The exposure of the larvae to the respective compound is limited to 10 min, while they are kept for 24 h in

Figure 5.2 Schematic depiction of the most commonly used bioassays for *in vitro* identification of cattle tick resistance to acaricides. (a). AIT: Adult engorged female ticks are immersed in defined dilutions of acaricide. The ticks are dried, placed in Petri dishes and kept at 27 to 28 °C and 80–95% relative humidity (RH). After 2 weeks, eggs are weighed, transferred to tubes, and incubated at the before-mentioned climatic conditions for ~3 additional weeks to evaluate the larval hatching. In a simplified protocol, oviposition is evaluated approximately 1 week after incubation, without further evaluation of the larval hatching. (b). LPT: Tick larvae are incubated at 27 to 28 °C and 85–95% RH for 24 h in filter paper pouches that were impregnated with acaricides. The pouches are opened to visually determine the viability of the larvae. (c). LIT: The tick larvae are immersed for ~10 min in defined concentrations of acaricide. This is followed by a transfer into filter paper pouches (no impregnation with acaricide) in which the larvae are incubated for 24 h at 27 to 28 °C and 80–90% RH. The viability of the larvae is determined directly in the pouch by removing the clips, analogous to the LPT. (d). LTT: Tick eggs are placed into 96-well flat bottom plates which were pre-coated with defined concentrations of acaricides. The plates are sealed and incubated at 27 to 29 °C and 70–80% RH. Two weeks after egg hatching the larval mortality is determined in each well.

dry filter paper pouches similar to those described in the LPT. The LIT is currently mainly used for the detection of resistance to ivermectin and fipronil and was shown to perform more reliably than the LPT for the detection of resistance to these two compounds [5, 6].

Larval Tarsal Test

The LTT was first described by Lovis *et al.* [4]. In contrast to the larval tests described, it uses a microtiter-plate format to expose the ticks to serial dilutions

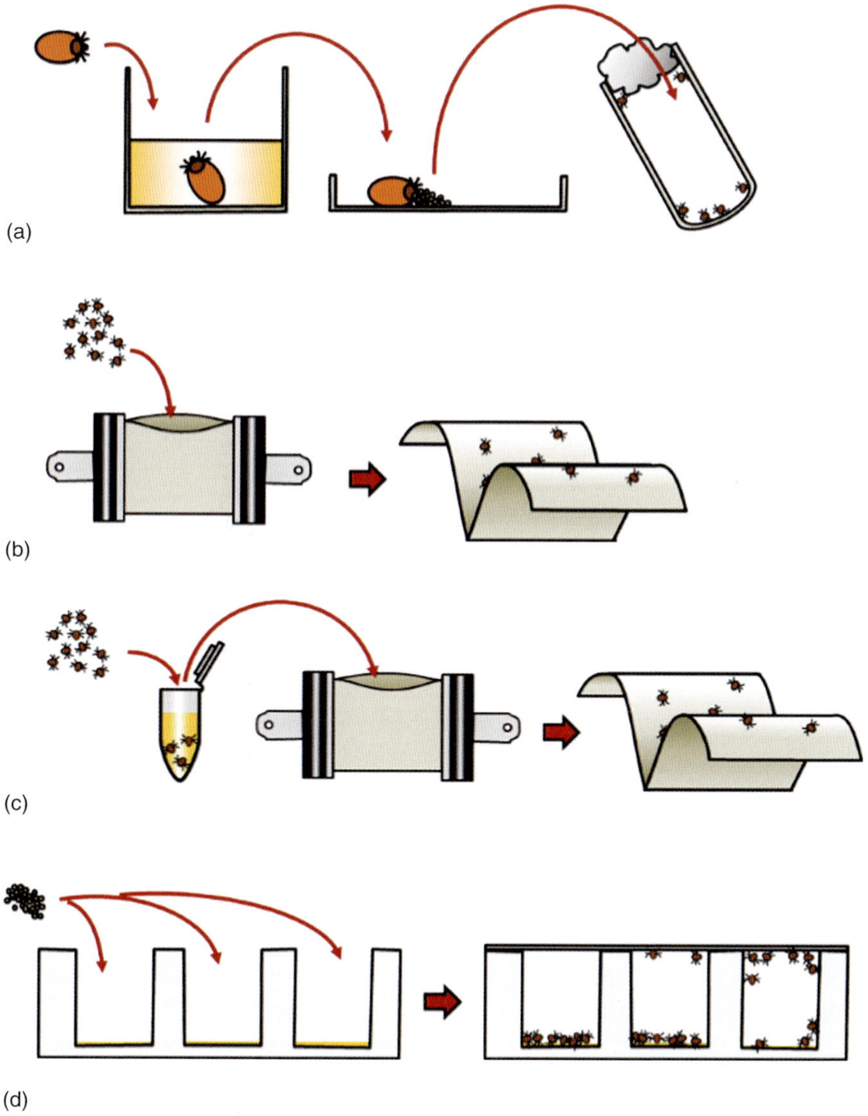

of acaricides and to evaluate viability (Figure 5.2d). The LTT allows testing of a high number of compounds at various doses with a minimum number of ticks. An important advantage of this system is the easy and safe handling, since no manipulations of larvae are required. The distribution of ticks is done with the eggs, and the plates are sealed afterwards. In addition, the concentrations of the acaricides used are considerably lower than in the LPT. One of the reasons preventing the test from a wider use is the required laboratory infrastructure for coating the plates with the compounds. Most diagnostic laboratories in Latin American countries do not have the required dry blocks for the evaporation of the solvents and thus far no commercial provider of pre-coated plates has emerged.

Biochemical Tools

Metabolic resistance can be diagnosed by measuring the activity of enzymes capable of contributing to known drug resistance mechanisms. Esterases have been shown to play an important role in the resistance to OPs and SPs [12–16]. Two categories of esterases are usually studied for potential OP- and SP-hydrolytic activity, namely, the carboxylesterases and the acetylcholinesterases. Potential involvement of esterases in metabolic resistance is investigated through studies measuring hydrolytic activity of esterases. To do so, total soluble proteins are extracted from larvae. The esterase activity can be quantified by different means such as gel electrophoresis or colorimetry [12, 16–18]. In addition, inhibitors can be used to determine the level of activity attributable to acetylcholinesterases or carboxylesterases separately [16, 19].

Molecular Tools

There are two prerequisites for the development of molecular tools to diagnose target-site resistance. First of all, the target site (receptor or enzyme) of the corresponding compound must be known. Secondly, mutations conferring resistance to the compound must have been identified in the target. In *R. (B.) microplus*, these conditions are met only for SPs, where four single nucleotide substitutions have been identified in the voltage-gated sodium channel gene of *R. (B.) microplus* (Figure 5.3) [20–23]. Polymerase chain reaction (PCR) assays have been developed to detect the presence of these mutations: Guerrero *et al.* [24] designed a PCR assay for the detection of the domain III mutation identified by He *et al.* [23]; and Morgan *et al.* [21] developed a PCR assay for the domain II mutation described in 2009. More recently, a multiplex PCR using allele-specific primers to amplify wild-type or mutated genotypes of all three previously described mutations was developed by Lovis *et al.* [25]. Finally, Stone *et al.* [20] used a quantitative PCR to identify resistant genotypes, including the latest (fourth) mutation they identified in domain II.

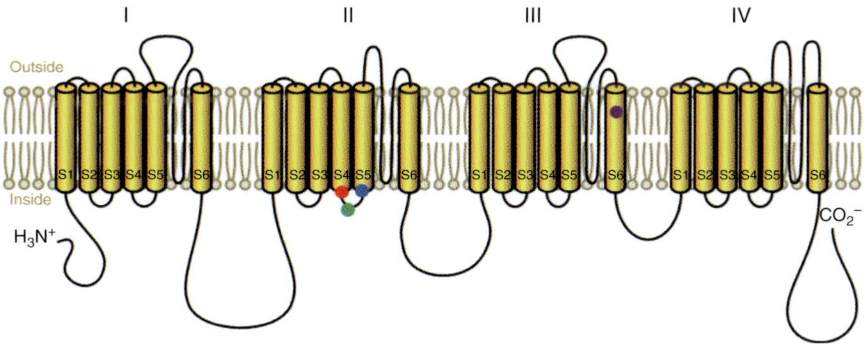

Figure 5.3 Position of the so far identified mutations in the voltage-gated sodium channel of *R. (B.) microplus,* which lead to resistance to synthetic pyrethroids. Three mutations lead to changes in the S4-S5 linker of domain II: T170C (red) [20], C190A (green) [21], and G215T (blue) [22]; and one in the S6-region of domain III: T2134A (violet) [23].

The availability of *in vitro* methods and of molecular tools to determine the resistance status of ticks is of great importance, from the scientific as well as from the economic viewpoint. However, although the available bioassays are very helpful (and some of them are even recommended by the FAO), each of them has limitations which restrict the number of compounds and doses that can be tested. The molecular tests, on the other hand, are very specific and are currently not or only rarely used for routine diagnosis, especially in developing markets where laboratory infrastructure is limited. Taking this into consideration, there is a need to develop new tests that are easy to perform, require only basic infrastructure, and provide a maximum of information on acaricide resistance with a limited number of ticks.

Use of Resistance Bioassays in the Discovery of New Acaricides

The requirements on a bioassay for identification of new acaricide candidates differ considerably from those used for identification of acaricide resistance in cattle ticks isolated in the field. While for the latter the constraints are given by the number of collected engorged females and the laboratory infrastructure, the former often use large numbers of ticks produced under standardized conditions and the tests are run in an automated way. Since most of the compounds are down-titrated, tests similar to the LTT seem most suitable for such an approach. Nevertheless, in this setup only the topical activity of a compound is evaluated. If a systemic activity is expected, artificial blood feeding systems would be required or the compound would need to be injected into the tick; this would most likely be done in engorged females.

For pivotal studies testing the efficacy of new acaricide candidates it is recommended – and requested by most authorities – to use recent field isolates for artificial infestations of cattle [26]. No such requirement exists for ticks used in bioassays for the discovery of new acaricides. However, given the widespread presence of acaricide resistance, it would make sense to include ticks with confirmed acaricide resistance at an early stage. This would help identify drug candidates with similar modes of action or modes of resistance (cross-resistance) before testing them *in vivo*. The problem of including recent field isolates in automated test systems is the need of recalibration of each test in case of changing parameters. This is labor intensive and is normally kept at a minimum. A feasible approach would be to use the (high-throughput) standard test using the lab-isolate for screening of new compounds and only select positive candidates for further evaluation, which includes testing the activity against (recent) resistant tick isolates. A characterization of the isolates with bioassays is a prerequisite. Especially due to the fact that not all tests are equally sensitive for the known acaricide classes, one should not refer to a single method. More information on the use of bioassays in ectoparasiticide discovery can be found in Chapter 8.

Resistance Management

The availability of tools to determine acaricide resistance in ticks is a prerequisite for giving recommendations to farmers for successfully treating or protecting cattle in the short term and on reducing the risk of further development of resistance in the long term. This applies for the current acaricides as well as for any new one. When a new class of compounds is introduced in the market, it is only a question of time before resistance appears. An additional problem is the persistence of resistance to a compound class once it is established in a population. Today's assumption is that reversion of acaricide resistance in ticks is unlikely to occur. To ensure the sustainable use of the acaricidal compounds that are currently available and still effective and to extend their life span, information for farmers on ticks, tick control, tick resistance, and integrated management programs is essential. This is discussed in more detail elsewhere in this volume.

References

1. FAO (2004) Ticks: acaricide resistance: diagnosis, management and prevention, in Resistance Management and Integrated Parasite Control in Ruminants: Guidelines, FAO Animal Production and Health Division, Rome, pp. 25–77. http://www.fao.org/tempref/docrep/fao/010/ag014e (accessed 04 November 2016).
2. Kemp, D.H., Thullner, R., Gale, W.G., Nari, A., *et al.* (1998) Acaricide resistance in the cattle-ticks *Boophilus microplus* and *B. decoloratus*: review of resistance data; standardisation of resistance tests and recommendations for integrated parasite control to delay resistance. Report to the Animal Health Services, AGAH, FAO.
3. Shaw, R.D. (1966) Culture of an organophosphorus-resistant strain of *Boophilus microplus* (Can.) and an assessment of its resistance spectrum. *Bull. Entomol. Res.*, **56**, 389–405.
4. Lovis, L., Perret, J.L., Bouvier, J., Fellay, J.M. *et al.* (2011) A new *in vitro* test to evaluate the resistance level against acaricides of the cattle tick,

Rhipicephalus (*Boophilus*) *microplus*. *Vet. Parasitol.*, **182**, 269–280.
5. Castro-Janer, E., Rifran, L., Piaggio, J., Gil, A. *et al.* (2009) *In vitro* tests to establish LC50 and discriminating concentrations for fipronil against *Rhipicephalus* (*Boophilus*) *microplus* (Acari: Ixodidae) and their standardization. *Vet. Parasitol.*, **162**, 120–128.
6. Klafke, G.M., Castro-Janer, E., Mendes, M.C., Namindome, A. *et al.* (2012) Applicability of *in vitro* bioassays for the diagnosis of ivermectin resistance in *Rhipicephalus microplus* (Acari: Ixodidae). *Vet. Parasitol.*, **184**, 212–220.
7. Lovis, L., Mendes, M.C., Perret, J.L., Martins, J.R. *et al.* (2013) Use of the larval tarsal test to determine acaricide resistance in *Rhipicephalus (Boophilus) microplus* Brazilian field populations. *Vet. Parasitol.*, **191**, 323–331.
8. Lovis, L., Reggi, J., Berggoetz, M., Betschart, B. *et al.* (2013) Determination of acaricide resistance in *Rhipicephalus (Boophilus) microplus* (Acari: Ixodidae) field populations of Argentina, South Africa, and Australia with the larval tarsal test. *J. Med. Entomol.*, **50**, 326–335.
9. Drummond, R.O., Ernst, S.E., Trevino, J.L., Gladney, W.J. *et al.* (1973) *Boophilus annulatus* and *B. microplus*: laboratory tests of insecticides. *J. Econ. Entomol.*, **66**, 130–133.
10. Stone, B.F. and Haydock, P. (1962) A method for measuring the acaricide susceptibility of the cattle tick *Boophilus microplus*. *Bull. Entomol. Res.*, **53**, 563–578.
11. Sabatini, G.A., Kemp, D.H., Hughes, S., Nari, A. *et al.* (2001) Tests to determine LC50 and discriminating doses for macrocyclic lactones against the cattle tick, *Boophilus microplus*. *Vet. Parasitol.*, **95**, 53–62.
12. Jamroz, R.C., Guerrero, F.D., Pruett, J.H., Oehler, D.D. *et al.* (2000) Molecular and biochemical survey of acaricide resistance mechanisms in larvae from Mexican strains of the southern cattle tick, *Boophilus microplus*. *J. Insect Physiol.*, **46**, 685–695.
13. Pruett, J.H., Guerrero, F.D., and Hernandez, R. (2002) Isolation and identification of an esterase from a Mexican strain of *Boophilus microplus* (Acari: Ixodidae). *J. Econ. Entomol.*, **95**, 1001–1007.
14. Villarino, M.A., Waghela, S.D., and Wagner, G.G. (2003) Biochemical detection of esterases in the adult female integument of organophosphate-resistant *Boophilus microplus* (Acari: Ixodidae). *J. Med. Entomol.*, **40**, 52–57.
15. Baffi, M.A., de Souza, G.R., de Sousa, C.S., Ceron, C.R. *et al.* (2008) Esterase enzymes involved in pyrethroid and organophosphate resistance in a Brazilian population of *Rhipicephalus* (*Boophilus*) *microplus* (Acari, Ixodidae). *Mol. Biochem. Parasitol.*, **160**, 70–73.
16. Baffi, M.A., de Souza, G.R., Vieira, C.U., de Sousa, C.S. *et al.* (2007) Identification of point mutations in a putative carboxylesterase and their association with acaricide resistance in *Rhipicephalus* (*Boophilus*) *microplus* (Acari: Ixodidae). *Vet. Parasitol.*, **148**, 301–309.
17. Rosario-Cruz, R., Guerrero, F.D., Miller, R.J., Rodriguez-Vivas, R.I. *et al.* (2005) Roles played by esterase activity and by a sodium channel mutation involved in pyrethroid resistance in populations of *Boophilus microplus* (Acari: Ixodidae) collected from Yucatan, Mexico. *J. Med. Entomol.*, **42**, 1020–1025.
18. Soberanes-Céspedes, N., Rosario-Cruz, R., Santamaria-Vargas, M., and Garcia-Vazquez, Z. (2005) General esterase activity variation in the cattle tick *Boophilus microplus* and its relationship with organophosphate resistance. *Téc. Pecu. Méx.*, **43**, 239–246.
19. Miller, R.J., Davey, R.J., and George, J.E. (1999) Characterization of pyrethroid resistance and susceptibility to coumaphos in Mexican *Boophilus microplus* (Acari: Ixodidae). *J. Med. Entomol.*, **36**, 533–538.
20. Stone, N.E., Olafson, P.U., Davey, R.B., Buckmeier, G. *et al.* (2014) Multiple mutations in the para-sodium channel gene are associated with pyrethroid

resistance in *Rhipicephalus microplus* from the United States and Mexico. *Parasit. Vectors*, **7**, 456–469.

21. Morgan, J.A., Corley, S.W., Jackson, L.A., Lew-Tabor, A.E. *et al.* (2009) Identification of a mutation in the para sodium channel gene of the cattle tick *Rhipicephalus* (*Boophilus*) *microplus* associated with resistance to synthetic pyrethroid acaricides. *Int. J. Parasitol.*, **39**, 775–779.
22. Jonsson, N.N., Cutulle, C., Corley, S.W., and Seddon, J.M. (2010) Identification of a mutation in the para-sodium channel gene of the cattle tick *Rhipicephalus microplus* associated with resistance to flumethrin but not to cypermethrin. *Int. J. Parasitol.*, **40**, 1659–1664.
23. He, H., Chen, A.C., Davey, R.B., Ivie, G.W. *et al.* (1999) Identification of a point mutation in the para-type sodium channel gene from a pyrethroid-resistant cattle tick. *Biochem. Biophys. Res. Commun.*, **261**, 558–561.
24. Guerrero, F.D., Davey, R.B., and Miller, R.J. (2001) Use of an allele-specific polymerase chain reaction assay to genotype pyrethroid resistant strains of *Boophilus microplus* (Acari: Ixodidae). *J. Med. Entomol.*, **38**, 44–50.
25. Lovis, L., Guerrero, F.D., Miller, R.J., Bodine, D.M. *et al.* (2012) Distribution patterns of three sodium channel mutations associated with pyrethroid resistance in *Rhipicephalus* (*Boophilus*) *microplus* populations from North and South America, South Africa and Australia. *Int. J. Parasitol. Drugs Drug Resist.*, **2**, 216–224.
26. Holdsworth, P.A., Kemp, D., Green, P., Peter, R.J. *et al.* (2006) World Association for the Advancement of Veterinary Parasitology (W.A.A.V.P.) guidelines for evaluating the efficacy of acaricides against ticks (Ixodidae) on ruminants. *Vet. Parasitol.*, **136**, 29–43.

6
New Developments in the Control of Human Lice

*John M. Clark**

Abstract

Human head and body lice represent one of the longest and most prevalent parasitic infestations of humans. While head lice are an economic and social concern, body lice pose a more serious public health concern. In the past 10 years, significant progress has been made in the study of human lice and the information generated has led to a number of new developments for their control. The *in vitro* rearing of head and body lice has allowed the establishment of insecticide-susceptible and -resistant reference strains, which have allowed more formal descriptions of pediculicide resistance, its underlying mechanisms, and the detection and monitoring of resistance. Likewise, the availability of inbred strains has allowed the efficient sequencing, assembly, and annotation of the genomes and transcriptomes of both lice. This information enabled the use of functional genomics and reverse genetics to study the genes involved in the evolution of resistance, the odorant response to attractants and repellents, and the discovery of novel target sites for the development of new pediculicidal/ovicidal chemicals and metabolic synergists. With these tools and techniques, we are now poised for the first time to make substantial advancements in providing society with effective treatments for pediculosis in a sustainable resistance management format.

Introduction

Both the human head louse (*Pediculus humanus capitis*) and the body louse (*Pediculus humanus humanus*) are obligatory ectoparasites that feed only on human blood and together represent one of the longest ectoparasitic relationships associated with mankind [1]. Body lice diverged from head lice when humans began wearing clothing ~40 000–70 000 years ago [2]. Genomic and transcriptomic comparisons show that these two lice are highly related genetically

*Corresponding author.

Ectoparasites: Drug Discovery Against Moving Targets, First Edition.
Edited by Charles Q. Meng and Ann E. Sluder

[3, 4] and can interbreed under laboratory conditions [5, 6]. Nevertheless, there are significant differences between head and body lice, and these differences have and continue to have significant impacts on humans.

Head and body lice have different living and feeding behaviors, morphologies, and disease vectoring abilities. Head lice live only on the human scalp. Females attach their eggs (ova) to the base of hair shafts, eggs hatch leaving an empty egg case (nit), and the first instar larva undergo three successive molts, resulting in a reproductive adult. All stages feed numerous times per day using their piecing-sucking mouth parts. In contrast, body lice feed on the non-head regions of the human body and the females secure eggs to clothing, where they usually reside, only coming onto the human body intermittently to feed. Body lice are generally larger than head lice as a result of taking larger but less frequent blood meals [7, 8]. Head lice do not transmit human diseases, whereas body lice are competent vectors of a number of bacterial diseases to humans.

In this chapter, new developments that have occurred over the past 10 years are summarized in terms of the impact that they have had on the control of human lice. This time frame was chosen because it encompasses the genomic sequencing of both the head and body louse genomes, the development of an *in vitro* louse-rearing system, the use of functional genomics and reverse genetics to study pediculicidal resistance, the discovery of new target sites for pediculicidal action and repellency, and the marketing of new, effective, and novel acting pediculicides with the possibility of implementing sustainable resistance management strategies. To accomplish this in an efficient manner, the information in a number of excellent recent reviews is used to guide our summarization [9–14].

Pediculosis and Medical Importance of Head and Body Lice

Pediculosis is the infestation of humans by lice and is one of the most prevalent parasitic infestations of humans. The costs of pediculosis in the United States are estimated at >$367 million USD/yr [12] and infestation rates range from 6 to 12 million cases annually, with 2.6 million households affected and 8% of all schoolchildren infested [15].

Although not a vector of disease, head lice represent a major economic and social concern worldwide [16]. Infestations can cause intense itching, which can injure the skin allowing secondary infections [17]. Unlike head lice, body lice pose a serious public health threat because they transmit several bacteria (*Rickettsia prowazekii, Borrelia recurrentis,* and *Bartonella quintana*) that cause human diseases (epidemic typhus, louse-borne relapsing fever, and trench fever, respectively) [18]. Since the advent of antibiotics, outbreaks are sporadic but do occur particularly during times of war, famine, and social unrest; and the body louse still serves as an important vector of reemerging diseases in developed countries [19].

Most people find lice intolerable and repeatedly and prophylactically apply pediculicides (insecticides) without realizing their harm and possible lethality. Misapplications affect children in particular due to their small size and higher sensitivity.

There are two ways to combat pediculosis: (i) proactive prevention or (ii) post-infestation treatment. Emphasis is increasingly on prevention (education) and physical removal (combing or shaving) because a crisis exists in the chemical management of pediculosis. The pediculicide arsenal is limited and shrinking due to insecticide resistance. Effective management information is hard to find and few alternatives exist when standard treatments fail. Thus, there is a critical need for biological, biochemical, and molecular information that can be used to implement sustainable novel lice control strategies, including the understanding of mechanisms of pediculicide resistance and monitoring of resistance, the use of genetic information to identify new and unique target sites, and the development of new and novel acting pediculicides.

Pediculicides and Resistance

Over the past 70 years, the control of pediculosis has been largely dependent upon the availability of natural and synthetic insecticides starting with dichlorodiphenyltrichloroethane (DDT) (1943), natural pyrethrins (1945), the organochlorine lindane (1960), organophosphorous insecticides (malathion, 1971), carbamates (carbaryl, 1977), and synthetic pyrethroids (permethrin, phenothrin, 1992) [20].

In the United States, the pyrethrins/pyrethroids have dominated the over-the-counter (OTC) market, followed by the prescription-only malathion-containing formulations, such as Ovide®. The pyrethrins/pyrethroids share a common target site in the nervous system, the voltage-sensitive sodium channel (VSSC), and act as agonistic neuroexcitants by increasing inward sodium current, leading to nerve depolarization and hyperexcitation, followed by neuromuscular paralysis and death. Malathion is a phosphorodithioate-type organophosphorous insecticide, which is an indirect nerve toxin that acts as a competitive irreversible inhibitor of acetylcholinesterase associated with the cholinergic nervous system. When inhibited, acetylcholinesterase cannot efficiently hydrolyze the neurotransmitter, acetylcholine, allowing overstimulation of post-synaptic effector organs, including muscle, leading to paralysis and death.

Insecticide resistance to currently used pediculicides, including permethrin, synergized pyrethrins, and malathion, has occurred worldwide, is increasing [21–24], and is certainly contributing to increased incidences of pediculosis.

Both clinical and parasitological pyrethroid resistance to *d*-phenothrin was first reported in France in 1994 [24] with additional reports of clinical control failures following: permethrin (2001) in the United States [25], phenothrin (1995) in the United Kingdom [26], and permethrin (2005) in the United Kingdom [27]. Also, parasitological resistance has been reported in the Czech Republic [28], the United Kingdom [26], Denmark [29], Israel [30], the United States [31], Argentina [32], Japan [33], and Australia [34].

Malathion resistance was first reported in France in 1995 [35], followed by the United Kingdom in 1999 [36], Australia in 2003 [34], and Denmark in 2006 [29].

The lack of extensive resistance in the United States is likely due to the use of the Ovide® formulation, which also includes pediculicidal terpenes likely resulting in a mixture of active compounds that has redundant killing action on multiple target sites [37].

Current control and resistance problems underscore the need to understand the molecular mechanisms of insecticide resistance in lice. The identification of resistance mechanisms and novel target sites may allow the development of resistance-breaking compounds and specific nontoxic synergists useful in the implementation of novel control and resistance management strategies.

Development of the *In Vitro* Rearing System: Maintenance of Insecticide-susceptible and -resistant Strains and Determination of Resistance

An improved *in vitro* rearing system was developed on the basis of modifications to a manual prototype, which allowed for the first time the sustainable maintenance of head and body lice without human infestation [38]. The improved system, based on a silicone-reinforced Parafilm® membrane, human hair tufts, and reconstituted human blood, enabled the large-scale rearing of pediculicide-susceptible and -resistant strains of lice in a semi-sterile condition [39] (Figure 6.1).

The efficacies of three commercially available OTC formulations (Nix®, Rid®, Proto® Plus) were assessed by applying the products directly to the hair tufts with

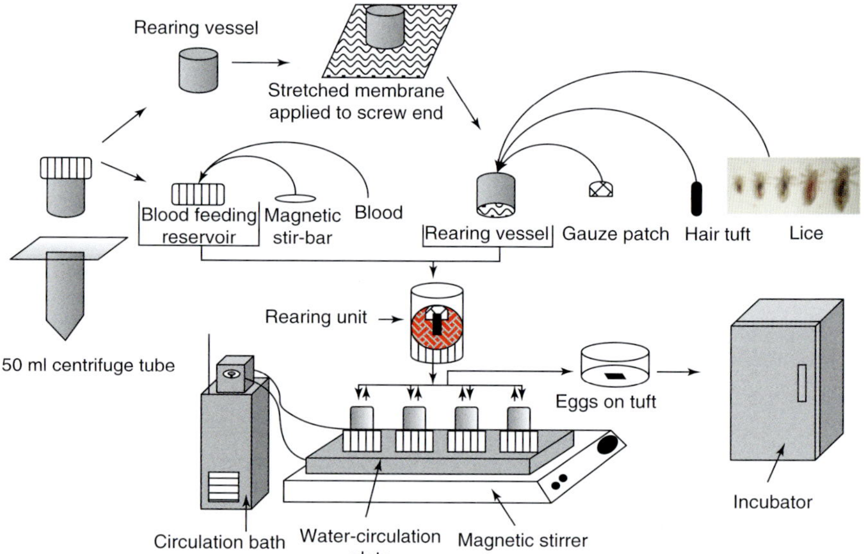

Figure 6.1 Assembly of the *in vitro* rearing system for lice. (Reproduced with permission from Yoon *et al.* [39]. Copyright 2006, Elsevier.)

lice attached following the manufacturers' instructions. All products were highly effective (100% mortality) in the pediculicide-susceptible strain but differentially efficacious (62–84% mortality) in the pediculicide-resistant strain, validating previous anecdotal reports of resistance to permethrin- and pyrethrin-based pediculicide formulations.

Sequencing of the Human Louse Genomes and Transcriptomes

Estimation of nuclear deoxyribonucleic acid (DNA) content by flow cytometry indicated that both head and body lice had small genomes compared to other insects, spanning only 108 Mb [40]. Sequencing of the body louse genome validated this finding and revealed that despite its small size the genome retained a remarkably complete basal insect repertoire of 10775 predicted genes [41]. Evolutionary reduction of the genome size relative to other insects was accomplished by removing intergenic DNA, reducing the size and number of introns within genes, and reducing the number of genes within large gene families, particularly those involved in environmental sensing and response, including odorant and gustatory receptor, detoxification enzyme, and innate immune response genes [41–43].

Comparison of the transcriptional profiles of body and head lice using expressed sequence tags identified 10771 body and 10770 head louse transcripts [4]. Illumina sequence reads were mapped to the 10775 body louse gene models and identified nine presence/absence differences between the two transcript sets. Only one gene difference between the two transcriptomes was determined, a hypothetical protein with no function, indicating that these two organisms share virtually the same genome and are likely ecotypes of the same species. Interestingly, the numbers of detoxification genes involved in xenobiotic metabolism (e.g., cytochrome P450 monooxygenases, glutathione-S-transferases, esterases) were dramatically reduced in both head and body lice compared with other insects, indicating that the decreased number of detoxification genes and small genome size would make human lice an efficient model to study insecticide resistance [42].

Recently, the whole genome sequence of the head louse was determined by next-generation sequencing methods and compared to the reference genome sequences of the body louse [3]. Results indicated that the number of genes in the body louse genome may be an underestimation and that both genomes appear to possess at least 10790 genes. In addition, the nucleotide diversity between the head and body louse genomes was determined to be larger than that reported initially using a transcriptome comparison [4], suggesting that head and body lice are evolving to separate species from their status as con-species [44] or ecotypes of the same species [45].

With this information in hand, it became apparent that human lice could serve as an efficient model system to study (i) the molecular mechanisms of insecticide resistance and use of this information in monitoring of resistance; (ii) how xenobiotic metabolism, which is involved in insecticide resistance, is induced and how this information might be used in 'proactive' resistance monitoring; and (iii) selective new target sites for the development of novel acting pediculicides, repellants/deterrents, and metabolic synergists.

Head Louse Resistance to Pyrethrins, Pyrethroids, and Malathion

Three Point Mutations in the α-subunit Gene of the VSSC Cause Knockdown Resistance (*kdr*) and can be used for Monitoring the Extent and Magnitude of Resistance

Lee *et al.* [46] first reported that head lice were resistant to a pyrethroid, permethrin, and exhibited *in vivo* responses in behavioral bioassays that were consistent with *kdr*. *Kdr* is a heritable trait associated with nerve insensitivity to DDT, the pyrethrins, and the pyrethroids [47] and point mutations in *VSSC* genes are functionally responsible for the *kdr*, *kdr*-type, and super-*kdr* traits [48].

Three point mutations located in the domain IIS1-2 extracellular loop (M815I) and in the domain IIS5 transmembrane segment (T917I and L920F) of VSSC α-subunit (numbered according to the head louse amino acid sequence) were identified in permethrin-resistant head lice [49]. Yoon *et al.* [50] inserted the three mutations associated with pyrethroid resistance in the head louse (MI, TI, and LF) in all possible combinations into the corresponding positions of the house fly $Vssc1^{WT}$ sequence, expressed wild-type and specifically mutated channels along with the house fly Vsscβ auxiliary subunit in *Xenopus* oocytes, and employed the two electrode voltage-clamp technique to electrophysiologically assess the impact of these mutations on permethrin sensitivity of the expressed channels (Figure 6.2). In the absence of the three mutations and their corresponding amino acid replacements, a dose-dependent increase in the late current seen during inactivation and a prolongation of the tail current seen during deactivation were apparent at increasing concentrations of permethrin. In the presence of the three amino acid replacements, superimposed current traces obtained at increasing concentrations of permethrin were indistinguishable from dimethyl sulfoxide (DMSO) control traces, confirming that the MITILF haplotype results in target-site insensitivity of the VSSC and contributes to permethrin resistance in the head louse.

Monitoring the Allele Frequency of *kdr* in North America

The extent and frequency of a *kdr*-type resistance allele in North American populations of head lice were initially determined from lice collected from 32 locations in Canada and the United States [51]. Using the serial invasive signal amplification (SISAR) technique to detect the frequency of the *kdr*-type T917I mutation (TI), it was found that TI occurs at high levels in North American lice (94.1%). The TI frequency in the US lice from 1999 to 2009 was 84.4%, increased to 99.6% from 2007 to 2009, and was 97.1% in Canadian lice in 2008. The authors of the study cautioned, however, that their results were preliminary (based only on the

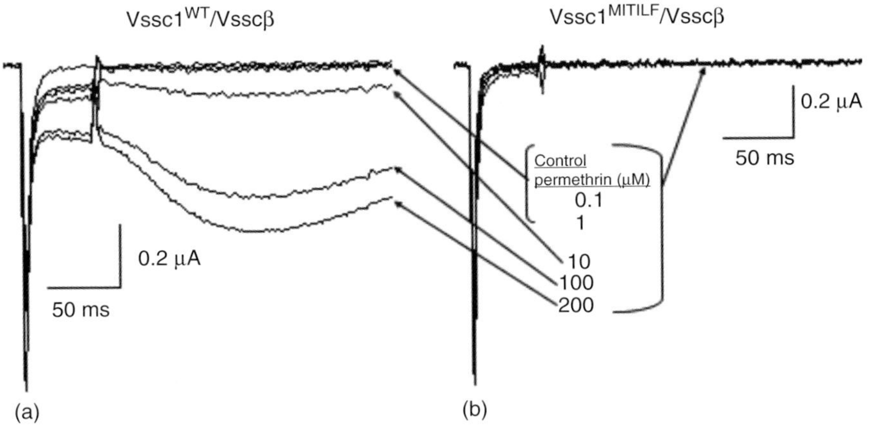

Figure 6.2 Comparative sodium current traces from the house fly VSSC variants with and without head louse mutations expressed in *Xenopus* oocytes before and after exposure to increasing concentrations of permethrin. (Reproduced with permission from Yoon *et al.* [50]. Copyright 2008, Elsevier.)

TI mutation) and perhaps biased due to the small number of lice analyzed and because most of the lice were collected from metropolitan and urban collection sites.

In response, a subsequent study expanded on the existing *kdr*-map [52]. It utilized three quantitative sequencing (QS) reactions to determine the *kdr*-type mutation frequency at each of the three alleles (MI, TI, and LF). Lice from 138 geographical collection sites, ranging from rural to metropolitan areas, were collected from 48 US states. Mean percent resistance allele frequency values across the three mutation loci (mean % RAF) were determined from each collection site. The overall mean % RAF (±S.D.) for all analyzed lice was 98.3 ± 10%. Of the 138 sites, 132 (95.6%) had a mean % RAF of 100%, 5 sites (3.7%) had intermediate values, and only a single site had no mutations (0.0%). Forty-two states (88%) had mean % RAF of 100%. The frequencies of *kdr*-type mutations did not differ regardless of the size of the human population from which the lice were collected, indicating a uniformly high level of resistant alleles. The loss of efficacy of the Nix® formulation from 1998 to 2013 was correlated to the increase in *kdr*-type mutations. These data provide a plausible reason for the decrease in the effectiveness of permethrin in the Nix® formulation, which is the parallel increase of *kdr*-type mutations in lice over time.

Thus, the frequency of *kdr*-type alleles in North American head louse populations was determined to be uniformly high, apparently due to the high selection pressure from the intensive and widespread use of the pyrethrins/pyrethroid-based pediculicides over many years, and is likely the main cause of increased pediculosis and failure of pyrethrins/permethrin-based products in Canada and the United States.

Malathion Resistance Is Due to Enhanced Hydrolytic Ester Cleavage by Malathion Carboxylesterase

Enhanced malathion carboxylesterase (MCE) activity was previously reported to be involved in malathion resistance in the head louse [23]. To identify the MCEs involved, the transcriptional profiles of five catalytically active esterases were determined and compared between the malathion-resistant (BR-HL) and malathion-susceptible (KR-HL) strains [53]. Only one esterase gene, *HLCbE3*, exhibited a significantly higher transcription level (5.4-fold) in the resistant BR-HL strain. Comparison of the entire complementary deoxyribonucleic acid (cDNA) sequences of *HLCbE3* revealed no sequence differences between the BR-HL and KR-HL strains. However, two copies of the *HLCbE3* gene were found in BR-HL, implying that over-transcription of *HLCbE3* is due to the combination of a gene duplication and upregulated transcription. Knockdown of HLCbE3 expression by RNA interference (RNAi) in the BR-HL strain caused increased malathion susceptibility, confirming the identity of HLCbE3 as the MCE responsible for malathion resistance in the head louse.

Optimization of the Noninvasive Induction Assay to Identify Detoxification Genes Involved in Insecticide Tolerance as a Proactive Resistance Monitoring Approach

Identifying xenobiotic detoxification genes based on insecticide-induced transcript profiles of insects has been suggested as a means of identifying metabolic pathways involved in insecticide resistance [42]. Initial pilot studies using *Drosophila melanogaster* did identify a number of detoxification genes, but most were not involved in insecticide metabolism [54].

The ability to identify detoxification genes that metabolize insecticides during the process of induced tolerance, prior to resistance evolving, would be a major step forward in resistance management because the expression of such genes could then be used proactively to monitor for metabolic resistance. In a proof-of-principle experiment, the transcriptional profiling results using an "optimized" noninvasive induction assay [short exposure intervals (2–5 h) to sublethal amounts of insecticides ($<LD_3$ at 24 h) administered by stress reducing means (contact vs. immersion screen) and with induction assessed in a time frame when tolerance is still present ($\sim LC_{90}$ in 2–4 h)] efficiently identified ivermectin-induced detoxification genes from body lice [55]. The cytochrome P450 monooxygenases (*CYP6CJ1*, *CYP9AG1*, and *CYP9AG2*) and ATP-binding cassette transporter (*PhABCC4*) genes were the most significantly over-expressed, had high basal expression levels, and were most closely related to genes from other organisms that metabolized insecticides, including ivermectin. Injection of dsRNAs against either *CYP9AG2* or *PhABCC4* into non-induced female lice reduced their respective transcript level and resulted in increased sensitivity to ivermectin, indicating that these two genes are involved in the xenobiotic

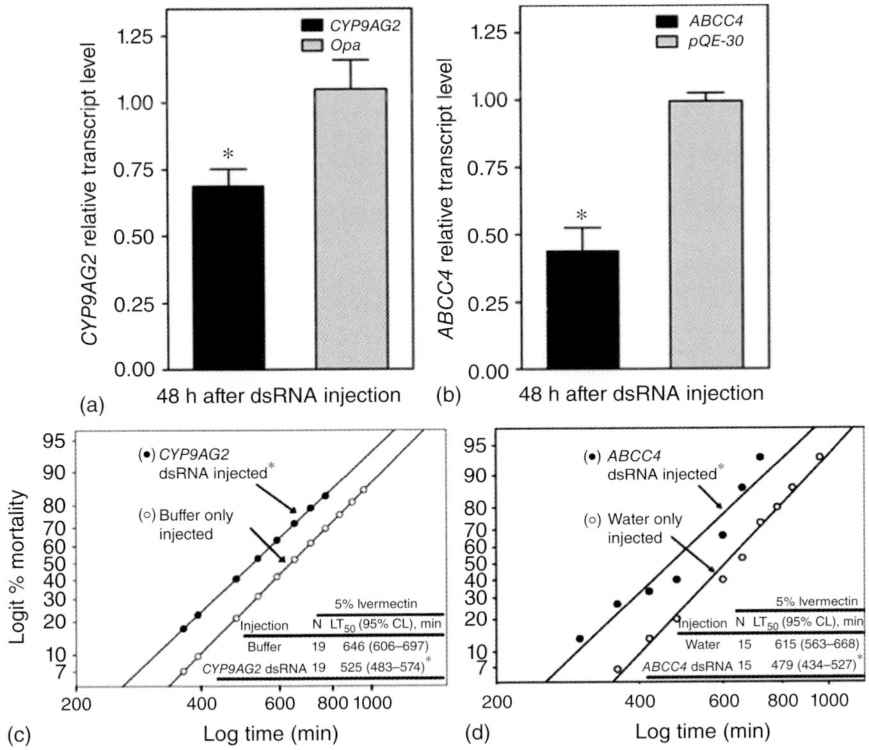

Figure 6.3 Relative transcript levels (panel a and b) and mortality responses (panel c and d) of body louse females to a lethal contact amount of ivermectin (5% IVM) following injection of dsRNA targeting either louse *CYP9AG2* or *ABCC4*. Lice were also injected with either dsRNA of the odd-paired gene, *opa*, (*GeneBank* accession # S78339) for P450 silencing or with dsRNA of the *E. coli* plasmid, pQE30, for ABC transporter silencing as sham injected controls. Asterisks (*) in panels a and b indicate that *CYP9AG2 and ABCC4* dsRNA significantly suppress the levels of *CYP9AG2 and ABCC4* transcripts, respectively (Student's *t*-test, $P < 0.05$). In panel c, the bioassay was started 48 h after *CYP9AG2* dsRNA injection. In panel d, the bioassay was started 12 h after *ABCC4* dsRNA injection. Asterisks (*) in panels c and d indicate that the mortality responses of lice injected with dsRNAs were significantly different from their respective controls (buffer or water only injected, maximum log-likelihood ratio test, $P < 0.05$). (Reproduced with permission from Yoon *et al.* [55]. Copyright 2011, John Wiley & Sons.)

metabolism of ivermectin and in the production of tolerance (Figure 6.3). Once identified, they can be used in proactive resistance monitoring schemes (e.g., quantitative, real-time, polymerase chain reaction) and in the construction of metabolic maps to establish cross- and negative cross-expression patterns during the acquisition of tolerance following induction. Such information is critical in establishing effective mixtures to be used in proactive resistance management of pediculosis. Certainly, the substantially reduced number of xenobiotic detoxification genes in the louse genome facilitated this process.

New Pediculicides, Infestation Deterrents/Repellents, and Metabolic Synergists with Novel Modes of Action

New pediculicides need to be safe, rapidly eliminate live lice and viable eggs, show no cross-resistance to other products, and be easily used and affordable [12]. The following list of formulations includes commercially available pediculicidal treatments that have been recently registered either in the United States or elsewhere, or are currently being registered for use. They all have unique chemicals as their active ingredients, the target sites of which do not overlap with those of the OTC- or malathion-containing formulations discussed earlier, indicating that cross-resistance is not likely. Only products that have had their active ingredient identified are discussed here.

Benzyl Alcohol-containing Formulations

A topical 5% benzyl alcohol formulation (Ulesfia®; Concordia, Pharmaceuticals, Inc., Bridgetown, Barbados) was registered in the United States in 2007 as a prescription treatment of children older than 6 months [12]. Although intravenous exposure to benzyl alcohol has been associated with neonatal gasping syndrome, this issue is unlikely given the product's labeled use protocols [14]. Ulesfia® is not neurotoxic to lice but appears to affect the spiracles controlling air movement in the louse respiratory system [56]. The product is not ovicidal so a second application 7 to 10 days following the first is needed for full efficacy.

Dimethicone-based Formulations

There has been a trend, primarily in Europe, toward the development of physical means to control head lice because of increasing instances of resistance, particularly to the neurotoxic pediculicides, and the increased scrutiny of the use of such products on children. The dimethicone-based anti-louse products (silicone oils) are of interest due to their low mammalian toxicity, novel modes of action (not neurotoxic), and the possibility that they will have a low potential for the development of resistance. Dimethicones are linear polydimethylsiloxanes ($CH_3SiO[SiO(CH_3)_2]_nSi(CH_3)_2$), where n is the number of repeating monomers $[SiO(CH_3)_2]$ in chains of varying length; chain length influences viscosity and spreading characteristics. The toxic action for the dimethicones is not definitively known (and may differ between products of differing chain lengths, see subsequent text), but they have been shown to be effective pediculicides [57, 58]. Two dimethicone-based products are better characterized scientifically in terms of their effectiveness and probable modes of action, and are discussed here.

Hedrin® 4% lotion (Thornton & Ross Ltd, Huddersfield, UK) is a 4% dimethicone lotion in 96% (w/w) decamethylcyclopentasiloxane (cyclomethicone D5). Treated head lice are rapidly immobilized, but small movements in their extremities over several hours indicate that death is delayed. Scanning electron microscopy coupled with X-ray microanalysis revealed that Hedrin® 4% lotion was found in the spiracles, in some cases blocking the opening, and penetrated into the outer aspects of the tracheae [57]. Asphyxia is unlikely given the slow onset of mortality. The inability of the louse to excrete the excess water acquired during blood feeding by transpiration out of the spiracles has been suggested as a toxic action [58].

The second dimeticone-based anti-louse product (NYDA®, G. Pohl-Boskamp GmbH & Co., Hohenlockstedt, Germany) contains a mixture of two dimethicones, one of low and the other of higher viscosity, at a final total concentration of dimeticones of 92% (w/w). Medium-chain-length triglycerides, jojoba wax, and two fragrances make up the remaining constituents. NYDA® rapidly enters the tracheal system due to its high spreading ability [59]. Death occurs rapidly and appears to be due to asphyxia. NYDA® is also an effective ovicide [60].

In the United States, the OTC product, Lice MD®, is currently available from Reckitt-Benckiser, Slough, England, and contains dimethicone as an emollient [12].

Ivermectin-based Formulations

Ivermectin is a macrocyclic lactone produced fermentatively by *Streptomyces avermitilis* followed by a chemical modification, and it is a widely-used oral anthelmintic agent for both humans and animals. In addition to muscles used in motility, ivermectin also acts to paralyze the muscles associated with the nematode pharyngeal pump, inhibiting the pumping action needed for feeding and attachment [61, 62]. The concentration of ivermectin needed to cause paralysis of the pharyngeal pump is 10- to 100-fold lower than the concentration needed to cause mortality [63].

Ivermectin increases chloride ion permeability in insect [64] and nematode [65] neurons and muscle membranes through binding to glutamate- and gamma-aminobutyric acid (GABA)-gated chloride ion channels. These channels are highly expressed in the neuromuscular system of the pharyngeal pump in the mouthparts of the free living nematode, *Caenorhabditis elegans*, which has been shown to be highly sensitive to ivermectin. During deworming, ivermectin paralyzes the mouthparts of the nematode, causing it to detach from the mammalian gut and be excreted. A similar mode of action in head lice, however, has not been directly characterized.

Recently, oral ivermectin was used to treat hard-to-control head louse infestations [66]. Successive treatments were necessary to kill nymphs that emerge from eggs present at the time of the initial treatment, indicating an absence of a direct ovicidal effect of oral ivermectin.

Ivermectin is also formulated as a topically applied pediculicide in a 0.5% ivermectin cream (Sklice®, Arbor Pharmaceutical, Atlanta, GA) that was approved by the U.S. Food and Drug Administration (FDA) in 2012 as a prescription treatment in patient 6 months or older [12]. Sklice® killed permethrin-resistant head lice [67] but was not directly ovicidal to treated eggs, as hatchability was not decreased [68]. Nevertheless, the percentage of hatched lice from treated eggs that took a blood meal significantly decreased (80–95%) compared to lice that hatched from untreated eggs, and all treated lice died within 48 h of hatching. Lice that hatched from eggs treated with nonlethal dilutions of Sklice® also fed significantly less than lice from untreated eggs. Using [^{3}H] inulin uptake as a means to measure blood feeding, lice from eggs treated with nonlethal dilutions of Sklice® ingested significantly less blood than lice from untreated eggs. Thus, the failure of hatched instars to take a blood meal following egg treatments with Sklice® is likely responsible for its action as a post-eclosion nymphicide [68].

Spinosad-based Formulation

Spinosad is a macrocyclic lactone insecticide produced fermentatively by a soil actinomycete bacterium, *Saccharopolyspora spinosa*. It has two active ingredients, spinosyn A and spinosyn D, in a 5 : 1 ratio. Spinosad is a neurotoxic agonist at the nicotinic acetylcholine receptor of the cholinergic nervous system where it selectively modifies the non-desensitizing aspect of the current flowing through this ligand-gated channel, causing prolonged excitability and then paralysis [69]. Spinosad is both pediculicidal and ovicidal [70].

Spinosad has been commercially formulated as a 0.9% viscous topical suspension and was approved by the FDA in 2011 as a prescription treatment for the treatment of pediculosis in patients 6 months or older (Natroba®, ParaPRO, LLC, Carmel, IN) (12).

Abametapir-based Formulation

Recently, an alternate approach investigated egg hatching [71] as a pathway to identify new sites of action for the development of novel ovicidal compounds. A number of proteases were identified as metalloproteases and their presence from newly hatched louse eggs indicated that they may play a role in egg hatching. Subsequently, eggs treated with known metalloproteinase inhibitors failed to hatch, indicating that metalloproteinase may function as novel ovicidal targets in lice.

To better understand this phenomenon, 5,5′-dimethyl-2,2′-bipyridyl (a bipyridine metal chelating ligand formerly termed Ha44, now referred to as abametapir) was used to determine its ovicidal, larvicidal, and adulticidal action on *D. melanogaster* as a model insect [72]. Although toxic to both larvae and adults, abametapir was particularly potent on eggs, providing further evidence of its ovicidal efficacy in insects.

Very recently, abametapir in isopropanol and as a formulated lotion (Xeglyze™, 0.74% abametapir) was determined to be 100% ovicidal on both head and body louse eggs [73]. Registration processes are currently under way.

Repellents/Deterrents and Odorant Receptor Discovery

Due to pyrethroid resistance and the lack of efficient resistance management, there is considerable interest in the protection of uninfested people and prevention of reinfestation by disrupting lice transfer. Recently, *in vitro* and *in vivo* models were used to determine the efficacy of the infestation deterrents, *Elimax lotion*® and/or *Elimax shampoo*® (oligodecene oil, sesame oil, and acrylate, Oystershell Laboratories NV, Drongen, Belgium), against human head or poultry chewing lice, respectively [74]. Head lice exhibited significantly higher ovipositional avoidance (~100% over 72 h) to both formulations when compared to lice on control hair tufts and both formulations were determined to be competent infestation deterrents (% avoidance >50%) in a competitive avoidance test against a known attractant (head louse feces extract). *Elimax shampoo®* was also an efficacious deterrent against poultry chewing lice. These results validate that human lice have both an aggregation response, which was found many years ago [75], and now a repellent response.

There is a drastic reduction of the chemosensory gene repertoires in the body louse genome when compared to other insects [41]. With this finding, seven putative full-length odorant receptors (ORs), in addition to the odorant receptor co-receptor (Orco), were identified and four of them expressed in the *Xenopus* oocyte system [76]. PhumOR2 responded electrophysiologically to a narrow set of compounds: 2,3-dimethylphenol >4-methylcyclohexanol >1-phenylethanol. At the behavior level, both head and body lice were repelled by these physiologically active chemicals in the same order that they were electrophysiologically active. This study presents the first evidence that an OR pathway is functional in lice and identifies PhumOR2 as a sensitive receptor of natural repellents, which could be used to develop novel efficient molecules for the control of these insects.

Metabolic Synergists

Pyrethroids are the preferred class of insecticides when human exposure is likely. Unfortunately, the pyrethroids share a common genetic resistance mechanism, *kdr*, with DDT and prior extensive use of DDT has predisposed the pyrethroids to cross-resistance via *kdr* and metabolic mechanisms. Thus, the use of metabolic synergists with pyrethroids is considered prudent to guard against the selection of insects with multiple mechanisms of resistance.

Recently, 3-phenoxybenzyl hexanoate (PBH) was synthesized as a multifunctional pyrethroid synergist that, besides being a surrogate substrate that sequesters hydrolytic carboxylesterases, also functions by out-competing pyrethroids in oxidative xenobiotic metabolism pathways [77]. Addition of PBH to permethrin-treated mosquitoes, *Culex pipiens quinquefasciatus,* resulted in a threefold increase in the synergistic ratio (LT_{95} of permethrin-treated mosquito/ LT_{95} of permethrin-treated + PBH mosquito). Similarly, PBH synergized the action of deltamethrin by sixfold on the common bed bug, *Cimex lectularius,* and was 2.8-fold more synergistic than piperonyl butoxide (PBO), the synergist in many pyrethrin-based pediculicides. Thus, PBH synergized the action of both type I and II pyrethroids in a range of blood-feeding insects. PBH has residual properties similar to permethrin and is itself nontoxic, unlike PBO, and therefore should be compatible with existing pyrethroid formulations, including the OTC pediculicides, perhaps extending their usefulness.

Sustainable Resistance Management

Resistance management entails processes that reduce resistant allele frequencies, dominance, and fitness of the resistant genotypes [78]. Many nonchemical processes used to delay resistance in agricultural settings (e.g., natural enemies, insect disease, and host–plant resistance) are not applicable for human lice and limit the operational choices to a chemical management format coupled with nit removal. Insecticides are applied to manage resistance by moderation, saturation, and multiple attack strategies [79]. Low tolerance of infestations and the "No-Nit" policy eliminate most moderation approaches. Saturation schemes that involve high concentrations of insecticides are not appropriate when treating children. Thus, only multiple attacks (e.g., mixtures, mosaics, and rotations of pediculicides) are available strategies. The use of mosaics on a human head is impractical and rotations have not worked in the United Kingdom [80], leaving only mixtures as a viable application strategy. Central to the use of mixtures as a resistance management scheme is the idea of "redundant killing." Simply put, all active ingredients in the mixture must have unique target sites and metabolism so as the insect acquires mutations that result in insecticide resistance; the other compounds in the mixture will act on their unaltered sites, thereby removing the insect and its resistance-yielding mutations. To this point, a mixture of α-terpineol, terpine-4-ol, and 0.5% malathion, found in the Ovide® formulation (Medicis Corp., Scottsdale, AZ), was effective in controlling a malathion- and permethrin-resistant head louse strain from the United Kingdom [37]. The successful application of insecticide mixtures, used in a resistance management format, shows that: (i) these approaches suppress resistance and (ii) they are likely to be effective in field situations when used in conjunction with efficient monitoring. Thus, as new pediculicides are introduced into the market place, it is imperative that we understand how lice may develop resistance or cross-resistance to these compounds. Such knowledge will have practical applications, in terms of recommending mixtures of compounds, where lice may develop very different forms of resistance to the two separate compounds.

In summary, resistance to traditional pediculicides has developed, leading to clinical failures, so management strategies are necessary. However, these resistance management strategies need to be built on sound scientific knowledge *before* resistance evolves. It is also imperative that the molecular mechanisms mediating resistance be identified for the selective targeting of novel compounds and proper formulation of mixtures aimed at controlling these insects. The head and body louse genome projects [3, 41] have now provided the necessary core information for finding novel target sites for improved louse control, the means of identifying genes responsible for resistance and establishing cross- and negative cross-resistance relationships, and the tools for effective and affordable resistance monitoring, as summarized in this chapter.

References

1. Pennisi, E. (2004) Human origins. Louse DNA suggests close contact between early humans. *Science*, **306**, 210.
2. Kittler, R. (2003) Molecular evolution of *Pediculus humanus* and the origins of clothing. *Curr. Biol.*, **13**, 1414–1417.
3. Kang, J.S., Cho, Y.J., Kim, J.H., Kim, S.H. *et al.* (2015) Comparison of the genome profiles between head and body lice. *J. Asia Pac. Entomol.*, **18**, 337–382.
4. Olds, B.P., Coates, B.S., Steele, L.D., Sun, W. *et al.* (2012) Comparison of the

transcriptional profiles of head and body lice. *Insect Mol. Biol.*, **21**, 257–268.

5. Bacot, A. (1917) A contribution to the bionomics of *Pediculus humanus* (vestimenti) and *Pediculus capitis*. *Parasitology*, **9**, 228–258.
6. Mullen, C. and Durden, L.A. (2009) Medical and Veterinary Entomology, 2nd edn, Academic Press, San Diego.
7. Light, J.E., Toups, M.A., and Reed, D.L. (2008) What's in a name: the taxonomic status of human head and body lice. *Mol. Phylogenet. Evol.*, **47**, 1203–1216.
8. Bonilla, D.I., Kabeya, H., Henn, J., Kramer, V.I. *et al.* (2009) *Bartonella quintana* in body lice and head lice from homeless persons, San Francisco, California, USA. *Emerg. Infect. Dis.*, **15**, 912–915.
9. Clark, J.M., Yoon, K.S., Lee, S.H., and Pittendrigh, B.R. (2013) Human lice: past present and future control. *Pestic. Biochem. Physiol.*, **106**, 162–171.
10. Pittendrigh, B.R., Olds, B.P., Yoon, K.S., Lee, S.H. *et al.* (2013) The genomics of human lice: from the genome to the potential for future control strategies. *Pestic. Biochem. Physiol.*, **106**, 172–176.
11. Clark, J.M., Yoon, K.S., Kim, J.H., Lee, S.H. *et al.* (2015) Utilization of the human louse genome to study insecticide resistance and innate immune response. *Pestic. Biochem. Physiol.*, **120**, 125–132.
12. Devore, C.D. and Schutze, G.E. (2015) Head lice. *Pediatrics*, **135**, e1356–e1361.
13. Pittendrigh, B.R., Clark, J.M., Lee, S.H., Yoon, K.S. *et al.* (2015) Body lice: from genome project to functional genomics and reverse genetics, in Short Views on Insect Genomics and Proteomics, Entomology in Focus 3, (eds C. Raman, M.R. Goldsmith, and T.A. Agunbiade), Springer International Publishing, Switzerland, pp. 1–18. ISBN: 10.1007/978-3-319-24235-4_1
14. Koch, E., Clark, J.M., Cohen, B., Meinking, T.L. *et al.* (2016) Management of head louse infestations in the United States – a literature review. *Pediatr. Dermatol.*, **33**, 466–472.
15. Gratz, N.G. (1999) Emerging and resurging vector-borne diseases. *Annu. Rev. Entomol.*, **44**, 51–75.
16. Williams, L.K., Reichart, A., MacKenzie, W.R., Hightower, A.W. *et al.* (2001) Lice, nits, and school policy. *Pediatrics*, **107**, 1011–1015.
17. Jones, D.S., Wache, S., and Chhokar, V. (1996) Toxins produced by arthropod parasites: salivary gland proteins of human body lice and venom protein of chelonine wasps. *Toxicon*, **34**, 1421–1429.
18. Kelly, D.J., Richards, A.L., Temenak, J., Strickman, J.D. *et al.* (2002) The past and present threat of rickettsial diseases to military medicine and international public health. *Clin. Infect. Dis.*, **34** (Suppl. 4), S245–S169.
19. Weiss, K. (1988) The role of rickettsioses in history, in Biology of Rickettsial Diseases, vol. **1** (ed. D.H. Walker), CRC, Boca Raton, FL, pp. 1–14.
20. Durand, R., Bouviesse, S., Berdjane, Z., Izri, A. *et al.* (2012) Insecticide resistance in head lice: clinical, parasitological and genetic aspects. *Clin. Microbiol. Infect.*, **18**, 338–344.
21. Hodgdon, H.E., Yoon, K.S., Previte, D.J., Kim, H.J. *et al.* (2010) Serial invasive signal amplification reaction for the determination of kdr frequencies in global human head louse populations for efficient resistance monitoring. *Pest Manag. Sci.*, **66**, 1031–1040.
22. Marcoux, D., Palma, K., Kaul, N., Hodgdon, H. *et al.* (2010) Pyrethroid pediculicide resistance of head lice in Canada evaluated by serial invasive signal amplification reaction. *J. Cutan. Med. Surg.*, **14**, 115–118.
23. Gao, J.-R., Yoon, K.S., Frisbie, R.K., Coles, G.C. *et al.* (2006) Esterase-mediated malathion resistance in the human head louse, *Pediculus capitis* (Anoplura: Pediculidae). *Pestic. Biochem. Physiol.*, **85**, 28–37.
24. Chosidow, O., Chastang, C., Brue, C., Bouvet, E. *et al.* (1994) Controlled study of malathion and d-phenothrin lotions for *Pediculus humanus* var *capitis*-infested schoolchildren. *Lancet*, **344**, 1724–1727.
25. Hipolito, R.B., Mallorca, F.G., Zuniga-Macaraig, Z.O., Apolinario, P.C. *et al.* (2001) Head lice infestation: single drug

versus combination therapy with one percent permethrin and trimethoprim/sulfamethoxazole. *Pediatrics*, **107** (3), E30.

26. Burgess, I.F., Brown, C.M., Peock, S., and Kaufman, J. (1995) Head lice resistant to pyrethroid insecticides in Britain. *BMJ*, **311**, 752.
27. Hill, N., Moor, G., Cameron, M.M., Butlin, A. *et al.* (2005) Single blind, randomised, comparative study of the Bug Buster kit and over the counter pediculicide treatments against head lice in the United Kingdom. *BMJ*, **331**, 384–387.
28. Rupes, V., Moravec, J., Chmela, J., Ledvinka, J., and Zelenkova, J. (1995) A resistance of head lice (*Pediculus capitis*) to permethrin in Czech Republic. *Cent. Eur. J. Public Health*, **3**, 30–32.
29. Kristensen, M., Knorr, M., Rasmussen, A.M., and Jespersen, J.B. (2006) Survey of permethrin and malathion resistance in human head lice populations from Denmark. *J. Med. Entomol.*, **43**, 533–538.
30. Mumcuoglu, K.Y., Hemingway, J., Miller, J., Ioffe-Uspensky, I. *et al.* (1995) Permethrin resistance in the head louse *Pediculus capitis* from Israel. *Med. Vet. Entomol.*, **9**, 427–432.
31. Meinking, T.L., Entzel, P., Villar, M.E., Vicaria, M. *et al.* (2001) Comparative efficacy of treatments for pediculosis capitis infestations: update 2000. *Arch. Dermatol.*, **137**, 287–292.
32. Picollo, M.I., Vassena, C.V., Casadio, A.A., Massimo, J. *et al.* (1998) Laboratory studies of susceptibility and resistance to insecticides in *Pediculus capitis* (Anoplura; Pediculidae). *J. Med. Entomol.*, **35**, 814–817.
33. Tomita, T., Yaguchi, N., Mihara, M., Takahashi, M. *et al.* (2003) Molecular analysis of a para sodium channel gene from pyrethroid-resistant head lice, *Pediculus humanus capitis* (Anoplura: Pediculidae). *J. Med. Entomol.*, **40**, 468–474.
34. Hunter, J.A. and Barker, S.C. (2003) Susceptibility of head lice (*Pediculus humanus capitis*) to pediculicides in Australia. *Parasitol. Res.*, **90**, 476–478.
35. Izri, M.A. and Briere, C. (1995) First cases of resistance of *Pediculus capitis* Linne 1758 to malathion in France. *Presse. Med.*, **24**, 1444.
36. Downs, A.M., Stafford, K.A., Harvey, I., and Coles, G.C. (1999) Evidence for double resistance to permethrin and malathion in head lice. *Br. J. Dermatol.*, **141**, 508–511.
37. Yoon, K.S., Gao, J.-R., Lee, S.H., Clark, J.M. *et al.* (2003) Permethrin-resistant human head lice, *Pediculus capitis*, and their treatment. *Arch. Dermatol.*, **139**, 994–1000.
38. Takano-Lee, M., Velten, R.K., Edman, J.D., Mullens, B.A. *et al.* (2003) An automated feeding apparatus for *in vitro* maintenance of the human head louse, *Pediculus humanus capitis*, (Anoplura: Pediculidae). *J. Med. Entomol.*, **40**, 795–799.
39. Yoon, K.S., Strycharz, J.P., Gao, J.R., Takano-Lee, M. *et al.* (2006) An improved *in vitro* rearing system for the human head louse allows the determination of resistance to formulated pediculicides. *Pestic. Biochem. Physiol.*, **86**, 195–202.
40. Johnson, J.S., Yoon, K.S., Strycharz, J.P., Pittendrigh, B.R. *et al.* (2007) Body and head lice (Phthiraptera: Pediculidae) have the smallest genomes of any hemimetabolous insect reported to date. *J. Med. Entomol.*, **44**, 1009–1012.
41. Kirkness, E.F., Haas, B.J., Sun, W., Braig, H.R. *et al.* (2010) Genome sequences of the human body louse and its endosymbiont provide insights into the permanent parasitic lifestyle. *Proc. Natl. Acad. Sci. U S A*, **107**, 12168–12173.
42. Lee, S.H., Min, J.S., Yoon, K.S., Strycharz, J.P. *et al.* (2010) Decreased detoxification genes and genome size makes the human body louse an efficient model to study xenobiotic metabolism. *Insect Mol. Biol.*, **19**, 599–615.
43. Kim, J.H., Min, J.S., Kang, J.S., and Kwon, D.H. (2011) Comparison of the humoral and cellular responses between

body and head lice following immune challenge. *Insect Biochem. Mol. Biol.*, **41**, 333–339.

44. Leo, N.P., Campbell, N.H.J., Yang, X., Mumcuoglu, K. *et al.* (2002) Evidence from mitochondrial DNA that head and body lice of humans (Phthiraptera: Pediculidae) are conspecific. *J. Med. Entomol.*, **39**, 663–666.

45. Li, W., Ortiz, G., Fournier, P., Gimenez, G. *et al.* (2010) Genotyping of human lice suggests multiple emergences of body lice from local head louse populations. *PloS Negl. Trop. Dis.*, **4**, e641, 10.1371/journal.pntd.0000641.

46. Lee, S.H., Yoon, K.S., Williamson, M.S., Goodson, S.J. *et al.* (2000) Molecular analysis of kdr-type resistance in permethrin-resistant strains of head lice, *Pediculus capitis. Pestic. Biochem. Physiol.*, **66**, 130–143.

47. Busvine, J.R. (1951) Mechanism of resistance to insecticide in houseflies. *Nature*, **168**, 193–195.

48. Soderlund, D.M. and Knipple, D.C. (2003) The molecular biology of knockdown resistance to pyrethroid insecticides. *Insect Biochem. Mol. Biol.*, **33**, 563–577.

49. Lee, S.H., Gao, J.-R., Yoon, K.S., Mumcuoglu, K.Y. *et al.* (2003) Sodium channel mutations associated with knockdown resistance in the human head louse, *Pediculus capitis*, (De Geer). *Pestic. Biochem. Physiol.*, **75**, 79–91.

50. Yoon, K.S., Symington, S.B., Lee, S.H., Soderlund, D.M. *et al.* (2008) Three mutations identified in the voltage-sensitive sodium channel alpha-subunit gene of permethrin-resistant human head lice abolish permethrin sensitivity of the house fly Vssc1 expressed in *Xenopus* oocytes. *Insect Biochem. Mol. Biol.*, **38**, 296–306.

51. Yoon, K.S., Previte, D.J., Hodgdon, H.E., Poole, B.C. *et al.* (2014) Knockdown resistance allele frequencies in North American head louse populations. *J. Med. Entomol.*, **51**, 450–457.

52. Gellatly, K.J., Krim, S., Palenchar, D.J., Shepherd, K. *et al.* (2016) Expansion of the knockdown resistance frequency map for human head lice (Phthiraptera: Pediculidae) in the United States using quantitative sequencing. *J. Med. Entomol.*, **53** (3), 653–659. doi: 10.1093/jme/tjw023

53. Kwon, D.K., Kim, J.H., Kim, Y.H., Yoon, K.S. *et al.* (2014) Identification and characterization of the esterase involved in malathion resistance in the head louse, *Pediculus humanus capitis. Pestic. Biochem. Physiol.*, **112**, 13–18.

54. Willoughby, L., Chung, H., Lumb, C., Robin, C. *et al.* (2006) A comparison of *Drosophila melanogaster* detoxification gene induction responses for six insecticides, caffeine and phenobarbital. *Insect Biochem. Mol. Biol.*, **36**, 934–942.

55. Yoon, K.S., Strycharz, J.P., Baek, J.H., Sun, W. *et al.* (2011) Brief exposures of human body lice to sub-lethal amounts of ivermectin over transcribe genes involved in tolerance. *Insect Mol. Biol.*, **20**, 687–699.

56. Meinking, T., Villar, M.E., Vicaria, M. *et al.* (2010) The clinical trials supporting benzyl alcohol lotion 5% (Ulesfia): a safe and effective topical treatment for head lice (*Pediculosis humanus capitis. Pediatr. Dermatol.*, **27**, 19–24.

57. Burgess, I.F. (2009) The mode of action of dimeticone 4% lotion against head lice, *Pediculus capitis. BMC Pharmacol.*, **9**, 3.

58. Burgess, I.F., Brunton, E.R., and Burgess, N.A. (2013) Single application of 4% dimeticone liquid gel versus two applications of 1% permethrin crème rinse for treatment of head louse infestations; a randomised controlled trial. *BMC Dermatol.*, **13**, 5.

59. Richling, I. and Bockele, W. (2008) Lethal effects of treatment with a special dimeticone formula on head lice and house crickets (Orthoptera, Ensifera: Acheta domestica and Anoplura, Phthiraptera: *Pediculus humanus*)-insights into physical mechanisms. *Arzneimittelforschung*, **58**, 248–254.

60. Strycharz, J.P., Lao, A.R., Alves, A.M., and Clark, J.M. (2012) Ovicidal response of NYDA formulations on the human head louse (Anoplura, Pediculidae) using a hair tuft bioassay. *J. Med. Entomol.*, **49**, 336–342.
61. Fisher, M.H. and Mrozik, H. (1984) The avermectin family of macrolide-like antibiotics, in Macrolide Antibiotics (ed. A. Omura), Academic Press, New York, NY, pp. 553–606.
62. Brownlee, D.J., Holde-Dye, A.L., and Walker, R.J. (1997) Actions of the anthelminic ivermectin on the pharyngeal muscle of the parasitic nematode, *Ascaris suum*. *Parasitology*, **115**, 553–561.
63. Gill, J.H., Redwin, J.M., Van Wyk, J.A., and Lacey, E. (1995) Avermectin inhibition of larval development in *Haomonchus contortus* effects ivermectin resistance. *Int. J. Parasitol.*, **25**, 463–470.
64. Duce, I.C. and Scott, R.H. (1985) Actions of dihydroavermectin B1a on insect muscle. *Br. J. Pharmacol.*, **85**, 395–401.
65. Dent, J.A., Davis, M.W., and Avery, L. (1997) avr-15 encodes a chloride channel subunit that mediates inhibitory glutamatergic neurotransmission and ivermectin sensitivity in *Caenorhabditis elegans*. *EMBO J.*, **16**, 5867–5879.
66. Chosidow, O., Giraudeau, B., Cottrell, J., Ozri, A. *et al.* (2010) Oral ivermectin versus malathion lotion for difficult-to-treat head lice. *N. Engl. J. Med.*, **362**, 896–905.
67. Strycharz, J.P., Yoon, K.S., and Clark, J.M. (2008) A new ivermectin formulation topically kills permethrin-resistant human head lice (Anoplura: Pediculidae). *J. Med. Entomol.*, **45**, 75–81.
68. Strycharz, J.P., Berge, N.M., Alves, A.M., and Clark, J.M. (2011) Ivermectin acts as a post-eclosion nymphicide by reducing blood feeding of human head lice (Anoplura, Pediculidae) that hatched from treated eggs. *J. Med. Entomol.*, **48**, 1174–1182.
69. Salgado, V. and Saar, R. (2004) Desensitizing and non-desensitizing subtypes of alpha-bungarotoxin-sensitive nicotinic acetylcholine receptors in cockroach neurons. *J. Insect Physiol.*, **50**, 867–879.
70. Villegas, S.C. and Breitzka, R.L. (2012) Head lice and the use of spinosad. *Clin. Ther.*, **34**, 14–23.
71. Bowles, V.M., Young, A.R., and Barker, S.C. (2008) Metalloproteases and egg-hatching in *Pediculus humanus*, the body (clothes) louse of humans (Phthiraptera: Insecta). *Parasitology*, **135**, 125–130.
72. Van Hiel, M.B., Breugelmans, B., Pagel, C.N., Williams, A.K. *et al.* (2012) The ovicidal, larvicidal and adulticidal properties of 5,5'-dimethyl-2,2'-bipyridyl against *Drosophila melanogaster*. *PLoS One*, **7**, e49961.
73. Bowles, V.M., Yoon, K.S., Barker, S.C., Tran, C. *et al.* (2016) Ovicidal efficacy of abametapir against eggs of human head and body lice (Anoplura: Pediculidae). *J. Med. Entomol.*, **54**, 167–172.
74. Yoon, K.S., Ketzis, J.K., Andrewes, S.W., Wu, C.S. *et al.* (2015) *In vitro* and *in vivo* evaluation of infestation deterrents against lice. *J. Med. Entomol.*, **52**, 970–978.
75. Mumcuoglu, Y., Galun, R., and Ikan, R. (1986) The aggregation response of human body louse (*Pediculus humanus*) (Insecta: Anoplura) to its excretory products. *Insect Sci. Appl.*, **7**, 629–632.
76. Pelletier, J., Xu, P., Yoon, K.S., Clark, J.M. *et al.* (2015) Odorant receptor-based discovery of natural repellents of human lice. *Insect Biochem. Mol. Biol.*, **66**, 103–109.
77. Hardstone, M., Strycharz, J.P., Kim, J.H., Park, I.K. *et al.* (2015) Development of multi-functional metabolic synergists to suppress the evolution of resistance against pyrethroids in insects that blood feed on humans. *Pest Manag. Sci.*, **71**, 842–849.
78. Leeper, J.R., Rough, J.R., and Reynolds, H.T. (1986) Preventing or managing resistance in arthropods, in Pesticide Resistance: Strategies and Tactics for Management, National Academy Press, Washington, D.C., pp. 335–346.

79. Georghiou, G.P. (1983) Management of resistance in arthropods, in Pest Resistance to Pesticides (eds G.P. Georghiou and T. Saito), Plenum Press, New York, NY, pp. 769–792.

80. Downs, A.M., Stafford, K.A., and Coles, G.C. (1999) Head lice: prevalence in schoolchildren and insecticide resistance. *Parasitol. Today*, **15**, 1–4.

Part Two
Screens & Models

Ectoparasites: Drug Discovery Against Moving Targets, First Edition.
Edited by Charles Q. Meng and Ann E. Sluder

7
Molecular Targets to Impair Blood Meal Processing in Ticks

Petr Kopáček, Jan Perner, Daniel Sojka, Radek Šíma, and Ondřej Hajdušek*

Abstract

Feeding and digestion of host blood are key physiological processes providing essential nutrients for the development and fecundity of ticks. Ingested host blood, which exceeds the weight of unfed females by more than one hundred times, is concentrated and stored in the tick gut lumen, gradually being taken up by digestive cells, and intracellularly digested by a multi-enzyme network of acidic aspartic and cysteine endo- and exo-peptidases. Digestion of hemoglobin, the major protein component of blood, results in the release of a vast excess of potentially toxic heme. In most eukaryotic cells, heme and iron homeostasis is based on a balanced flux between heme biosynthesis and heme degradation, mediated by heme oxygenase. In contrast, ticks are not capable of synthesizing or degrading heme. Therefore, ticks have evolved specific molecular mechanisms of heme and iron acquisition, detoxification, intracellular trafficking, and inter-tissue transport. This chapter reviews current knowledge on the molecular mechanisms of these processes and discusses their potential as targets for anti-tick interventions.

Introduction

Ticks (Acari, Ixodida) are ectoparasites that have adapted to an obligate blood-feeding lifestyle (hematophagy) during their evolution from ancestral mites. About 900 species have been described and these are divided into two major families – the hard ticks (Ixodidae, ~700 species) and soft ticks (Argasidae, ~200 species) [1, 2]. Of special interest is a monotypic family, Nuttalliellidae, represented by a single species *Nuttalliella namaqua* that is hypothesized to represent an evolutionary link between these two families [3, 4]. Ticks of both families are dangerous vectors of a wide variety of pathogens, causing severe infectious diseases of humans as well as wild and domestic animals [5]. Although hard and soft ticks may have followed independent paths toward hematophagy [6] and differ in many

*Corresponding author.

Ectoparasites: Drug Discovery Against Moving Targets, First Edition.
Edited by Charles Q. Meng and Ann E. Sluder

physiological and developmental aspects [7], they still share several common traits that offer targets for rational and specific tick control. Major physiological differences in feeding strategies of soft and hard ticks can be summarized as follows: In soft ticks, the larvae, several nymphal stages, and adults of both sexes take their blood meal rapidly by feeding on the host for up to 1 h. In adult females, feeding and oviposition are cyclic processes, and feeding performance is not necessarily related to mating status. In contrast, hard ticks feed only once per life stage (larva, one nymphal stage, and adult) for long periods of several days. An adult, mated female can engorge a huge amount of host blood that exceeds more than one hundred times the weight of the unfed tick. The substantial amount of imbibed blood is then digested and converted into a large number (several thousands) of laid eggs, following which the hard tick female dies.

What all ticks noticeably have in common is the process of blood meal digestion that clearly distinguishes ticks from their vertebrate hosts as well as from blood-feeding insects that have the digestive apparatus based on neutral or alkaline serine proteases. In contrast, ticks digest blood intracellularly in the acidic lysosome-like vesicles of digestive cells lining the midgut epithelium [7]. The multi-enzymatic machinery of acidic aspartic and cysteine peptidases involved in blood meal processing in ticks resembles the digestive system in other, evolutionary distant blood-feeding parasites, such as the malaria-causing *Plasmodium*, flatworms, or nematodes [8].

The blood meal, the ultimate source of nutrients for all tick developmental stages, is a rather unhealthy diet that required ticks to evolve specific adaptations to cope with excessive amounts of blood. Hemoglobin, the major protein constituent (accounting for about 60% of total blood proteins), has been assumed to be the main source of amino acids for tick proteosynthesis and vitellogenesis. Digestion of hemoglobin brings about the release of its prosthetic group – heme, which, when unbound, is toxic through its involvement in the formation of free oxygen radicals via the Fenton reaction [9]. One adult hard tick female, such as *Ixodes ricinus*, can imbibe and concentrate up to 1 ml of host blood containing ~150 mg of hemoglobin, out of which heme comprises 6 mg. In the case of complete degradation of acquired heme, about 500 μg of ferrous iron would be released. These extremely large amounts of pro-oxidative molecules would clearly represent a lethal burden for a ~2 mg organism, as is the case of the unfed female tick. This simple consideration indicates how important in tick physiology are efficient "waste management" strategies for superfluous blood meal components.

In the majority of eukaryotic organisms, heme and iron homeostasis is based on balancing the flux between heme biosynthesis and its degradation, mediated by heme oxygenase (HO). An earlier biochemical study on the cattle tick *Rhipicephalus* (*Boophilus*) *microplus* [10] and a recent genome-wide analysis of *Ixodes scapularis* genome [11] revealed that at least hard ticks lack a functional pathway for biosynthesis of endogenous heme [12]. Ticks have retained only the last three mitochondrial enzymes (coproporphyrinogen oxidase, protoporphyrinogen oxidase, and ferrochelatase) out of eight enzymes that make up the canonical heme-biosynthetic pathway that was present in tick mite ancestors [12]. The most likely explanation for the loss of heme biosynthesis during tick evolution is their exposure to a vast surplus of heme originating from digested host hemoglobin. On the other hand, the inability to synthesize endogenous heme makes ticks completely dependent on the acquisition of host-derived heme for use as a prosthetic group for proteosynthesis of their own hemoproteins. Besides lacking heme biosynthesis, ticks are also unable to degrade heme as the gene coding for the key enzyme, HO, is apparently not present in their genome [11]. This trait, however, does not seem to have evolved as an adaptation to hematophagy, since HO is also absent from the genomes of other non-hematophagous mites and possibly also chelicerates [12].

Taken together, the physiological processes of ticks associated with digestion of huge amounts of blood meal, elimination of waste products and excessive water, lack of heme biosynthesis and catabolism, and unique mechanisms for acquisition and inter-tissue distribution of heme or iron, all constitute a major departure from their vertebrate hosts (Figure 7.1). There are basically two ways to effectively target tick Achilles' heels: "anti-tick" vaccines and acaricides. The development of effective "anti-tick" vaccines or highly selective and environmentally

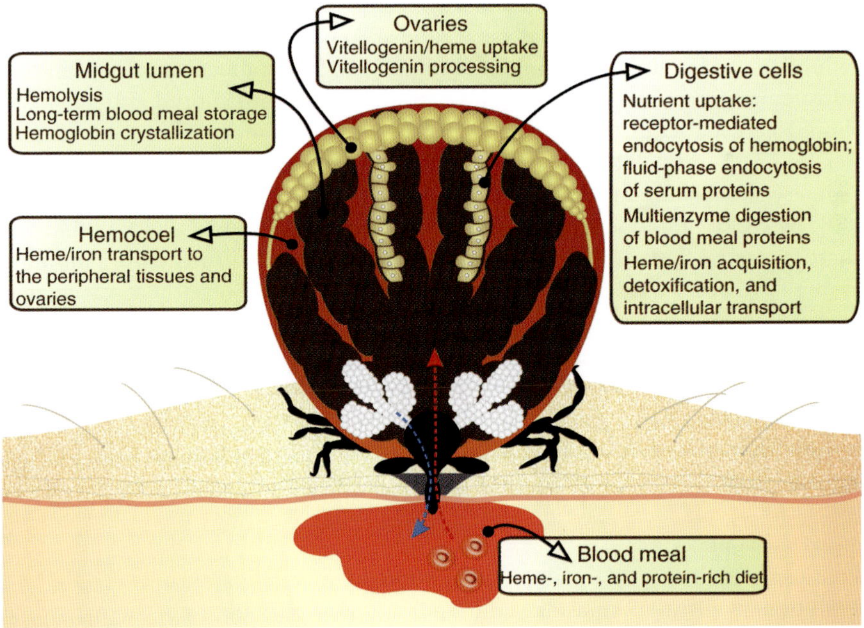

Figure 7.1 Physiological processes associated with blood meal processing in ticks. A schematic overview of tick tissues and related processes of blood meal uptake, digestion, heme and iron metabolism, detoxification and inter-tissue transport that may serve as rational targets for "anti-tick" intervention. The red and blue arrows indicate blood meal uptake and reverse water secretion, respectively.

friendly acaricides depends on the discovery of suitable protective antigens or distinctive traits in tick physiology. This effort comprises a variety of strategies including the target-based approach (the primary focus of this chapter) as well as high-throughput approaches such as functional genomics, vaccinomics and tick–host–pathogen interactomics [13–15]; anti-tick vaccines are discussed further in Chapter 2 of this volume.

Processes Associated with Blood Feeding and Digestion

Attachment and On-host Feeding

Blocking of tick attachment to the host and impairing their long undisturbed feeding is intuitively the most straightforward "anti-tick" strategy. Therefore, a focus on the tick–host interface and the role of tick saliva components in modulation of host hemostasis, inflammation, and innate (complement-based) or acquired (antibody-based) immunity represents the largest and most explored field of tick research. However, it is out of the scope and aim of this chapter to cover the complexity of bioactive factors present in tick saliva and we can only refer to several recent comprehensive reviews that deal with this extensive area of research [14, 16–18].

Nevertheless, identification of suitable targets in the rich cocktail of compounds secreted by the tick salivary glands is quite problematic given the high redundancy of multi-genic protein families [14]. The molecules present in tick saliva come into contact with the host immune system (exposed antigens) and should be capable of eliciting an antibody response. Yet the counteraction of tick immune modulators usually suppresses the host's ability to prevent or reduce tick attachment and feeding during repeated exposure to tick infestation [19]. By contrast, blood meal digestion and associated processes located inside the tick are mediated by molecules that are not exposed to the host immune system (concealed antigens). Therefore, tick infestation on a host vaccinated with a recombinant concealed antigen might have a protective effect if the specific antibodies present in ingested blood meal block a physiologically important mechanism. The feasibility of targeting concealed antigens has been successfully tested by the use of the only existing "anti-tick" commercial vaccine against the cattle tick *R. microplus*, based on the midgut membrane protein Bm-86 [20, 21].

Blood Meal Uptake

Processes of blood meal uptake and digestion differ significantly between the two evolutionary distinct Argasidae and Ixodidae, as well as between the individual developmental stages of hard ticks [7]. However, morphological, tissue, cellular, and molecular backgrounds of these processes are best explored for female hard ticks, which are thus mainly discussed in this chapter.

Following attachment to the host, blood meal uptake occurs in two phases: (i) a *slow feeding period* (taking about 6 days post-attachment in the case of *Ixodes* spp. ticks), during which time about one-third of total blood meal is ingested and the tick body size continuously grows; (ii) *rapid engorgement*, occurring 12–24 h prior to detachment, in which the main portion (about 2/3) of the host blood is imbibed [8, 22]. Rapid engorgement is conditioned by female fertilization that might occur during feeding in the presence of the male on the host. After the fully engorged female drops off the host, the ingested blood meal is gradually digested and the pool of free amino acids is mainly utilized for synthesis of yolk proteins, – vitellogenins (Vgs), that takes place mainly in the fat body and midgut. Following transport to ovaries, Vgs are proteolytically processed to vitellins (Vns), the major storage proteins for embryonic development and larval stages [7, 12].

Mating Factors

Rapid engorgement of females, conditioned by mating, is triggered by the factor *voraxin*, which is passed from males to females during copulation [23]. *Amblyomma hebraeum* voraxin is composed of two independent peptides, AHEFα and AHEFβ, of MW 16.1, and 11.6 kDa, respectively. Interestingly, rabbits vaccinated with a mixture of recombinant AHEFα and AHEFβ were highly protected against infestation with mated *A. hebraeum* females as only 25% were capable of full engorgement [23]. The vaccination potential of voraxin was also successfully tested against *Rhipicephalus appendiculatus* infested on rabbits immunized with a recombinant homolog of voraxinα from this tick species. Fully engorged *R. appendiculatus* females were reduced in weight by ~40% and also the efficacies of subsequent oviposition and larval hatching were markedly lowered [24].

Secretion of Blood Meal Water

The imbibed blood meal is concentrated and stored in the gut lumen and excessive water is excreted via mechanisms that substantially differ between *N. namaqua*, soft ticks, and hard ticks, providing another piece of evidence in support of independent evolution of hematophagy among these tick families [3, 4]. The ancestral mode of secreting excessive water of blood meal origin via the Malpighian tubules was described in the "living fossil" tick *N. namaqua* [4]. In soft ticks, about 40% of blood meal water is secreted via the coxal glands that are, by contrast, absent in hard ticks [7]. Hard ticks are capable of removing 60–70% of blood meal water by salivation during the rapid engorgement phase. The water is transported from the gut contents via the hemolymph to the salivary glands and its secretion to the host is dependent on the presence of prostaglandin E_2, synthesis of which is specific for the salivary glands of hard ticks [25, 26]. The water channels that allow this massive reverse water flow through the hydrophobic lipidic membranes are named *aquaporins* [27]. The potential of aquaporins as anti-tick molecular targets was recently demonstrated for several tick species [28–31]. A vaccine based on the recombinant aquaporin 1 from the cattle tick *R. microplus* (RmAQP1) exerted high protection against this species as only about one-third of ticks infested on vaccinated cattle completed feeding [30].

Digestion of Blood Meal in the Tick Gut

The entire process of blood digestion takes place in the tick gut, the largest organ (>80%) of the tick body (Figure 7.1). Blood digestion occurs in the central part of the gut (midgut) that is branched into individual protuberances (caeca). In contrast to blood-feeding insects, the tick gut lumen serves mainly as a storage organ and the host blood is degraded intracellularly in the digestive cells of the midgut epithelium [7, 32]. Blood digestion in ticks is a slow and gradual process that allows survival of immature as well as adult ticks for long periods of time (months to years) of starvation. The blood-filled midgut lumen is surrounded by a thin layer of histologically distinguishable epithelial cells and a thin outer layer of muscle fibers. The inner epithelial surface is covered by a peritrophic matrix (PM) composed of mucopolysaccharides underpinned by a chitin network. PM has been described in several hard and soft tick species and is repeatedly formed in each instar during blood meal ingestion [7]. Therefore, the endogenous tick *chitinase* likely plays an important role in inter-stage molting and turnover of the PM. Chitinase cloned and characterized from the hard tick *Haemaphysalis longicornis* is a protein of MW 116 kDa that contains one chitin-binding peritrophin A domain and two glycosyl hydrolase family 18 chitin-binding domains, further corroborating the role of the enzyme as a chitin hydrolase in the tick life cycle [33]. The potential of chitinase as a bioacaricide component was tested using a recombinant baculovirus expressing *H. longicornis* chitinase, giving promising results in increased tick mortality, especially in combination with the pyrethroid flumethrin [34].

The existing nomenclature of tick gut cells is very inconsistent (gastrointestinal, basal, secretory, and other cell types described differently in various tick species), when in reality this may only reflect one type of cell that undergoes asynchronous differentiation during blood meal uptake and digestion. The original basal cells of the intestinal epithelium change to digestive cells that first multiply their proteosynthetic apparatus, and then secrete components of the PM on the surface and begin to digest hemoglobin [35]. During preparation for rapid engorgement (see subsequent text), the cell filled with condensed heme leaves the epithelial layer and migrates toward the inner intestinal lumen. This phenomenon is not apparent in fully engorged females where digestion occurs in all intestinal cells [22].

The imbibed host blood is stored in the gut lumen and the major protein component – hemoglobin, is gradually released from lysed red blood cells (RBCs). Hemolysis is not merely an osmotic or mechanical destruction of erythrocytes, but more likely a complex biochemical process involving extracellular hemolysins. Hemolytic activity of unspecified origin was described in the midgut of *Ixodes dammini* (nowadays *I. scapularis*) [36]. Later, a multi-domain cubulin-like serine protease from *H. longicornis*, tagged as HlSP, was reported to exert hemolytic activity in both *in vivo* and *in vitro* experiments, suggesting its role in initiation of the whole hemoglobinolytic process [37, 38]. Hemoglobin released from the RBCs of some vertebrates tends to form relatively large protein crystals in the tick gut lumen, most likely as a result of a high protein concentration that follows blood meal water secretion [39]. A physiological rationale for this fascinating phenomenon remains unclear, but could possibly be long-term preservation of hemoglobin as a nutrient source and/or physiological protection against potential risks from excess hemoglobin and released free heme.

The process of uptake and digestion of blood meal proteins, termed *heterophagy*, in ticks consists of at least two parallel endocytic mechanisms (Figure 7.2), as elegantly demonstrated by Lara et al. who monitored the intracellular fate of fluorescently labeled hemoglobin and albumin in a primary culture of intestinal cells from *R. microplus* [40]. Serum albumin (and possibly other serum proteins) is nonspecifically transported into the population of small acidic vesicles within tick digestive cells by fluid phase endocytosis (FPE), whereas hemoglobin seems to be recognized specifically by as yet unidentified cell surface receptor(s) and transported via clathrin-coated pits to large endosomal vesicles (receptor-mediated endocytosis, RME) [40, 41]. Intracellular transport and degradation of these two major host blood proteins thus occurs separately (Figure 7.2). The need for specific recognition and endocytosis of hemoglobin evolved most likely as a response to the toxicity of free heme released upon the enzymatic cleavage of hemoglobin. Tick digestive cells adopted a mechanism of heme detoxification by heme accumulation and aggregation within specialized

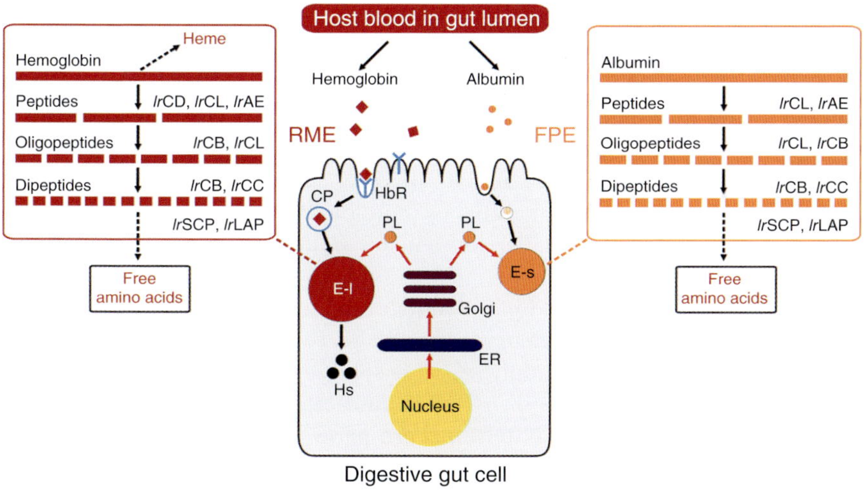

Figure 7.2 The current model of uptake and digestion of major host blood proteins by *Ixodes ricinus* digestive cells. Red arrows – expressional and secretory pathway of hydrolases involved in protein digestion. Black arrows – endocytic pathways of hemoglobin and serum albumin. RME – receptor-mediated endocytosis of hemoglobin; FPE – fluid-phase endocytosis of albumin (and other serum proteins). HbR – hemoglobin receptor (putative, yet not identified); CP – coated pits, E-l – large endosomal vesicles containing hemoglobin; E-s – small endosomal vesicles containing dissolved serum albumin [40]; Hs – hemosomes containing condensed heme [42]; Golgi – Golgi apparatus; ER – endoplasmic reticulum; PL – primary lysosomes. *Ir*AE – *I. ricinus* legumain/AE; *Ir*CD – *I. ricinus* cathepsin D; *Ir*CB – *I. ricinus* cathepsin B; *Ir*CC – *I. ricinus* cathepsin C; *Ir*SCP – *I. ricinus* serine carboxypeptidase; *Ir*LAP – *I. ricinus* leucine aminopeptidase. For details, see the text.

membrane organelles – hemosomes [42] (Figure 7.2). Although hemoglobin seems to be a replaceable source of amino acids, its absence in the tick diet leads to defects in tick embryogenesis [12]. Hence, molecular identification and targeting of the gut surface-localized hemoglobin receptor by a recombinant vaccine has great potential to control tick reproduction.

The previous concept of blood digestion in tick digestive cells by a single "hemoglobinase" was replaced by the current model of multienzyme system that comprises a complex of acidic cysteine and aspartic peptidases involved in processing of host hemoglobin [8, 43, 44] and albumin [41]. Complexes comprising clan CA (cathepsin B, L, and C), clan CD (asparaginyl endopeptidase – legumain), and clan AA (cathepsin D) peptidases are also present in the unicellular malarial agent, *Plasmodium* spp., as well as in the intestines of parasitic platyhelminthes and nematodes (for review, see Ref. [8]). Blood digestion based on acidic lysosomal proteases apparently predates the evolution of the pancreas and serine-based extracellular digestion used in most blood-feeding insects such as mosquitoes, fleas, lice, sandflies, or tse-tse flies.

Despite spatial separation, digestion of hemoglobin and albumin seems to be performed by the same cadre of peptidases [8, 41, 43]. The initial cleavage of hemoglobin is performed by three endopeptidases, namely, cathepsin D (CatD) and cathepsin L (CatL) with a supportive (possibly activating) role of asparaginyl endopeptidase (AE) [43]. By contrast, CatD does not seem to be necessary for the initial cleavage of albumin [41]. The large protein fragments are further cleaved by the endo-peptidolytic activity of cathepsin B (CatB), the most abundant peptidase in the system. The dipeptides are cleaved from the small fragments by the exo-peptidolytic activity of CatB and cathepsin C (CatC). The free amino acid residues are liberated by monopeptidases, comprising a class of serine carboxypeptidases (SCPs) and leucine aminopeptidases (LAPs) [8, 43] (Figure 7.2).

Individual enzymes involved in blood digestion were characterized in several tick species. The hard tick *I. ricinus* possesses three CatD isoenzymes tagged as *Ir*CD1, 2, and 3, out of which only *Ir*CD1 is solely expressed in the gut tissue of partially engorged females [45]. *Ir*CD1 is auto-catalytically activated from its zymogen upon cleavage of its N-terminal pro-part. The activated enzyme preferentially cleaves hemoglobin as well as synthetic substrates between large hydrophobic amino acids in the P1 and P1′ position [45]. RNAi silencing of *ircd1* gene expression resulted in a substantial reduction of CatD activity in the midgut of semi-engorged *I. ricinus* females. The *Ir*CD2 isoform that is expressed in the gut of fully engorged females (post rapid engorgement) is phylogenetically more related to tick aspartic peptidases *Bm*AP from *R. microplus* [46] and longepsin from *H. longicornis* [47]. *Bm*AP was found to be responsible for generation of heme-derived antimicrobial fragments (hemocidins) that were isolated from the gut of fully engorged *R. microplus* females [46, 48].

AEs (legumains) were shown to have a supportive role in the initial phases of hemoglobinolysis and albuminolysis [41, 43]. The *I. scapularis* genome contains nine genes encoding AEs, out of which four seem to be exclusively expressed in the midgut [49]. *Ir*AE1 from *I. ricinus*, the first ever identified and characterized invertebrate AE, is localized intracellularly in digestive cells as well as extracellularly within the PM covering the midgut epithelium of semi-engorged females [50]. *Ir*AE1 has a strict cleavage specificity for asparagine at the P1 position and is irreversibly inactivated at pH > 6.0. Functional characterization of two gut-associated legumains, *Hl*Lgm and *Hl*Lgm2 from *H. longicornis* by RNAi displayed a phenotype in gut cell remodeling and reduced tick post-engorgement weight, oviposition, and hatching rate [51–53].

A cysteine endopeptidase of *CatL*-type is involved in the initial cleavage of blood meal proteins and is capable of substituting for the activity of *Ir*CD1 in hemoglobin digestion after its specific inhibition [43]. At least three isoenzymes of CatL could be identified in the *I. scapularis* genome [11]. The enzyme responsible for CatL activity in the midgut of *I. ricinus* (*Ir*CL1) is markedly upregulated during the slow feeding phase, has a very low pH optimum (pH 3–4), and undergoes autocatalytic activation [54], features described also for *Hl*CPL-A, the CatL from the *H. longicornis* midgut [55]. Expression and activity of *Ir*CL1 is markedly reduced in fully engorged females [22, 54] and its role is most likely substituted by expression of the isoform *Ir*CL3 during the off-host digestive phase [56]. The *Ir*CL1 ortholog in *R. microplus* (*Bm*CL1), together with *Bm*AP, was reported to be involved in production of hemoglobin-derived hemocidins [46].

CatB of the papain family of cysteine peptidases is capable of both endo- and exopeptidase activity. *Ir*CB1 is the most abundant component of the *I. ricinus* digestive apparatus [22, 43, 44]. Western blotting analysis of midgut homogenates using *Ir*CB1-specific antibodies detected the zymogen, prevailing intermediates, and active enzyme of ~38, ~33, and ~31 kDa, respectively [22]. The enzyme activity of the native CatB in the *I. ricinus* midgut extracts has a pH optimum at 5.5–6.0, significantly higher than the pH optima of the initial peptidases *Ir*CD1, *Ir*CL1, and *Ir*AE1. A search of the *I. scapularis* genome, as well as in the rich transcriptomic data from *I. ricinus*, revealed the existence of another two isoforms of CatB, tagged as *Ir*CB2 and *Ir*CB3. Transcripts encoding *Ir*CB1and *Ir*CB2 are expressed more highly in the course of feeding than in the fully engorged females [57]. *Ir*CB3 is orthologous to the *H. longicornis* CatB, termed longipain, which was reported to be involved in blood processing and was shown to exert a babesiacidal effect by killing the midgut stage of *Babesia* parasites in *H. longicornis* [58]. By contrast, expression of *ircb3* mRNA in the *I. ricinus* midgut is marginal and, therefore, the role of this isoenzyme in *I. ricinus* remains unclear [57].

A papain family cysteine peptidase, *CatC* (aka dipeptidyl-peptidase I), is involved in the terminal phase of blood protein digestion by cleavage of dipeptides from fragments produced by the upstream endopeptidases. Only a single gene encoding CatC is present in the *I. scapularis* genome, and accordingly only one type of transcript could be found in *I. ricinus* midgut transcriptomes [57]. The gene encoding *I. ricinus* CatC (IrCC) zymogen (~50 kDa) is mainly expressed in the tick gut but transcripts are also present in other tissues [44]. The pH optimum of native CatC activity in *I. ricinus* gut extracts is similar to that of CatB, pointing to a pH shift toward neutral values along with the blood meal processing [43].

Monopeptidase activities of *SCPs* and *LAPs* types have been detected in *I. ricinus* midgut homogenates [43]. The SCP *Hl*SCP1 was shown to liberate free amino acids from the blood meal–derived peptides in the digestive vesicles of *H. longicornis.* The enzyme is induced by blood meal and is active over a broad range of acidic and neutral pHs [59]. The LAP, termed *Hl*LAP from the same tick species, a member of the M17 family of cytosolic aminopeptidases, was found to be mainly expressed in the cytosol of midgut epithelial cells [60]. The transcription of its encoding gene peaks during the post-feeding period [61]. A follow-up study revealed that *Hl*LAP is also localized to the ovarian cells, indicating its role in the supply of free amino acids for the developing oocytes [62]. The pH optimum and cytosolic localization of *Hl*LAP thus support the concept that at least part of the blood meal processing (cleavage of dipeptides to free amino acids) takes place in the cytosol of digestive gut cells.

Except for the abovementioned RNAi silencing of *hllgm*, *hllgm2*, and *longepsin* in *H. longicornis* [52, 58], other attempts to silence genes encoding individual components of tick digestive machinery by RNAi usually did not substantially affect tick fitness and fecundity, despite a clear reduction in transcription, protein content, and activity of the targeted enzymes in gut tissue extracts from partially engorged females [45, 54]. Vaccination of laboratory animals with individual recombinant digestive enzymes, as well as with a mixed cocktail of recombinant antigens, did not exert any significant protective effect against *I. ricinus* infestation [13]. The limited potential of targeting the tick digestive apparatus may be explained by redundancies in the system, which was also demonstrated *in vitro* by the specific inhibition of individual enzymes that did not prevent completion of hemoglobinolysis or albuminolysis [41, 43].

Heme Detoxification and Intracellular Transport

Ticks detoxify the majority of heme liberated from digested hemoglobin in the digestive vesicles via its accumulation in specialized, membrane-delimited organelles called hemosomes that were first described from the gut cells of the cattle tick *R. microplus* [40, 42]. Heme detoxification via formation of hemosomes (also tagged as residual bodies in soft ticks [7]) is a process functionally analogous to hemozoin formation in other hematophagous parasites such as the malarial *Plasmodium*, triatominae bug *Rhodnius prolixus*, or the flatworm *Schistosoma mansoni* [42]. In contrast to hemozoin, which consists of a crystalline form of heme, the tick hemosomes contain noncrystalline heme aggregates [42].

The mechanism of intracellular heme trafficking within the tick digestive cell is unknown and its depiction in Figure 7.3 is only putative. In the model heme auxotrophic nematode *Caenorhabditis elegans*, heme acquisition was shown to be mediated by the heme-responsive gene (*hrg-1*) [63]. Proteins homologous to HRG-1 were also described in the unicellular, heme auxotrophic parasites of the genus *Trypanosoma* and *Leishmania* [64, 65] which, similar to ticks, exploit a mechanism of acquisition of exogenous heme via RME and lysosomal digestion

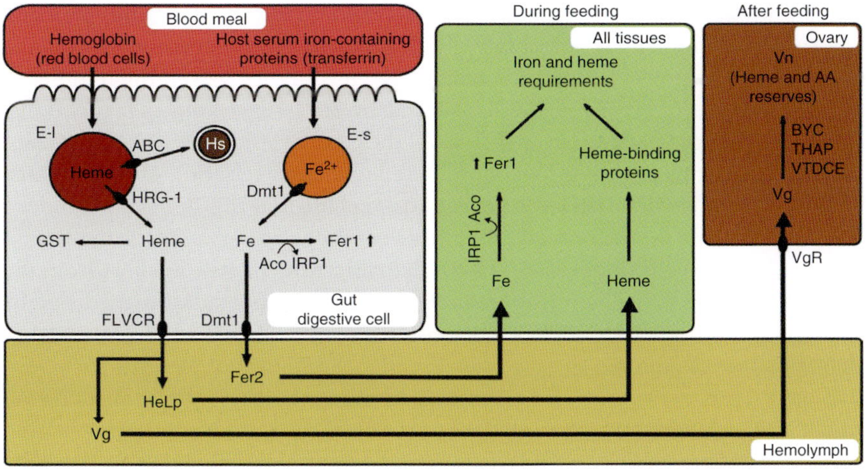

Figure 7.3 A model of putative tick iron and heme metabolic pathways. Iron and heme pathways in ticks are independent as ticks are not capable of heme degradation given the absence of heme oxygenase. Iron is most likely acquired from the host serum transferrin digested in the acidic environment of small endosomal vesicles (E-s) of midgut digestive cells. Upon reduction, ferrous iron is transported from the lysosome by the divalent metal transporter 1 (Dmt1). Once in the cytoplasm, Fe^{2+} ions are scavenged by intracellular ferritin 1 (Fer1), whose translation is strictly regulated by the cytoplasmic Aconitase/Iron responsive protein1 (Aco/IRP1), which senses the iron cellular level. Iron destined to be delivered to the peripheral tissues is transported from digestive cells via DMT1 to the hemolymph and bound to the iron secreted ferritin 2 (Fer2) that functions as an iron transporter. The excessive iron in peripheral tissues is scavenged and stored in Fer1. Heme released from the digested host hemoglobin in the large endosomal vesicles (E-l) is transferred to the cytoplasm via the heme responsive gene 1 (HRG-1) transporter. Ticks detoxify the majority of acquired heme by an ABC transporter (ABC)-mediated transport to hemosomes. Glutathione-S-transferase(s) (GST) serve as an intracellular scavenger of free heme. A small portion of acquired heme required for proteosynthesis of endogenous hemoproteins is exported from the digestive cells to the hemocoel by FLVCR transporter. In hemolymph, heme is bound by the abundant carrier protein(s), heme-lipo-glycoprotein (HeLp), which serves in all developmental stages both as a scavenger of excessive heme and transporter into peripheral tissues. In the post-repletion period of fully engorged females, most of the heme is bound to vitellogenins (Vg) and transported to the ovaries to supply heme metabolic demands of developing embryos and larvae. After entry into the developing oocytes via vitellogenin receptor (VgR), Vg is proteolytically processed to vitellins (Vns) by aspartic proteases BYC and THAP, and cathepsin L-like activity of VTDCE. For details, see the text.

of hemoglobin. These protozoan parasites transport heme required for synthesis of their endogenous hemoproteins from the endolysosomes to the cytosol via HRG transporters; these present very promising targets for rational development of antiparasitic drugs [64]. One gene related to *hrg-1* (ISCW001847) was identified in the *I. scapularis* genome [11] and its corresponding ortholog in *I. ricinus* (Gen Bank GEFM01005533) was found to be expressed in ovaries and the midgut of adult females, in all developmental stages [66]. The gene encodes a protein of

204 amino acids, with no signal sequence and one HRG-superfamily domain. RNAi-mediated silencing of putative tick *hrg-1* had no marked effect on the number and weight of females that completed feeding. However, the number of females capable of laying eggs was reduced by 30%, suggesting that functional HRG-1 is involved in tick reproduction. As the conservation of HRG-1 molecules among metazoans is rather low, more detailed characterization and functional studies of putative tick HRG-1 is needed to conclude that this molecule is indeed the transporter of heme from digestive vesicles (endolysosomes) to the cytoplasm of tick digestive cells (Figure 7.3). Heme detoxification via its transport from the digestive vesicles to the hemosomes has been recently reported to be mediated by the *ATP-binding cassette* (ABC) transporter (Figure 7.3) that is also involved in detoxification of amitraz, the acaricide used for the control of the cattle tick population [67]. It remains to be examined whether the tick ABC transporter(s) might also be involved in extracellular heme export, as was recently demonstrated in *C. elegans* for MRP-5, a multidrug resistance protein belonging to the ABC transporter family [68].

In mammalian macrophages, heme originating from degradation of RBC is exported via *FLVCR* (the cell surface receptor for feline leukemia virus, subgroup C) [69, 70]. The gene (ISCW022805) coding for a protein related to FLVCR was also identified in the genome of *I. scapularis*. The putative tick FLVCR is a protein of 413 amino acids that clearly belongs to the large and diverse group of secondary transporters of the major facilitator superfamily (MSF), and contains a signal sequence peptide, ten transmembrane motifs, and predicted target sequences for localization to the cellular plasma membrane or Golgi apparatus vesicles [66]. The BLAST (Basic Local Alignment Search Tool) search at https://blast.ncbi.nlm.nih.gov/Blast.cgi displayed up to 50% sequence identity to FLVCR-related proteins of other organisms such as the horseshoe crab or *Drosophila* sp. Expression of a corresponding *flvcr* ortholog in *I. ricinus* developmental stages or tissues of adult female was undetectable by quantitative reverse transcription polymerase chain reaction (qRT-PCR) analysis, and RNAi-mediated silencing of *flvcr* in this species revealed no phenotype. However, it is possible that the RNAi experiment did not hit the right target as transcriptomes from *I. ricinus* midgut or *I. scapularis* synganglion contain transcripts coding for different FLVCR-related proteins of the MSF family (Gen Bank GEFM01002811, GANP01002798, and GBBN01005999, respectively).

Glutathione S-transferases (GSTs) form a family of enzymes that catalyze the reaction of xenobiotics as well as endogenous molecules with reduced glutathione and thus facilitate their solubility and biological detoxification [66, 71]. GSTs were proposed to bind heme in several blood-feeding parasites, the nematodes *Haemonchus contortus* [72] and *Ancylostoma caninum* [73], or the malarial parasite *Plasmodium falciparum* [74]. As such, GSTs have long been in focus as promising antiparasitic targets [73, 75] and GST, together with the digestive CatD-like peptidase Na-ASP1, eventually became a leading vaccine candidate in vaccine development against the human hookworm *Necator americanus* [76]. In ticks, the GSTs were characterized in *H. longicornis* and *R. appendiculatus* [77].

Vaccination of cattle with recombinant GST from *H. longicornis* exerted a high (about 50%) protection against infestation with *R. microplus*, given the cross-reactivity of anti GST-antibodies in both species [78]. On the other hand, identification of the best "anti-tick" target among the tick GST family might be a quite demanding task, as the *I. scapularis* genome, for example, contains more than 40 genes annotated as putative *gst* [11]. It remains to be investigated whether any tick GSTs participate in heme binding and/or metabolism in the tick gut.

Inter-tissue Transport of Heme

Despite heme auxotrophy of ticks, more than 200 gene encoding enzymes that utilize heme as a cofactor were identified in the genome of *I. scapularis* [12]. Among the hemoproteins that are vitally important for basic metabolism of ticks are respiratory chain cytochromes, catalase, and a large number of genes encoding members of the cytochrome P450 family. Certainly, to fulfill the heme demands for synthesis of endogenous hemoproteins in all tick developmental stages, effective transport and distribution of heme from the site of hemoglobin digestion to the peripheral tissues has to be secured throughout the tick life cycle. Moreover, it was recently demonstrated by artificial membrane feeding of ticks on hemoglobin-depleted serum that the presence of heme in developing ovaries is absolutely necessary for successful embryogenesis [12]. Therefore, efficiently targeting heme transport to the ovaries of mated, fully engorged females can substantially reduce tick populations, a goal of particular importance for one-host tick species, the cattle tick *R. microplus*. The inter-tissue transport or scavenging of heme in all tick stages is facilitated by the abundant *hemolymph carrier proteins* (CPs) also termed HeLp (for hemo-lipo-glycoprotein) [79], whereas heme transport to the ovaries of fully engorged females seems instead to be mediated by *heme-binding* Vgs [12, 80–82] (Figure 7.3). Tick CPs and Vgs belong to the same family of large lipid transfer proteins known to facilitate the circulation of hydrophobic molecules across bodies of vertebrate as well as invertebrate animals [83]. To clearly distinguish between genes encoding tick CPs from those encoding Vgs, their expression profile has to be known. While HeLp/CPs are expressed ubiquitously in all tick stages including males, Vgs are expressed only in fertilized, fully engorged females [12]. The genome of *I. scapularis* contains at least five carrier proteins (*cp1–5*) [11]. In *I. ricinus*, the *cp3* ortholog, tagged as *ircp3*, was identified and sequenced (GenBank KP663716). It encodes a protein of 1537 amino acid residues including the signal peptide with a theoretical MW of about 175 kDa. In accordance with aforementioned criteria, the *ircp3* is expressed in all developmental stages and is upregulated by blood feeding. In adult females, *ircp3* is mainly expressed by the fat body associated with trachea, and to a lesser extent also in salivary glands and ovaries. Silencing of *ircp3* by RNAi reduced the amount of the protein in hemolymph by about 80% and correspondingly lowered the concentration of associated heme [12]. On the other hand, silencing of *ircp3* did not markedly reduce the amount of heme present in tick ovaries (Perner, unpublished data). The *I. scapularis* genome encodes two Vg molecules *Is*Vg1 (ISCW013727) and *Is*Vg2 (ISCW021228) that differ by the absence of the DUF1943 domain in the latter. The corresponding *I. ricinus* orthologs are preferentially synthesized in the tick gut and fat body. Their silencing by RNAi revealed that heme transport to developing ovaries occurs mainly during the off-host digestive phase and is dependent on Vgs [12]. Preliminary results further suggest that *Ir*Vg1 has a higher affinity for heme binding than *Ir*Vg2 (Perner, unpublished data). Whether the DUF1943 domain that is also present in *Ir*CP3 is responsible for heme-binding capacity remains to be examined. It is also possible that in the native state, *Ir*Vg1 and *Ir*Vg2 form a functional heterodimer capable of heme binding and transport.

The crucial function in Vg and most likely also Vg-bound heme uptake to tick ovaries is mediated by the *vitellogenin receptor* (VgR), which was characterized in three hard-tick species, the American dog tick *Dermacentor variabilis* [84], the Asian hard tick *H. longicornis* [85], and in the African bont tick *A. hebraeum* [86]. Tick VgRs are large membrane proteins, of about MW 200 kDa, sharing the common multi-domain architecture comprising two ligand-binding sites, two epidermal growth factor (EGF)-precursor domains, an O-sugar-binding domain, a transmembrane domain and a cytoplasmic C-terminal tail [84–86]. In all these studies, the function of VgR was examined by RNA interference, clearly demonstrating that VgR knockdown led to impaired development of heme-depleted oocytes, finally resulting in limited egg production. These findings present tick VgRs as a good target capable of reducing tick reproduction. However, the key to the rational control of vitellogenesis and Vg/heme uptake by tick oocytes seems to be rather in targeting hormonal regulation of Vg and VgR expression, which is still inadequately understood [87, 88].

Embryogenesis and Fertility

Rational targeting of oogenesis and embryogenesis would be specifically important for the control of the cattle tick *R. microplus* population. To date, three enzymes processing Vgs in *R. microplus* eggs have been characterized: (i) An aspartic peptidase, BYC (Boophilus yolk pro-cathepsin D) was isolated from tick eggs [89] and its molecular cloning revealed that BYC lacks the highly conserved second catalytic Asp residue that is essential for CatD-type endoproteolytic activity [90]. Despite this, the recombinant as well as isolated native BYC exerted limited proteolytic activity, which is actually a desired feature for the slow degradation of Vn in the course of embryonal development [90]. (ii) *THAP*, a tick-heme-binding aspartic peptidase isolated and cloned from *R. microplus* eggs, has conserved both catalytic Asp residues and is specifically active against hemoproteins [91]. The authors conclude that THAP uses heme bound to Vns as a docking site to increase the specificity of degradation of its physiological substrate – Vn and regulation of gradual heme supply for the developing embryos [91]. (iii) Another enzyme proposed to be involved in Vn processing in *R. microplus* eggs has been described as CatL-like vitellogenin degrading cysteine endopeptidase (VTDCE) [92]. This enzyme was purified from tick eggs and characterized as CatL based on its substrate/inhibitor specificity. However, the molecular mass of the purified protein did not match the size of CatL-type peptidases and recent molecular cloning revealed that VTDCE is more related to tick antimicrobial peptides of microplusin and/or hebraein types [93]. Vaccination of cattle with native or recombinant BYC and isolated VTDCE conferred only limited protection (some 25%) in overall efficacy against *R. microplus* ticks [94–96].

Iron Acquisition and Metabolism

Iron is an essential element that acts as an electron donor/acceptor involved in vitally important physiological processes across the whole animal kingdom. Iron or iron–sulfur (Fe–S) clusters are core components of many enzymes functioning, for instance, in the respiratory chain of mitochondria, DNA biosynthesis, and energy metabolism [97]. The major source of iron for most known animals, including hematophagous insects, originates from heme degradation that is catalyzed by HO [98]. As mentioned, this enzyme is, however, absent from the tick genome [11]. Artificial membrane feeding *I. ricinus* females on hemoglobin-depleted serum revealed that the amount of iron in tick tissues does not depend on hemoglobin in the diet, experimentally proving that hemoglobin-derived heme is not a source of iron for ticks [12]. The lack of heme degradation thus raises the question of the dietary source of iron for ticks, which is most likely explained by acquisition of a sufficient amount of nonheme iron from the host serum transferrin [99–101] (Figure 7.3). Mammalian cells uptake iron from circulating transferrin via RME of the transferrin/transferrin receptor complex, followed by iron release in the mildly acidic environment of endosomes [102, 103]. A similar mechanism of iron acquisition from host transferrin was also described for the blood-stream form of the sleeping sickness agent, *Trypanosoma brucei.* However, the transferrin receptor of this protozoan parasite is structurally completely different from its mammalian counterpart [104, 105]. No protein related either to mammalian or trypanosomal transferrin receptor could be found by BLAST searches of the available tick genome and/or transcriptome databases. It is possible that the release of iron from the host transferrin occurs in the acidic environment of digestive vesicles along with digestion of other serum proteins. Released iron must be first reduced to Fe^{2+} before its transport from the lysosome to the cytoplasm, mediated in other organisms by the *divalent metal-transporter* (Dmt1), also tagged as malvolio in *Drosophila* [106]. A gene encoding a putative *dmt1/malvolio* homologue was identified in the *I. ricinus* midgut transcriptomes (GenBank GANP01004329 or GEFM01002799). The tick *dmt1* gene was shown to be expressed in all developmental stages and tissues. However, its silencing by RNAi did not result in any obvious phenotype [66].

Free iron is potentially toxic for all living cells as it participates in the formation of free oxygen radicals and its intracellular levels must therefore be strictly maintained at low levels [107]. This function is carried out by the intracellular iron storage protein ferritin, referred to here as *ferritin1* (Fer1) (Figure 7.3). Fer1, first characterized in the hard tick *I. ricinus* and the soft tick *Ornithodoros moubata*, shared high-sequence similarities and were closely related to the mammalian heavy-chain ferritins, including the typically conserved motifs for ferroxidase center [108]. Tick Fer1 are proteins of MW about 20 kDa, which, in the native state, form homopolymers of MW about 500 kDa, most likely composed of 24 subunits as in vertebrates. The 5′-untranslated regions of tick *fer1* mRNA contains a stem-loop structure of an iron-responsive element (IRE) (the only IRE found in the tick genome), which points to the regulation of Fer1 at the translational level by an *iron-responsive protein1* (IRP1) (Figure 7.3). Increasing levels of intracellular iron allow insertion of newly synthesized Fe–S clusters into IRP, which then becomes an active cytoplasmic aconitase that subsequently detaches from *fer1* mRNA IRE thereby allowing its translation. This was experimentally proved by RNAi silencing of the *I. ricinus irp1* gene, which resulted in a marked increase in Fer1 protein levels in tick tissues [12, 100]. The *irp1* KD did not affect tick feeding, but exerted a clear impact on tick reproduction as larval hatching from laid eggs was significantly reduced [100]. Unlike vertebrates, ticks possess another form of heavy-chain-type ferritin, called *ferritin2* (Fer2) that is synthesized mainly in the tick gut and secreted into the hemolymph [100] (Figure 7.3). RNAi KD of *fer2* resulted in a substantial decrease in Fer1 levels in tick peripheral tissues, suggesting that Fer2 plays a role in iron inter-tissue transport [100]. It remains an unresolved issue whether iron is loaded into Fer2 inside the midgut cells or if iron is first secreted to the hemolymph and subsequently scavenged by Fer2. Impairment of iron storage and transport by RNAi KD of tick ferritins 1 and 2 severely affects tick development and reproduction as first demonstrated in *I. ricinus* [100] and later in *H. longicornis* [109, 110]. Fer2 possesses all the important attributes of a suitable concealed antigen as it has no counterpart in mammals, is encoded by a single gene, and is mainly expressed in the tick gut where it comes into direct contact with ingested host blood. This all makes Fer2 a promising candidate for development of an "anti-tick" vaccine. The concept of using recombinant tick Fer2 for vaccination of animals against tick infestation was successfully demonstrated on laboratory rabbits against *I. ricinus* [111] and *H. longicornis* [112]. More importantly, vaccination of cattle with recombinant Fer2 from *R. microplus* exerted a protective effect against this one-host tick species that was comparable with the commercial vaccine based on Bm-86 [111].

The *I. scapularis* genome contains a gene encoding a putative *transferrin* (GenBank XM_002400404), but a phylogenetic analysis of this gene revealed that it is most closely related to the insect type 2 transferrins (Tf2) (also termed melanotransferrins) [66]. The function of Tf2 remained obscure until a study showed that *Drosophila* Tf2, capable of binding iron, is a component of epithelial septate junctions and apparently does not play a role in inter-tissue iron transport [113]. In line with this finding, RNAi KD of the *tf2* ortholog in *I. ricinus* did not affect iron supply into tick tissues, as monitored by the levels of Fer1 [66].

Conclusions

Although our understanding of blood meal digestion and associated physiological processes in the tick gut has progressed remarkably during the past decade, it nevertheless remains limited mainly to adult hard-tick females during their on-host feeding phase. In contrast, we know almost nothing about blood digestion in fully engorged females following their detachment from the host. An in-depth knowledge of molecules and processes capable of transforming the huge amount of ingested blood into the imposing egg mass laid by the females would be especially useful for the control of tick reproduction and for reducing their population in the field. In order to protect hosts from tick infestation and transmission of tick-borne diseases, it would, however, be particularly important to effectively impair blood uptake and digestion during the early stages of feeding of both adult and immature ticks. Preliminary results suggest that the activities of digestive enzymes stay low in virgin females and are only upregulated in fertilized females [56]. Therefore, male factors such as voraxins [23], most likely linked with hormonal control, trigger the exponential upregulation of digestive enzymes during the slow-feeding period in fertilized females [22]. But a similar upregulation of the digestive apparatus, which is obviously independent of mating status, also occurs in the nymphal stage (Konvičková, unpublished results). What is then the triggering stimulus? Assuming that nutrient sensing is responsible, what blood meal component is important for that? These questions might be experimentally addressed using *in vitro* membrane feeding techniques (see Chapter 9 by Tyson and Nijhof). The necessary prerequisite for such experiments is developing a chemically defined diet, similar to that recently implemented for the mosquito *Aedes aegypti* [114]. The other possibility of sensing incoming blood could be via neuropeptidergic signaling, facilitated by the complex system of neurons present in the tick hindgut [115].

Blood-feeding arthropods are challenged not only by a surplus of heme and iron but also by a huge amount of amino acids originating from their protein-rich diet. Hence, functional catabolism of amino acids is also essential for their survival, as recently demonstrated for the triatominae bug *R. prolixus* by RNAi silencing and/or inhibition of the tyrosine degradative pathway in this insect [116]. Disabling tyrosine detoxification by specific inhibitors also caused the premature death of *A. aegypti* and of the cattle tick *R. microplus*, suggesting the potential of using these compounds to selectively target arthropod blood feeders [116].

In addition to the molecules and processes described in this chapter, there will definitely appear in the future many other targets that may eventually turn out to be the right keys to reach the ultimate goal – discovery of effective anti-tick vaccines and/or selective acaricides protecting us and our household animals against ticks and infectious diseases that they transmit.

Acknowledgments

This work was primarily supported by the Czech Science Foundation grant no. 13-11043S to P.K. and, additionally by grants nos. 14-33693S, 17-27386S, 17-27393S, and 18-01832S to D.S., O.H., R.S., and P.K., respectively. We thank

Martina Hajduskova (www.biographix.cz) for the graphical design and processing of the figures.

References

1. Guglielmone, A.A., Robbins, R.G., Apanaskevich, D.A., Petney, T.N., Estrada-Pena, A., Horak, I.G., Shao, R.F., and Barker, S.C. (2010) The Argasidae, Ixodidae and Nuttalliellidae (Acari: Ixodida) of the world: a list of valid species names. *Zootaxa*, **2528**, 1–28.
2. Nava, S., Guglielmone, A.A., and Mangold, A.J. (2009) An overview of systematics and evolution of ticks. *Front. Biosci. (Landmark Ed.)*, **14**, 2857–2877.
3. Mans, B.J., de Castro, M.H., Pienaar, R., de Klerk, D., Gaven, P., Genu, S., and Latif, A.A. (2016) Ancestral reconstruction of tick lineages. *Ticks Tick Borne Dis.*, **7**, 509–535.
4. Mans, B.J., de Klerk, D., Pienaar, R., and Latif, A.A. (2011) *Nuttalliella namaqua*: a living fossil and closest relative to the ancestral tick lineage: implications for the evolution of blood-feeding in ticks. *PLoS One*, **6**, e23675.
5. de la Fuente, J., Estrada-Pena, A., Venzal, J.M., Kocan, K.M., and Sonenshine, D.E. (2008) Overview: ticks as vectors of pathogens that cause disease in humans and animals. *Front. Biosci.*, **13**, 6938–6946.
6. Mans, B.J. and Neitz, A.W. (2004) Adaptation of ticks to a blood-feeding environment: evolution from a functional perspective. *Insect Biochem. Mol. Biol.*, **34**, 1–17.
7. Sonenshine, D.E. and Roe, R.M. (2014) Biology of Ticks, 2nd edn, vol. **1**, Oxford University Press, Oxford.
8. Sojka, D., Franta, Z., Horn, M., Caffrey, C.R., Mares, M., and Kopacek, P. (2013) New insights into the machinery of blood digestion by ticks. *Trends Parasitol.*, **29**, 276–285.
9. Graca-Souza, A.V., Maya-Monteiro, C., Paiva-Silva, G.O., Braz, G.R., Paes, M.C., Sorgine, M.H., Oliveira, M.F., and Oliveira, P.L. (2006) Adaptations against heme toxicity in blood-feeding arthropods. *Insect Biochem. Mol. Biol.*, **36**, 322–335.
10. Braz, G.R., Coelho, H.S., Masuda, H., and Oliveira, P.L. (1999) A missing metabolic pathway in the cattle tick *Boophilus microplus*. *Curr. Biol.*, **9**, 703–706.
11. Gulia-Nuss, M., Nuss, A.B., Meyer, J.M., Sonenshine, D.E., Roe, R.M., Waterhouse, R.M., Sattelle, D.B., de la Fuente, J., Ribeiro, J.M., Megy, K., Thimmapuram, J., Miller, J.R., Walenz, B.P., Koren, S., Hostetler, J.B., Thiagarajan, M., Joardar, V.S., Hannick, L.I., Bidwell, S., Hammond, M.P., Young, S., Zeng, Q., Abrudan, J.L., Almeida, F.C., Ayllon, N., Bhide, K., Bissinger, B.W., Bonzon-Kulichenko, E., Buckingham, S.D., Caffrey, D.R., Caimano, M.J., Croset, V., Driscoll, T., Gilbert, D., Gillespie, J.J., Giraldo-Calderon, G.I., Grabowski, J.M., Jiang, D., Khalil, S.M., Kim, D., Kocan, K.M., Koci, J., Kuhn, R.J., Kurtti, T.J., Lees, K., Lang, E.G., Kennedy, R.C., Kwon, H., Perera, R., Qi, Y., Radolf, J.D., Sakamoto, J.M., Sanchez-Gracia, A., Severo, M.S., Silverman, N., Simo, L., Tojo, M., Tornador, C., Van Zee, J.P., Vazquez, J., Vieira, F.G., Villar, M., Wespiser, A.R., Yang, Y., Zhu, J., Arensburger, P., Pietrantonio, P.V., Barker, S.C., Shao, R., Zdobnov, E.M., Hauser, F., Grimmelikhuijzen, C.J., Park, Y., Rozas, J., Benton, R., Pedra, J.H., Nelson, D.R., Unger, M.F., Tubio, J.M., Tu, Z., Robertson, H.M., Shumway, M., Sutton, G., Wortman, J.R., Lawson, D., Wikel, S.K., Nene, V.M., Fraser, C.M., Collins, F.H., Birren, B., Nelson, K.E., Caler, E., and Hill, C.A. (2016) Genomic insights into the *Ixodes scapularis* tick vector of Lyme disease. *Nat. Commun.*, **7**, 10507.

12. Perner, J., Sobotka, R., Sima, R., Konvickova, J., Sojka, D., Oliveira, P.L., Hajdusek, O., and Kopacek, P. (2016) Acquisition of exogenous haem is essential for tick reproduction. *eLife*, **5**, e12318.
13. de la Fuente, J., Kopacek, P., Lew-Tabor, A., and Maritz-Olivier, C. (2016) Strategies for new and improved vaccines against ticks and tick-borne diseases. *Parasite Immunol.* doi: 10.1111/pim.12339
14. Chmelar, J., Kotal, J., Kopecky, J., Pedra, J.H., and Kotsyfakis, M. (2016) All for one and one for all on the tick-host battlefield. *Trends Parasitol.*, **32**, 368–377.
15. Hajdusek, O., Sima, R., Ayllon, N., Jalovecka, M., Perner, J., de la Fuente, J., and Kopacek, P. (2013) Interaction of the tick immune system with transmitted pathogens. *Front. Cell. Infect. Microbiol.*, **3**, 26.
16. Francischetti, I.M., Sa-Nunes, A., Mans, B.J., Santos, I.M., and Ribeiro, J.M. (2009) The role of saliva in tick feeding. *Front. Biosci. (Landmark Ed.)*, **14**, 2051–2088.
17. Chmelar, J., Calvo, E., Pedra, J.H., Francischetti, I.M., and Kotsyfakis, M. (2012) Tick salivary secretion as a source of antihemostatics. *J. Proteomics*, **75**, 3842–3854.
18. Chmelar, J., Kotal, J., Karim, S., Kopacek, P., Francischetti, I.M., Pedra, J.H., and Kotsyfakis, M. (2016) Sialomes and mialomes: a systems-biology view of tick tissues and tick-host interactions. *Trends Parasitol.*, **32**, 242–254.
19. Nuttall, P.A., Trimnell, A.R., Kazimirova, M., and Labuda, M. (2006) Exposed and concealed antigens as vaccine targets for controlling ticks and tick-borne diseases. *Parasite Immunol.*, **28**, 155–163.
20. de la Fuente, J., Almazan, C., Canales, M., Perez de la Lastra, J.M., Kocan, K.M., and Willadsen, P. (2007) A ten-year review of commercial vaccine performance for control of tick infestations on cattle. *Anim. Health Res. Rev.*, **8**, 23–28.
21. Willadsen, P. and Kemp, D.H. (1988) Vaccination with 'concealed' antigens for tick control. *Parasitol. Today*, **4**, 196–198.
22. Franta, Z., Frantova, H., Konvickova, J., Horn, M., Sojka, D., Mares, M., and Kopacek, P. (2010) Dynamics of digestive proteolytic system during blood feeding of the hard tick *Ixodes ricinus*. *Parasit. Vectors*, **3**, 119.
23. Weiss, B.L. and Kaufman, W.R. (2004) Two feeding-induced proteins from the male gonad trigger engorgement of the female tick *Amblyomma hebraeum*. *Proc. Natl. Acad. Sci. U. S. A.*, **101**, 5874–5879.
24. Yamada, S., Konnai, S., Imamura, S., Ito, T., Onuma, M., and Ohashi, K. (2009) Cloning and characterization of *Rhipicephalus appendiculatus* voraxinalpha and its effect as anti-tick vaccine. *Vaccine*, **27**, 5989–5997.
25. Bowman, A.S., Dillwith, J.W., and Sauer, J.R. (1996) Tick salivary prostaglandins: presence, origin and significance. *Parasitol. Today*, **12**, 388–396.
26. Meredith, J. and Kaufman, W.R. (1973) A proposed site of fluid secretion in the salivary gland of the ixodid tick *Dermacentor andersoni*. *Parasitology*, **67**, 205–217.
27. Campbell, E.M., Ball, A., Hoppler, S., and Bowman, A.S. (2008) Invertebrate aquaporins: a review. *J. Comp. Physiol. B.*, **178**, 935–955.
28. Ball, A., Campbell, E.M., Jacob, J., Hoppler, S., and Bowman, A.S. (2009) Identification, functional characterization and expression patterns of a water-specific aquaporin in the brown dog tick, *Rhipicephalus sanguineus*. *Insect Biochem. Mol. Biol.*, **39**, 105–112.
29. Campbell, E.M., Burdin, M., Hoppler, S., and Bowman, A.S. (2010) Role of an aquaporin in the sheep tick *Ixodes ricinus*: assessment as a potential control target. *Int. J. Parasitol.*, **40**, 15–23.
30. Guerrero, F.D., Andreotti, R., Bendele, K.G., Cunha, R.C., Miller, R.J., Yeater, K., and Perez de Leon, A.A. (2014) *Rhipicephalus* (*Boophilus*) *microplus* aquaporin as an effective vaccine antigen to protect against cattle tick infestations. *Parasit. Vectors*, **7**, 475.

31. Hussein, H.E., Scoles, G.A., Ueti, M.W., Suarez, C.E., Adham, F.K., Guerrero, F.D., and Bastos, R.G. (2015) Targeted silencing of the Aquaporin 2 gene of *Rhipicephalus* (*Boophilus*) *microplus* reduces tick fitness. *Parasit. Vectors*, **8**, 618.
32. Coons, L.B., Rosell-Davis, R., and Tarnowski, B.I. (1986) Blood meal digestion in ticks, in Morphology, Physiology, and Behavioural Biology of Ticks (eds J.R. Sauer and J.A. Hair), Ellis Horwood Ltd./John Wiley & Sons, Inc., New York, pp. 248–279.
33. You, M., Xuan, X., Tsuji, N., Kamio, T., Taylor, D., Suzuki, N., and Fujisaki, K. (2003) Identification and molecular characterization of a chitinase from the hard tick *Haemaphysalis longicornis*. *J. Biol. Chem.*, **278**, 8556–8563.
34. Assenga, S.P., You, M., Shy, C.H., Yamagishi, J., Sakaguchi, T., Zhou, J., Kibe, M.K., Xuan, X., and Fujisaki, K. (2006) The use of a recombinant baculovirus expressing a chitinase from the hard tick *Haemaphysalis longicornis* and its potential application as a bioacaricide for tick control. *Parasitol. Res.*, **98**, 111–118.
35. Agyei, A.D. and Runham, N.W. (1995) Studies on the morphological changes in the midguts of two ixodid tick species *Boophilus microplus* and *Rhipicephalus appendiculatus* during digestion of the blood meal. *Int. J. Parasitol.*, **25**, 55–62.
36. Ribeiro, J.M. (1988) The midgut hemolysin of *Ixodes dammini* (Acari: Ixodidae). *J. Parasitol.*, **74**, 532–537.
37. Miyoshi, T., Tsuji, N., Islam, M.K., Huang, X., Motobu, M., Alim, M.A., and Fujisaki, K. (2007) Molecular and reverse genetic characterization of serine proteinase-induced hemolysis in the midgut of the ixodid tick *Haemaphysalis longicornis*. *J. Insect Physiol.*, **53**, 195–203.
38. Miyoshi, T., Tsuji, N., Islam, M.K., Kamio, T., and Fujisaki, K. (2004) Cloning and molecular characterization of a cubilin-related serine proteinase from the hard tick *Haemaphysalis longicornis*. *Insect Biochem. Mol. Biol.*, **34**, 799–808.
39. Smit, J.D., Grandjean, O., Guggenheim, R., and Winterhalter, K.H. (1977) Haemoglobin crystals in the midgut of the tick *Ornithodorus moubata* Murray. *Nature*, **266**, 536–538.
40. Lara, F.A., Lins, U., Bechara, G.H., and Oliveira, P.L. (2005) Tracing heme in a living cell: hemoglobin degradation and heme traffic in digest cells of the cattle tick *Boophilus microplus*. *J. Exp. Biol.*, **208**, 3093–3101.
41. Sojka, D., Pytelkova, J., Perner, J., Horn, M., Konvickova, J., Schrenkova, J., Mares, M., and Kopacek, P. (2016) Multienzyme degradation of host serum albumin in ticks. *Ticks Tick Borne Dis.*, **7**, 604–613.
42. Lara, F.A., Lins, U., Paiva-Silva, G., Almeida, I.C., Braga, C.M., Miguens, F.C., Oliveira, P.L., and Dansa-Petretski, M. (2003) A new intracellular pathway of haem detoxification in the midgut of the cattle tick *Boophilus microplus*: aggregation inside a specialized organelle, the hemosome. *J. Exp. Biol.*, **206**, 1707–1715.
43. Horn, M., Nussbaumerova, M., Sanda, M., Kovarova, Z., Srba, J., Franta, Z., Sojka, D., Bogyo, M., Caffrey, C.R., Kopacek, P., and Mares, M. (2009) Hemoglobin digestion in blood-feeding ticks: mapping a multipeptidase pathway by functional proteomics. *Chem. Biol.*, **16**, 1053–1063.
44. Sojka, D., Franta, Z., Horn, M., Hajdusek, O., Caffrey, C.R., Mares, M., and Kopacek, P. (2008) Profiling of proteolytic enzymes in the gut of the tick *Ixodes ricinus* reveals an evolutionarily conserved network of aspartic and cysteine peptidases. *Parasit. Vectors*, **1**, 7.
45. Sojka, D., Franta, Z., Frantova, H., Bartosova, P., Horn, M., Vachova, J., O'Donoghue, A.J., Eroy-Reveles, A.A., Craik, C.S., Knudsen, G.M., Caffrey, C.R., McKerrow, J.H., Mares, M., and Kopacek, P. (2012) Characterization of gut-associated cathepsin D hemoglobinase from tick *Ixodes ricinus* (IrCD1). *J. Biol. Chem.*, **287**, 21152–21163.
46. Cruz, C.E., Fogaca, A.C., Nakayasu, E.S., Angeli, C.B., Belmonte, R., Almeida,

I.C., Miranda, A., Miranda, M.T., Tanaka, A.S., Braz, G.R., Craik, C.S., Schneider, E., Caffrey, C.R., and Daffre, S. (2010) Characterization of proteinases from the midgut of *Rhipicephalus (Boophilus) microplus* involved in the generation of antimicrobial peptides. *Parasit. Vectors*, **3**, 63.

47. Boldbaatar, D., Sikalizyo Sikasunge, C., Battsetseg, B., Xuan, X., and Fujisaki, K. (2006) Molecular cloning and functional characterization of an aspartic protease from the hard tick *Haemaphysalis longicornis. Insect Biochem. Mol. Biol.*, **36**, 25–36.

48. Fogaca, A.C., da Silva, P.I. Jr., Miranda, M.T., Bianchi, A.G., Miranda, A., Ribolla, P.E., and Daffre, S. (1999) Antimicrobial activity of a bovine hemoglobin fragment in the tick *Boophilus microplus. J. Biol. Chem.*, **274**, 25330–25334.

49. Hartmann, D., Šíma, R., Konvičková, J., Perner, J., Kopáček, P., Sojka, D. (2018) Multiple legumain isoenzymes in ticks. *Int. J. Parasitol.*, **48**, 167–178.

50. Sojka, D., Hajdusek, O., Dvorak, J., Sajid, M., Franta, Z., Schneider, E.L., Craik, C.S., Vancova, M., Buresova, V., Bogyo, M., Sexton, K.B., McKerrow, J.H., Caffrey, C.R., and Kopacek, P. (2007) IrAE – an asparaginyl endopeptidase (legumain) in the gut of the hard tick *Ixodes ricinus. Int. J. Parasitol.*, **37**, 713–724.

51. Abdul Alim, M., Tsuji, N., Miyoshi, T., Khyrul Islam, M., Huang, X., Motobu, M., and Fujisaki, K. (2007) Characterization of asparaginyl endopeptidase, legumain induced by blood feeding in the ixodid tick *Haemaphysalis longicornis. Insect Biochem. Mol. Biol.*, **37**, 911–922.

52. Alim, M.A., Tsuji, N., Miyoshi, T., Islam, M.K., Hatta, T., and Fujisaki, K. (2009) Legumains from the hard tick *Haemaphysalis longicornis* play modulatory roles in blood feeding and gut cellular remodelling and impact on embryogenesis. *Int. J. Parasitol.*, **39**, 97–107.

53. Alim, M.A., Tsuji, N., Miyoshi, T., Islam, M.K., Huang, X., Hatta, T., and Fujisaki, K. (2008) HlLgm2, a member of asparaginyl endopeptidases/legumains in the midgut of the ixodid tick *Haemaphysalis longicornis*, is involved in blood-meal digestion. *J. Insect Physiol.*, **54**, 573–585.

54. Franta, Z., Sojka, D., Frantova, H., Dvorak, J., Horn, M., Srba, J., Talacko, P., Mares, M., Schneider, E., Craik, C.S., McKerrow, J.H., Caffrey, C.R., and Kopacek, P. (2011) IrCL1 – the haemoglobinolytic cathepsin L of the hard tick, *Ixodes ricinus. Int. J. Parasitol.*, **41**, 1253–1262.

55. Yamaji, K., Tsuji, N., Miyoshi, T., Islam, M.K., Hatta, T., Alim, M.A., Anisuzzaman, M., Takenaka, A., and Fujisaki, K. (2009) Hemoglobinase activity of a cysteine protease from the ixodid tick *Haemaphysalis longicornis. Parasitol. Int.*, **58**, 232–237.

56. Konvičková, J. (2015). Factors regulating the expression and activity of digestive enzymes in the tick *Ixodes ricinus*. Master thesis. Faculty of Sciences (Ceske Budejovice: University of South Bohemia).

57. Perner, J., Provaznik, J., Schrenkova, J., Urbanova, V., Ribeiro, J.M.C., and Kopacek, P. (2016) RNA-seq analyses of the midgut from blood- and serum-fed *Ixodes ricinus* ticks. *Sci. Rep.*, **6**, 36695.

58. Tsuji, N., Miyoshi, T., Battsetseg, B., Matsuo, T., Xuan, X., and Fujisaki, K. (2008) A cysteine protease is critical for *Babesia* spp. transmission in *Haemaphysalis ticks. PLoS Pathog.*, **4**, e1000062.

59. Motobu, M., Tsuji, N., Miyoshi, T., Huang, X., Islam, M.K., Alim, M.A., and Fujisaki, K. (2007) Molecular characterization of a blood-induced serine carboxypeptidase from the ixodid tick *Haemaphysalis longicornis. FEBS J.*, **274**, 3299–3312.

60. Hatta, T., Kazama, K., Miyoshi, T., Umemiya, R., Liao, M., Inoue, N., Xuan, X., Tsuji, N., and Fujisaki, K. (2006) Identification and characterisation of a leucine aminopeptidase from the hard tick *Haemaphysalis longicornis. Int. J. Parasitol.*, **36**, 1123–1132.

61. Hatta, T., Tsuji, N., Miyoshi, T., Alim, M.A., Islam, M.K., and Fujisaki, K.

(2009) Leucine aminopeptidase in the ixodid tick *Haemaphysalis longicornis*: endogenous expression profiles in midgut. *J. Vet. Med. Sci.*, **71**, 589–594.
62. Hatta, T., Tsuji, N., Miyoshi, T., Islam, M.K., Alim, M.A., Yamaji, K., Anisuzzaman, and Fujisaki, K. (2010) Leucine aminopeptidase, HlLAP, from the ixodid tick *Haemaphysalis longicornis*, plays vital roles in the development of oocytes. *Parasitol. Int.*, **59**, 286–289.
63. Rajagopal, A., Rao, A.U., Amigo, J., Tian, M., Upadhyay, S.K., Hall, C., Uhm, S., Mathew, M.K., Fleming, M.D., Paw, B.H., Krause, M., and Hamza, I. (2008) Haem homeostasis is regulated by the conserved and concerted functions of HRG-1 proteins. *Nature*, **453**, 1127–1131.
64. Cabello-Donayre, M., Malagarie-Cazenave, S., Campos-Salinas, J., Galvez, F.J., Rodriguez-Martinez, A., Pineda-Molina, E., Orrego, L.M., Martinez-Garcia, M., Sanchez-Canete, M.P., Estevez, A.M., and Perez-Victoria, J.M. (2016) Trypanosomatid parasites rescue heme from endocytosed hemoglobin through lysosomal HRG transporters. *Mol. Microbiol.*, **101** (6), 895–908.
65. Huynh, C., Yuan, X., Miguel, D.C., Renberg, R.L., Protchenko, O., Philpott, C.C., Hamza, I., and Andrews, N.W. (2012) Heme uptake by *Leishmania amazonensis* is mediated by the transmembrane protein LHR1. *PLoS Pathog.*, **8**, e1002795.
66. Hajdusek, O., Sima, R., Perner, J., Loosova, G., Harcubova, A., and Kopacek, P. (2016) Tick iron and heme metabolism – new target for an anti-tick intervention. *Ticks Tick Borne Dis.*, **7** (4), 565–572.
67. Lara, F.A., Pohl, P.C., Gandara, A.C., Ferreira Jda, S., Nascimento-Silva, M.C., Bechara, G.H., Sorgine, M.H., Almeida, I.C., Vaz Ida, S. Jr., and Oliveira, P.L. (2015) ATP binding cassette transporter mediates both heme and pesticide detoxification in tick midgut cells. *PLoS One*, **10**, e0134779.
68. Korolnek, T., Zhang, J., Beardsley, S., Scheffer, G.L., and Hamza, I. (2014) Control of metazoan heme homeostasis by a conserved multidrug resistance protein. *Cell Metab.*, **19**, 1008–1019.
69. Keel, S.B., Doty, R.T., Yang, Z., Quigley, J.G., Chen, J., Knoblaugh, S., Kingsley, P.D., De Domenico, I., Vaughn, M.B., Kaplan, J., Palis, J., and Abkowitz, J.L. (2008) A heme export protein is required for red blood cell differentiation and iron homeostasis. *Science*, **319**, 825–828.
70. Quigley, J.G., Yang, Z., Worthington, M.T., Phillips, J.D., Sabo, K.M., Sabath, D.E., Berg, C.L., Sassa, S., Wood, B.L., and Abkowitz, J.L. (2004) Identification of a human heme exporter that is essential for erythropoiesis. *Cell*, **118**, 757–766.
71. Sheehan, D., Meade, G., Foley, V.M., and Dowd, C.A. (2001) Structure, function and evolution of glutathione transferases: implications for classification of non-mammalian members of an ancient enzyme superfamily. *Biochem. J.*, **360**, 1–16.
72. van Rossum, A.J., Jefferies, J.R., Rijsewijk, F.A., LaCourse, E.J., Teesdale-Spittle, P., Barrett, J., Tait, A., and Brophy, P.M. (2004) Binding of hematin by a new class of glutathione transferase from the blood-feeding parasitic nematode *Haemonchus contortus*. *Infect. Immun.*, **72**, 2780–2790.
73. Zhan, B., Liu, S., Perally, S., Xue, J., Fujiwara, R., Brophy, P., Xiao, S., Liu, Y., Feng, J., Williamson, A., Wang, Y., Bueno, L.L., Mendez, S., Goud, G., Bethony, J.M., Hawdon, J.M., Loukas, A., Jones, K., and Hotez, P.J. (2005) Biochemical characterization and vaccine potential of a heme-binding glutathione transferase from the adult hookworm *Ancylostoma caninum*. *Infect. Immun.*, **73**, 6903–6911.
74. Hiller, N., Fritz-Wolf, K., Deponte, M., Wende, W., Zimmermann, H., and Becker, K. (2006) Plasmodium falciparum glutathione S-transferase – structural and mechanistic studies on ligand binding and enzyme inhibition. *Protein Sci.*, **15**, 281–289.
75. Sharp, P.J., Smith, D.R., Bach, W., Wagland, B.M., and Cobon, G.S.

(1991) Purified glutathione S-transferases from parasites as candidate protective antigens. *Int. J. Parasitol.*, **21**, 839–846.

76. Hotez, P.J., Diemert, D., Bacon, K.M., Beaumier, C., Bethony, J.M., Bottazzi, M.E., Brooker, S., Couto, A.R., Freire Mda, S., Homma, A., Lee, B.Y., Loukas, A., Loblack, M., Morel, C.M., Oliveira, R.C., and Russell, P.K. (2013) The human hookworm vaccine. *Vaccine*, **31** (Suppl 2), B227–B232.

77. da Silva Vaz Jnr, I., Imamura, S., Ohashi, K., and Onuma, M. (2004) Cloning, expression and partial characterization of a *Haemaphysalis longicornis* and a *Rhipicephalus appendiculatus* glutathione S-transferase. *Insect Mol. Biol.*, **13**, 329–335.

78. Parizi, L.F., Utiumi, K.U., Imamura, S., Onuma, M., Ohashi, K., Masuda, A., and da Silva Vaz, I. Jr. (2011) Cross immunity with *Haemaphysalis longicornis* glutathione S-transferase reduces an experimental *Rhipicephalus (Boophilus) microplus* infestation. *Exp. Parasitol.*, **127**, 113–118.

79. Maya-Monteiro, C.M., Daffre, S., Logullo, C., Lara, F.A., Alves, E.W., Capurro, M.L., Zingali, R., Almeida, I.C., and Oliveira, P.L. (2000) HeLp, a heme lipoprotein from the hemolymph of the cattle tick, *Boophilus microplus*. *J. Biol. Chem.*, **275**, 36584–36589.

80. Donohue, K.V., Khalil, S.M., Sonenshine, D.E., and Roe, R.M. (2009) Heme-binding storage proteins in the Chelicerata. *J. Insect Physiol.*, **55**, 287–296.

81. Logullo, C., Moraes, J., Dansa-Petretski, M., Vaz, I.S., Masuda, A., Sorgine, M.H., Braz, G.R., Masuda, H., and Oliveira, P.L. (2002) Binding and storage of heme by vitellin from the cattle tick, *Boophilus microplus*. *Insect Biochem. Mol. Biol.*, **32**, 1805–1811.

82. Smith, A.D. and Kaufman, W.R. (2014) Molecular characterization of two vitellogenin genes from the tick, *Amblyomma hebraeum* (Acari: Ixodidae). *Ticks Tick Borne Dis.*, **5**, 821–833.

83. Smolenaars, M.M., Madsen, O., Rodenburg, K.W., and Van der Horst, D.J. (2007) Molecular diversity and evolution of the large lipid transfer protein superfamily. *J. Lipid Res.*, **48**, 489–502.

84. Mitchell, R.D. 3rd, Ross, E., Osgood, C., Sonenshine, D.E., Donohue, K.V., Khalil, S.M., Thompson, D.M., and Michael Roe, R. (2007) Molecular characterization, tissue-specific expression and RNAi knockdown of the first vitellogenin receptor from a tick. *Insect Biochem. Mol. Biol.*, **37**, 375–388.

85. Boldbaatar, D., Battsetseg, B., Matsuo, T., Hatta, T., Umemiya-Shirafuji, R., Xuan, X., and Fujisaki, K. (2008) Tick vitellogenin receptor reveals critical role in oocyte development and transovarial transmission of *Babesia* parasite. *Biochem. Cell Biol.*, **86**, 331–344.

86. Smith, A.D. and Kaufman, W.R. (2013) Molecular characterization of the vitellogenin receptor from the tick, *Amblyomma hebraeum* (Acari: Ixodidae). *Insect Biochem. Mol. Biol.*, **43**, 1133–1141.

87. Friesen, K.J. and Kaufman, W.R. (2004) Effects of 20-hydroxyecdysone and other hormones on egg development, and identification of a vitellin-binding protein in the ovary of the tick, *Amblyomma hebraeum*. *J. Insect Physiol.*, **50**, 519–529.

88. Seixas, A., Friesen, K.J., and Kaufman, W.R. (2008) Effect of 20-hydroxyecdysone and haemolymph on oogenesis in the ixodid tick *Amblyomma hebraeum*. *J. Insect Physiol.*, **54**, 1175–1183.

89. Logullo, C., Vaz Ida, S., Sorgine, M.H., Paiva-Silva, G.O., Faria, F.S., Zingali, R.B., De Lima, M.F., Abreu, L., Oliveira, E.F., Alves, E.W., Masuda, H., Gonzales, J.C., Masuda, A., and Oliveira, P.L. (1998) Isolation of an aspartic proteinase precursor from the egg of a hard tick, *Boophilus microplus*. *Parasitology*, **116** (Pt 6), 525–532.

90. Nascimento-Silva, M.C., Leal, A.T., Daffre, S., Juliano, L., da Silva Vaz, I. Jr., Paiva-Silva Gde, O., Oliveira, P.L., and Sorgine, M.H. (2008) BYC, an atypical aspartic endopeptidase from

Rhipicephalus (*Boophilus*) *microplus* eggs. *Comp. Biochem. Physiol. B: Biochem. Mol. Biol.*, **149**, 599–607.

91. Sorgine, M.H., Logullo, C., Zingali, R.B., Paiva-Silva, G.O., Juliano, L., and Oliveira, P.L. (2000) A heme-binding aspartic proteinase from the eggs of the hard tick *Boophilus microplus*. *J. Biol. Chem.*, **275**, 28659–28665.
92. Seixas, A., Dos Santos, P.C., Velloso, F.F., Da Silva Vaz, I. Jr., Masuda, A., Horn, F., and Termignoni, C. (2003) A *Boophilus microplus* vitellin-degrading cysteine endopeptidase. *Parasitology*, **126**, 155–163.
93. Oldiges, D.P., Parizi, L.F., Zimmer, K.R., Lorenzini, D.M., Seixas, A., Masuda, A., da Silva Vaz, I. Jr., and Termignoni, C. (2012) A *Rhipicephalus* (*Boophilus*) *microplus* cathepsin with dual peptidase and antimicrobial activity. *Int. J. Parasitol.*, **42**, 635–645.
94. da Silva Vaz, I. Jr., Logullo, C., Sorgine, M., Velloso, F.F., Rosa de Lima, M.F., Gonzales, J.C., Masuda, H., Oliveira, P.L., and Masuda, A. (1998) Immunization of bovines with an aspartic proteinase precursor isolated from *Boophilus microplus* eggs. *Vet. Immunol. Immunopathol.*, **66**, 331–341.
95. Leal, A.T., Seixas, A., Pohl, P.C., Ferreira, C.A., Logullo, C., Oliveira, P.L., Farias, S.E., Termignoni, C., da Silva Vaz, I. Jr., and Masuda, A. (2006) Vaccination of bovines with recombinant *Boophilus* Yolk pro-Cathepsin. *Vet. Immunol. Immunopathol.*, **114**, 341–345.
96. Seixas, A., Leal, A.T., Nascimento-Silva, M.C., Masuda, A., Termignoni, C., and da Silva Vaz, I. Jr. (2008) Vaccine potential of a tick vitellin-degrading enzyme (VTDCE). *Vet. Immunol. Immunopathol.*, **124**, 332–340.
97. Dunn, L.L., Suryo Rahmanto, Y., and Richardson, D.R. (2007) Iron uptake and metabolism in the new millennium. *Trends Cell Biol.*, **17**, 93–100.
98. Zhou, G., Kohlhepp, P., Geiser, D., Frasquillo Mdel, C., Vazquez-Moreno, L., and Winzerling, J.J. (2007) Fate of blood meal iron in mosquitoes. *J. Insect Physiol.*, **53**, 1169–1178.
99. Galay, R.L., Umemiya-Shirafuji, R., Mochizuki, M., Fujisaki, K., and Tanaka, T. (2014) Iron metabolism in hard ticks (Acari: Ixodidae): the antidote to their toxic diet. *Parasitol. Int.*, **64**, 182–189.
100. Hajdusek, O., Sojka, D., Kopacek, P., Buresova, V., Franta, Z., Sauman, I., Winzerling, J., and Grubhoffer, L. (2009) Knockdown of proteins involved in iron metabolism limits tick reproduction and development. *Proc. Natl. Acad. Sci. U. S. A.*, **106**, 1033–1038.
101. Mori, H., Galay, R.L., Maeda, H., Matsuo, T., Umemiya-Shirafuji, R., Mochizuki, M., Fujisaki, K., and Tanaka, T. (2014) Host-derived transferrin is maintained and transferred from midgut to ovary in *Haemaphysalis longicornis* ticks. *Ticks Tick Borne Dis.*, **5**, 121–126.
102. Hentze, M.W., Muckenthaler, M.U., Galy, B., and Camaschella, C. (2010) Two to tango: regulation of Mammalian iron metabolism. *Cell*, **142**, 24–38.
103. Ponka, P. and Lok, C.N. (1999) The transferrin receptor: role in health and disease. *Int. J. Biochem. Cell Biol.*, **31**, 1111–1137.
104. Basu, S., Horakova, E., and Lukes, J. (2016) Iron-associated biology of *Trypanosoma brucei*. *Biochim. Biophys. Acta*, **1860**, 363–370.
105. Steverding, D. (2000) The transferrin receptor of *Trypanosoma brucei*. *Parasitol. Int.*, **48**, 191–198.
106. Rodrigues, V., Cheah, P.Y., Ray, K., and Chia, W. (1995) Malvolio, the *Drosophila* homologue of mouse NRAMP-1 (Bcg), is expressed in macrophages and in the nervous system and is required for normal taste behaviour. *EMBO J.*, **14**, 3007–3020.
107. Hentze, M.W., Muckenthaler, M.U., and Andrews, N.C. (2004) Balancing acts: molecular control of mammalian iron metabolism. *Cell*, **117**, 285–297.
108. Kopacek, P., Zdychova, J., Yoshiga, T., Weise, C., Rudenko, N., and Law, J.H. (2003) Molecular cloning, expression and isolation of ferritins from two tick

species – *Ornithodoros moubata* and *Ixodes ricinus*. *Insect Biochem. Mol. Biol.*, **33**, 103–113.

109. Galay, R.L., Aung, K.M., Umemiya-Shirafuji, R., Maeda, H., Matsuo, T., Kawaguchi, H., Miyoshi, N., Suzuki, H., Xuan, X., Mochizuki, M., Fujisaki, K., and Tanaka, T. (2013) Multiple ferritins are vital to successful blood feeding and reproduction of the hard tick *Haemaphysalis longicornis*. *J. Exp. Biol.*, **216**, 1905–1915.
110. Galay, R.L., Umemiya-Shirafuji, R., Bacolod, E.T., Maeda, H., Kusakisako, K., Koyama, J., Tsuji, N., Mochizuki, M., Fujisaki, K., and Tanaka, T. (2014) Two kinds of ferritin protect ixodid ticks from iron overload and consequent oxidative stress. *PLoS One*, **9**, e90661.
111. Hajdusek, O., Almazan, C., Loosova, G., Villar, M., Canales, M., Grubhoffer, L., Kopacek, P., and de la Fuente, J. (2010) Characterization of ferritin 2 for the control of tick infestations. *Vaccine*, **28**, 2993–2998.
112. Galay, R.L., Miyata, T., Umemiya-Shirafuji, R., Maeda, H., Kusakisako, K., Tsuji, N., Mochizuki, M., Fujisaki, K., and Tanaka, T. (2014) Evaluation and comparison of the potential of two ferritins as anti-tick vaccines against *Haemaphysalis longicornis*. *Parasit. Vectors*, **7**, 482.
113. Tiklova, K., Senti, K.A., Wang, S., Graslund, A., and Samakovlis, C. (2010) Epithelial septate junction assembly relies on melanotransferrin iron binding and endocytosis in *Drosophila*. *Nat. Cell Biol.*, **12**, 1071–1077.
114. Talyuli, O.A., Bottino-Rojas, V., Taracena, M.L., Soares, A.L., Oliveira, J.H., and Oliveira, P.L. (2015) The use of a chemically defined artificial diet as a tool to study *Aedes aegypti* physiology. *J. Insect Physiol.*, **83**, 1–7.
115. Simo, L. and Park, Y. (2014) Neuropeptidergic control of the hindgut in the black-legged tick *Ixodes scapularis*. *Int. J. Parasitol.*, **44**, 819–826.
116. Sterkel, M., Perdomo, H.D., Guizzo, M.G., Barletta, A.B., Nunes, R.D., Dias, F.A., Sorgine, M.H., and Oliveira, P.L. (2016) Tyrosine detoxification is an essential trait in the life history of blood-feeding arthropods. *Curr. Biol.*, **26**, 2188–2193.

8
Whole-organism Screens for Ectoparasites

Jeffrey N. Clark and Cedric J. Pearce*

Abstract

Whole-organism *ex vivo* assays that are key to the discovery process for new animal health ectoparasiticides and that form an essential bridge between *in vitro* target-based screens and animal models are described. The purpose and basic principles of conducting whole-organism assays are discussed. Typical workflows are outlined for the use of whole-organism screens in the discovery process assessment of pure compounds and natural products. The fundamentals and procedures of assays for mosquito larvae, adult fleas (contact and ingestion) and flea eggs (contact), ticks (contact and ingestion), and flies (contact) are described in detail along with activity information in these assays for many marketed ectoparasiticides. Assays using mites and lice are described in summary and with literature references. The relevance and importance of agricultural chemical insecticidal and acaricidal compounds and discovery programs are emphasized as a source of new animal health ectoparasiticides.

Purpose, Advantages, and Limitations of Whole-organism Screens

Whole-organism ectoparasite assays, or *ex vivo* screens, have been used for nearly 50 years at the early phase of novel ectoparasiticide discovery to determine compound activity directly on the target parasite in the laboratory rather than on the animal. These assays avoid several aspects of host physiology that can complicate early compound assessment, including compound bioavailability, metabolism, absorption across the relevant tissue barrier (skin, gastrointestinal tract, etc.), immune responses and other processes that could compromise the effectiveness, tissue or blood levels, and/or molecular stability of the drug. However, because this *ex vivo* screening does not take into account these host responses, a false prediction of the efficacy in the host can occur. It is not practical or desirable to test a large number of compounds using only animal models (see

*Corresponding author.

Ectoparasites: Drug Discovery Against Moving Targets, First Edition.
Edited by Charles Q. Meng and Ann E. Sluder

Chapters 10 and 11). Eventually, of course, during the drug development program, the effects of host physiology on drug exposure at the site(s) of parasite attachment must be taken into account. However, in the early ectoparasiticide discovery phase, it is paramount to determine if a compound is potent and effective in direct contact with the ectoparasite.

Whole-organism screening bridges the gap between *in vitro* molecular target-based assays, conducted using a validated drug target in the parasite, and testing in host animal parasite infestation models (see Chapter 11). *In vitro* screening focuses on a particular key enzyme, receptor, or other target that may be modulated by a compound in a manner detrimental to the parasite. However, it does not take into account those barriers that the whole-organism screening will encompass, namely, the metabolism of the drug by the living arthropod, such as the effect of mixed function oxidases on the integrity of pyrethroids, or tissue barriers, such as the cuticle of a tick or flea and other barriers that can prevent compound absorption or bioavailability.

While whole-organism screening does *not* allow for testing of ~1 million compounds as can be achieved with *in vitro* target-based screening, nevertheless, with the proper assay procedures, a few hundred compounds can be evaluated at a time. Furthermore, if liquid handling, automated parasite dispensing, and computerized efficacy determination programs are available, several thousand individual compounds can be evaluated in each screening campaign. Most whole-organism assays are conducted over 1–4 days, allowing more than one assay to be conducted in a week. Whole-organism screening also has the major advantage of requiring relatively little compound compared to animal model tests – usually less than 1 mg for several assays, including dose–response measurements or down titrations of interesting compounds, although *in vitro* target-based screening only requires a few micrograms.

General Workflow for Ectoparasiticide Discovery Screens

A typical screening paradigm for novel ectoparasiticides, used in our laboratories at Mycosynthetix, Inc. in Hillsborough, NC, is shown in the flow charts in Figure 8.1. When screening compounds or natural product extracts of unknown activity, we start with the mosquito larval assay (MLA) as a sensitive predictor of ectoparasiticide activity (Figure 8.1a). Pure compounds or natural product extracts are first evaluated in the MLA in single point (SP, one concentration), and active compounds are further assessed by a dose titration (DT) to determine the EC_{100} or EC_{50}. Compounds that are active at certain potency (e.g., EC_{50} of 1 ppm (μg/ml)) are thereafter evaluated in mammalian cytotoxicity assays, and those with a therapeutic index (TI, the ratio of activity against the mammalian cell compared to that toward the parasite) of ≥5X become possible program ectoparasiticides with appropriate potency and safety characteristics. Natural product crude extracts, following mammalian cytotoxicity assays, undergo a de-convolution/isolation process whereby the structure of the active compound is identified and determined

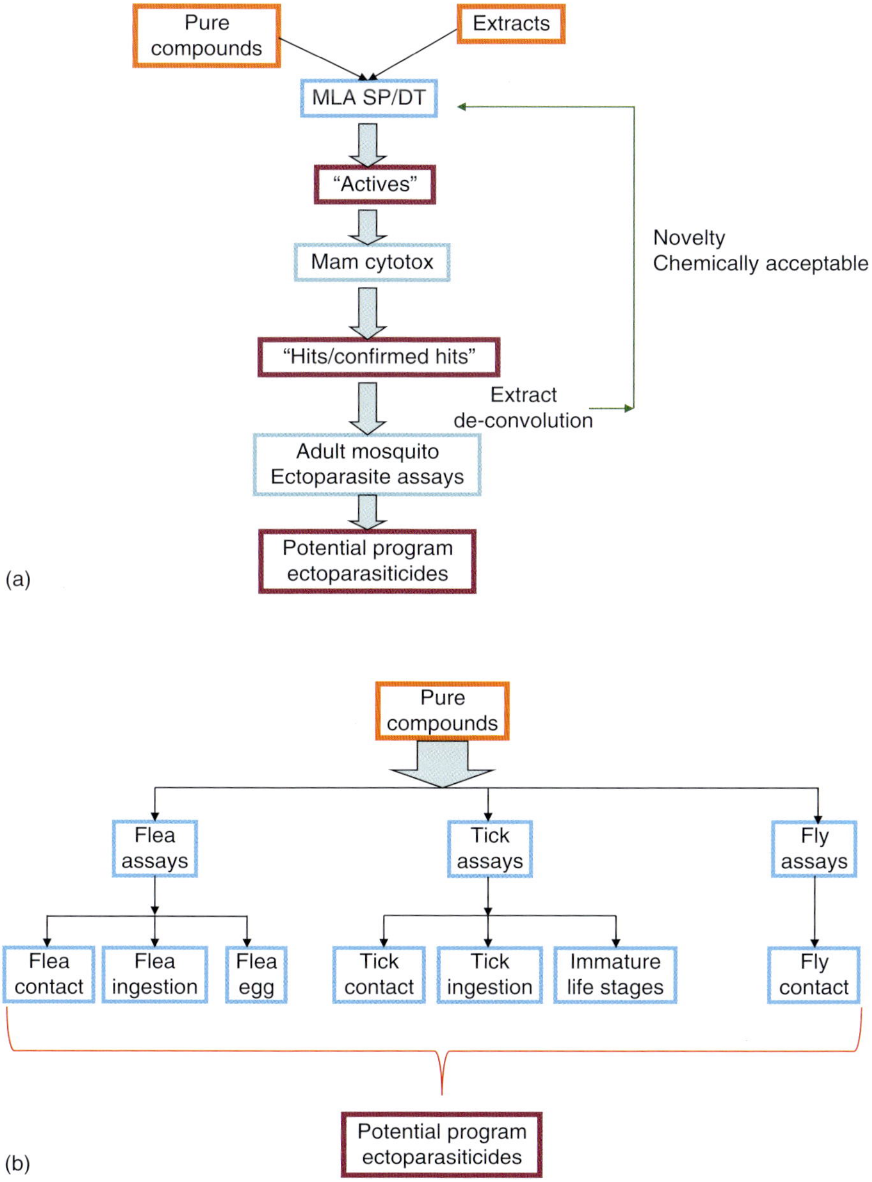

Figure 8.1 Ectoparasiticide discovery flow diagrams: (a) Part 1; (b) Part 2.

to be novel (or not), and then this structure undergoes a chemical review to determine if it represents a potentially "druggable" (drug candidate based on structure and physicochemical properties) entity. If not, it is discarded.

Candidate program insecticide leads are evaluated in a broad selection of ectoparasite assays to assess activity against fleas, ticks, and flies (Figure 8.1b). At this point, all samples will consist of single compounds usually with known structures. These pure compounds can be evaluated against several adult and immature stages and several species of fleas, ticks, and flies. Assays are conducted in SP tests (~100 compounds/test, generally performed in triplicate) followed by DT of active compounds.

Basic Principles in Ectoparasite Screen Design

Throughput is an important consideration in the design of assay methods. For example, the MLA is conducted in a 96-well plate-based system, so it is possible to evaluate a few hundred to as many as several thousand compounds over the course of a campaign. The MLA's throughput capacity is why it is used as a "primary ectoparasite assay" at Mycosynthetix, Inc. In contrast, the flea, tick, and fly assays are "secondary ectoparasite assays" and only allow for low throughput (~hundred compounds/week) due to the significant amount of skilled manual, as opposed to automated, labor required. However, the MLA can demonstrate that the compound is an ectoparasiticide and also can indicate its relative potency, allowing for prioritization in the secondary assays.

The lower throughput secondary assays largely encompass primary ectoparasite target species for companion animals (fleas and ticks) and some livestock animals (ticks, flies). Other ectoparasites can be utilized, however, to complete the spectrum for the host animal species of interest. Assays that are conducted for one species of canine tick might not suffice for that tick in cats or a related species in livestock. Therefore, it is best to utilize the specific tick that is applicable (e.g., *Rhipicephalus* (*Boophilus)* spp. for cattle). The life cycle of the parasite is important when determining the appropriate assay, as one stage (e.g., the adult) may be more important for the animal infestation in question than another stage (e.g., egg or larvae). In some cases, such as fleas, all parts of the life cycle are important, as control of all stages is required for complete flea control. This leads to the concept of a developmental assay, where insect growth regulators (IGRs) are important for such stages as the egg or larva, and where all the immature forms up to the adult can be present at some time during the course of the assay. For fleas, for example, developmental assays for the egg, larvae, pupae, and adult stages are well established.

For an effective screen, the ectoparasite must survive the rigors of the *ex vivo* assay in the absence of treatment or the actual efficacy of the test compound is in question. To demonstrate that survival is adequate to support a valid assay, either a negative (no treatment) or a solvent (that used to dissolve the compounds being tested) control is used. Solvent-only treated ectoparasites should maintain high viability, preferably ≥90% throughout the test period. As all solvents have some detrimental effect, the compounds should be dissolved in as small a quantity of solvent as possible. The volume and nature of the solvent at each concentration in each test must be consistent so that solvent effects are identical at all concentrations. Common solvents used are dimethyl sulfoxide (DMSO) and acetone, both of which have effects directly on the organism. In many assays, such as those for fleas, ticks, and flies, the solvent evaporates before introduction of the ectoparasites, thereby minimizing solvent effects.

Likewise, ectoparasites may require certain media or support systems to maintain viability in the control assays. For example, the MLA is conducted in an aqueous medium which contains some nutrients. Also, flies dehydrate rapidly, so they require sugar water or actual blood sustenance in the petri dish assay. Furthermore, some ectoparasites require temperature and humidity control (e.g., flea and tick), so these assays are conducted in an environmental chamber. Flea, tick, and fly assays also require a solid support, such as a filter paper, which allows the ectoparasite to come into contact with the compound. These support systems provide containment and allow constant exposure of the compound to the ectoparasite.

Considerations in Assay Implementation and Screen Execution

One or more known ectoparasiticide or IGR standards of different modes of action are utilized in every assay as positive controls. The common ectoparasiticides used are fipronil, pyrethrins/pyrethroids, organophosphates such as chlorpyrifos, ivermectin, or other macrocyclic lactones and amitraz. Methoprene is used as a representative IGR in immature flea assays. Occasionally, ectoparasiticide-resistant organisms are available. Novel compounds which remain active against these resistant parasites could indicate that they act by a different mode of action. The known ectoparasiticide or acaricide compound should be used at a concentration where it reaches 100% kill (EC_{100}) and can also be used in a DT. The latter is routinely done in the MLA, as throughput is not an issue; however, the EC_{100} alone may be used in the secondary assays such as flea, tick, and fly to limit the numbers of samples that need to be evaluated. When initiating a new test, the efficacy of various ectoparasiticide standards is evaluated, followed by a DT to the EC_{100}. Thereafter, this EC_{100} level is used in the secondary assays. Replicates are used in most assays to determine if variability or consistency in response is seen between identical treatments. Ideally, the results should be very similar within replicates. Otherwise the validity of the assay is in question.

Exposure of the ectoparasite to the compound can occur by various means. In the MLA, the compound is added to the medium in each well followed by mixing. The larvae are then added to the mixture, and compound contact occurs through the oral route. For flea, tick, and fly contact assays, the compound is exposed to the ectoparasite through the cuticle. For this reason, a detergent is often added to the solvent, at least for ticks, to assist in transporting the compound through the cuticle.

The concentration of the compound tested is subject to the type of assay being conducted (e.g., plate-based or direct application to filter paper in a Petri dish or scintillation vial). A useful starting SP concentration is 50 ppm (50 μg/ml) which is a relatively high concentration compared to the activity of the positive controls. If the test compound is not active at this level, it is most likely not of interest, as lead compounds must be much more active than this. A typical dose–response titration following the SP assay utilizes serial fourfold dilutions from the 50-ppm starting dose, yielding concentrations of 50, 12.5, 3.125, 0.781, and 0.195 ppm. For very potent compounds, the titration levels can go lower in a subsequent assay. For the fly assays described subsequently, solutions of the compound in an organic solvent are placed on filter papers contained in Petri dishes, and therefore the concentration rates used are in $\mu g/cm^2$ based on the area of the filter paper.

The duration of the assay varies between ectoparasite test species and should be at least as long as the time required to kill 100% of the ectoparasites in the positive controls. This can vary from a rapid kill, such as permethrin's effects in stable flies, where the flies are killed within 5 h, to a slower kill, such as fipronil killing of fleas or ticks which takes 1 to 2 days. This variation in speed of kill is another reason why multiple positive controls are used, and it could be important to the scientific or marketing interest in regard to the objective of the compound.

Ectoparasite assays can have different scoring end points depending on the species, life cycle stage, and testing medium, which document certain phenotypic responses. Most ectoparasites show decreased motility up to death when affected by the compound. Amitraz, which does not cause death in ticks up to 50 ppm, consistently creates intense hypermotility in ticks down to very low concentrations. While usually quiescent at the bottom of the scintillation vial unless disturbed or exposed to CO_2, the ticks in the presence of amitraz constantly move quickly around the vial and at the same time have rapid movement of their legs, head, and sensory appendages. Flies generally turn upside down when dead but can show slight to severe gait abnormalities in an upright or recumbent position until death occurs. For flies, fleas, and ticks, a scoring paradigm of dead/live/affected was developed. The "affected" score indicates that there were phenotypic changes that occurred that were due to the compound, but at the particular evaluation time, the ectoparasite was still alive. At times, the organism can recover. This "affected" score is documented as an early sign of the compound's effect.

Specific Whole-organism Ectoparasite Assays

This section discusses individual assays in more detail, and focuses primarily on contact and ingestion assays. The utility of specific feeding assays will be noted; feeding systems for hematophagous arthropods and the associated assays are covered in more detail in Chapter 9.

Mosquito Larval Assay

This assay uses first instar larvae of *Aedes aegytpi*, eggs of which can be obtained from Louisiana Biological Supply (Dr. Steve Sackett) or from Benzon Research, Inc.

A nutrient broth solution is prepared, and 180 μl of this broth is added to each well of a 96-well flat-bottom plate (Figure 8.2). Ectoparasiticide standards

		Test compounds (10–50 ppm)												
	Drug													
A														A
B														B
C														C
D														D
E														E
F	Ivermectin	Standards down titration												F
G	Fipronil													G
H	Control													H
		1	2	3	4	5	6	7	8	9	10	11	12	

Figure 8.2 Typical 96-well plate setup for mosquito larval assay.

are prepared in DMSO or acetone (depending on the solubility of the compound) and added to the standard wells, as noted in Figure 8.2, to prepare a titration regimen of 10, 2.5, 0.625, 0.156, and 0.039 ppm (μg/ml). Mosquito eggs are placed in a conical centrifuge tube, the broth is added, and the tube is shaken vigorously for 5 min. After settling, the upper layer containing the larvae is removed and placed in a separate beaker in preparation for adding to the 96-well plate. The concentration of larvae is adjusted to 10 larvae per 20 μl aliquot which is added to each well. Unknown compounds are initially tested in SP at 10 ppm, thereafter undergoing a DT, if they are 100% active at 10 ppm.

Mosquito larvae usually swim rapidly around the well in the 96-well plate, but the movement slows or stops as they die. Also, rotifers, microscopic pseudocoelomate animals which are a common contaminant of mosquito eggs, multiply greatly after the mosquito larvae die, as they are likely a food source for the larvae. The presence of large numbers of rotifers in the well is another indication that the mosquito larvae have died.

The results for ectoparasiticide standards are noted in Table 8.1. The organophosphate chlorpyrifos was found to be so potent that severe edge effects (activity

Table 8.1 Ectoparasiticide standards in the mosquito larval assay.

Standard	EC100 (ppm)
Ivermectin	0.125
Eprinomectin	0.031
Chlorpyrifos	<0.00195
Permethrin	0.031
Fipronil	0.031
Amitraz	>10

on the larvae in adjacent wells not containing chlorpyrifos) were seen, so this standard was dropped from subsequent tests. Ivermectin and fipronil, which did not show edge effects but have different modes of action (although on the same ion channel), were used as routine positive controls.

This assay was validated and was sufficiently robust to evaluate several thousand compounds over a few weeks. Several potent compounds and active extracts have been found using this assay.

Flea Contact Assay

The adult stage of *Ctenocephalidis felis*, the cat flea, which is the primary flea species infesting both cats and dogs, is used. Adult fleas can be obtained from several sources including EctoServices in North Carolina, TRS Laboratories in Georgia, and EL Labs in California.

The assay is conducted in 20 mL glass scintillation vials (Figure 8.3a). Test compounds are dissolved in a mixture of DMSO, acetone, and Triton X®100, a surfactant which facilitates entry of compound through the flea cuticle. Each compound is titrated from 50 ppm in fourfold dilution steps (50, 12.5, 3.1, 0.78, 0.195 ppm). A 1/8 inch hole is drilled into the top of the plastic cap of the scintillation vial to allow air to enter the vial, and a piece of 2.1 cm diameter circular filter paper is placed inside the cap. A 0.5 inch length of pipe cleaner is placed in the bottom of each vial (Figure 8.3a), and 0.75 μl of compound mixed in the solvent is put on each pipe cleaner, the vial cap is replaced loosely, and the vial is then air dried for at least 1 h. Infestation takes place the next day. Vials are protected from the fluorescent light in the laboratory to avoid photodegradation of the compound. Unfed adult fleas are released into a large (empty) aquarium of sufficient height to prevent the fleas from escaping (Figure 8.3b). Fleas are put into the previously treated scintillation vial using a vacuum system (Figure 8.3b,c). The vial is then removed from the system and quickly capped to prevent flea escape. The assay is maintained for 72 h in an incubator at about 25°C and 80% relative humidity.

Several ectoparasiticide standards (fipronil, permethrin, ivermectin, and chlorpyrifos) are used, and their activity in this assay along with their IRAC (Insecticide Resistance Action Committee) classification are noted in Table 8.2 below. Several of the compounds show poor activity in the flea contact assay but

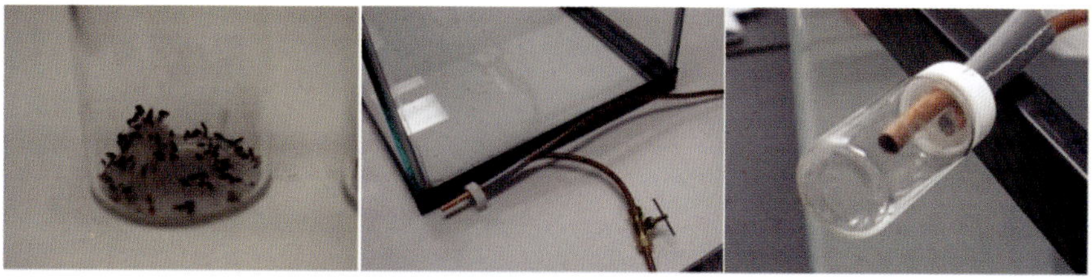

(a) (b) (c)

Figure 8.3 (a–c) Apparatus for flea contact assay.

Table 8.2 Activity of ectoparasiticide standards in the flea contact assay.

Standard	Chemical class	IRAC classification	EC50 (ppm)
Fipronil	Fiprole	2B	3.125
Pyrafluprole	Fiprole	2B	48.85
Pyraprole	Fiprole	2B	46.21
Fenpyroximate	METI	21	>50
Chlorfenapyr	METI	13	13.0
Carbosulfan	Carbamate	1A	6.25
Carbofuran	Carbamate	1A	14.06
Permethrin	Pyrethroid	3A	46.28
Cypermethrin	Pyrethroid	3A	20.97
Acetamiprid	Neonicotinoid	4A	44.32
Imidacloprid	Neonicotinoid	4A	11.65
Nitempyran	Neonicotinoid	4A	>50
Denotefuran	Neonicotinoid	4A	1.73
Indoxacarb	Semicarbazone	22A	>50
Metaflumizone	Semicarbazone	22B	>50
Ivermectin	Macrocyclic lactone	6	>50
Selamectin	Macrocyclic lactone	6	>100
Chlorpyrifos	Organophosphate	1B	3.125
Spinosad	Spinosyns	5	>50
Amitraz	*N*-Methyl-formamide	19	>50

are sold commercially to treat flea infestations, because they act through the oral route (see section titled "Flea Ingestion Assay"). The most potent compounds for contact killing of fleas include fipronil, the carbamates, the neonicotinoids imidacloprid and dinotefuran, the mitochondrial electron transport inhibitor (METI) chlorfenapyr, and the organophosphate chlorpyrifos. The pyrethroids (permethrin, cypermethrin) are moderately active. This assay is a very valuable tool to identify compounds that are active through the contact route.

Flea Ingestion Assay

This assay allows oral exposure to the flea of compounds that are mixed in with a blood meal. This was made possible by the "artificial dog," an ingenious invention of Dr. Jay Georgi, a noted parasitologist formerly of Cornell University Veterinary College, which is described in more detail by Nijhof and Tyson. His invention is covered in US Patent 5133289 [1].

Table 8.3 shows data from several known ectoparasiticides evaluated in this assay. There are several potent chemistries including the fiproles, the METIs, carbamates, neonicotinoids, macrocyclic lactones, and spinosad. Unlike its action in ticks, amitraz is not known to be effective in fleas, and this is borne out by the results of both the flea contact and flea ingestion assays.

Flea Egg/Larval Assays

An important consideration to control flea infestations is the killing of environmental immature stages such as the egg, larvae, and pupae. Due to the cocoon nature of the pupae, this stage is difficult to treat. However, the egg and larvae are the prime targets of IGRs such as methoprene, pyriproxyfen, and fenoxycarb.

A Petri dish/filter-paper-based system is used to evaluate the effect of IGRs on the hatchability of isolated flea eggs in a Petri dish/filter-paper-based system. Filter papers (47 mm) are placed into glass Petri dishes (60 mm) and then treated with various concentrations (0–0.020 $\mu g/cm^2$) of methoprene dissolved in

Table 8.3 Activity of ectoparasiticide standards in the flea ingestion assay.

Standard	Chemical class	IRAC classification	EC50 (ppm)
Fipronil	Fiprole	2B	0.048
Pyrafluprole	Fiprole	2B	1.503
Pyraprole	Fiprole	2B	0.196
Fenpyroximate	METI	21	12.28
Chlorfenapyr	METI	13	5.56
Carbosulfan	Carbamate	1A	6.25
Carbofuran	Carbamate	1A	3.98
Permethrin	Pyrethroid	3A	>50
Cypermethrin	Pyrethroid	3A	>400
Acetamiprid	Neonicotinoid	4A	10.58
Imidacloprid	Neonicotinoid	4A	0.14
Nitempyran	Neonicotinoid	4A	<0.195
Denotefuran	Neonicotinoid	4A	<1.563
Indoxacarb	Semicarbazone	22A	>50
Metaflumizone	Semicarbazone	22B	>50
Ivermectin	Macrocyclic lactone	6	3.88
Selamectin	Macrocyclic lactone	6	0.09
Chlorpyrifos	Organophosphate	1B	ND
Spinosad	Spinosyns	5	<0.195
Amitraz	*N*-Methyl-formamide	19	>400

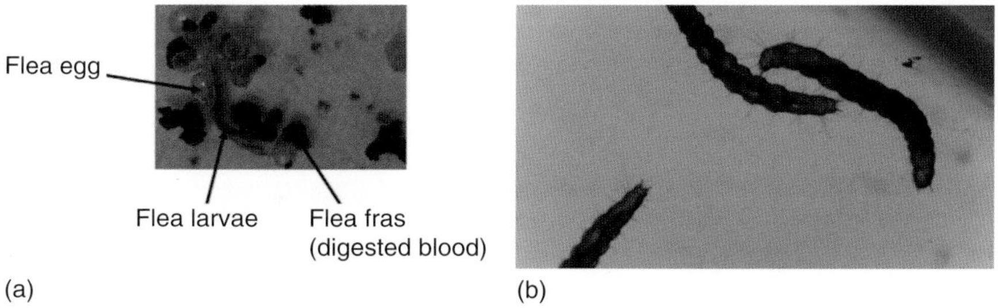

Figure 8.4 (a) Flea eggs, larvae, fras in Petri dish. (b) Flea larvae in Petri dish.

acetone, the latter of which is then evaporated. The isolated flea eggs are introduced onto the filter paper (~20 eggs/dish) and the dishes are maintained in an environmental chamber at 25–29°C and 85% relative humidity. Dishes remain in the chamber for 3 days after which they are evaluated for egg hatching and larvae number and viability by microscopic examination. Pictures of the eggs and larvae from these experiments are shown in Figure 8.4a and b.

Methoprene has a profound effect on egg hatching, resulting in inhibition from 30% (0.0105 $\mu g/cm^2$) to 60% (0.0210 $\mu g/cm^2$) when compared to solvent control. The effect of compounds on later immature flea stages up to the adult flea can thereafter be followed by providing the hatched larvae in the Petri dish with appropriate medium to develop a pupa and finally to pupate into the adult flea. This is a very useful assay to discover IGR effects on flea egg hatching and the development of larvae as well as the effect of insecticides on the larval stage.

Tick Contact Assay

Ticks are an important ectoparasite of companion animals and livestock and also are a vector for many zoonotic diseases. Dogs are infested by four primary ticks – *Rhipicephalus sanquineus, Dermacentor variabilis, Amblyomma americanum,* and *Ixodes* spp. Ticks can be obtained from several sources including EctoServices, EL Labs, and Nu-Era Farms. Ticks are released from their shipping containers (Figure 8.5a) on to a white paper (Figure 8.5b), sorted, and then placed in scintillation vials (Figure 8.6a and c) for the assay.

The tick contact assay is conducted using 20 mL glass scintillation vials using procedures similar to that for fleas. A 1/8 inch hole is drilled in the cap of the vial to supply air to the ticks. About 400 µl of compound solution is placed in the vial, and the vials are rotated horizontally on a hot dog roller at laboratory room temperature to allow compound spreading and evaporation of the solvent. The next day, one 2.1 cm filter paper is placed inside the cap and a second filter paper is placed in the bottom of the vial, 50 µl of compound solution is added, and the solvent is allowed to evaporate. Ten adult ticks up to 30 days old are introduced into the treated vial, and the vial is capped. Ticks are thereafter maintained in an environmental chamber at 24°C and 95% relative humidity, and efficacy is determined, based on the tick mobility at 24 and 48 h and occasionally at 72 h,

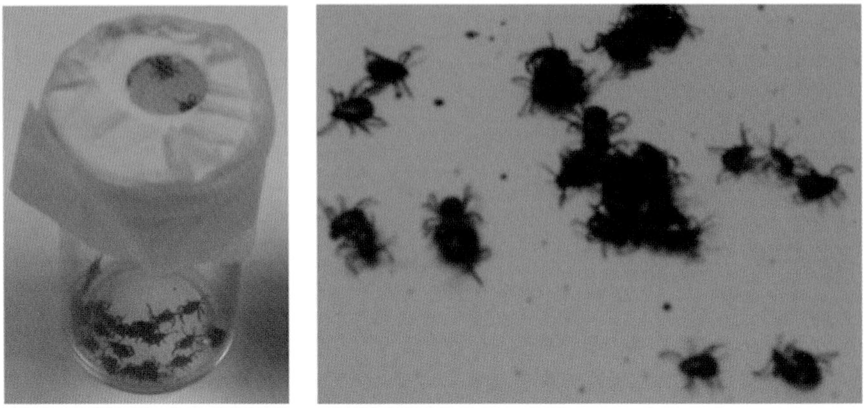

(a) (b)

Figure 8.5 (a) Ticks in shipping container. (b) Ticks on white paper.

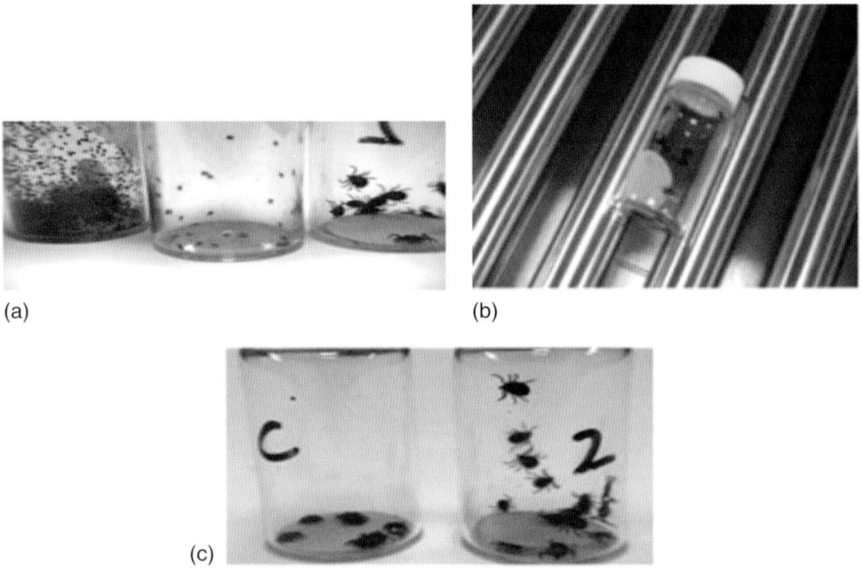

(a) (b) (c)

Figure 8.6 (a) Dv larvae nymph, adult. (b) Vial on hotdog roller. (c) Amitraz effect.

particularly for *D. variabilis*. Ticks become quiescent with no stimulation, so they may be stimulated to determine viability by rotating the vial on the hotdog roller or breathing through the top of the vial with the lid removed. Live ticks will start moving quickly; dead ticks, aside from being non-motile, are generally flat and desiccated in appearance. Pictures of the ticks, vial arrangement, and use of the hotdog roller are shown in Figure 8.6a–c.

Figure 8.6a shows three stages of the life cycle of the tick *D. variabilis*. Similar vial-based assays can be conducted for all three, although the size of the two immature stages makes containment interesting. Figure 8.6b shows the use of the hotdog roller. Figure 8.6c shows the effect of amitraz on the ticks. The control, diluent-only treated vial, labeled "C" in Figure 8.6c shows the normal quiescent nature of *D. variabilis* ticks when left undisturbed. However, amitraz treatment, without any other stimulation, causes the ticks to move quickly and continuously with rapid moving of their head, legs, and other appendages (Figure 8.6c treatment vial labeled "2"). This quite dramatic effect can be seen within 24 h and affects all ticks in the vial down to at least 0.195 ppm. The effects of various ectoparasiticide standards on ticks following exposure by the contact route are shown in the Table 8.4.

It is readily seen that the fiproles, pyrethroids, carbamates, and organophosphates have the greatest lethal effect on ticks by the contact route. The macrocyclic

Table 8.4 Activity of ectoparasiticide standards in the tick contact assay.

Standard	Chemical class	IRAC classification	EC50 (ppm)
Fipronil	Fiprole	2B	1.385
Pyrafluprole	Fiprole	2B	2.876
Pyraprole	Fiprole	2B	ND
Fenpyroximate	METI	21	19.64
Chlorfenapyr	METI	13	ND
Carbosulfan	Carbamate	1A	3.28
Carbofuran	Carbamate	1A	2.64
Permethrin	Pyrethroid	3A	10.73
Cypermethrin	Pyrethroid	3A	2.255
Acetamiprid	Neonicotinoid	4A	>50
Imidacloprid	Neonicotinoid	4A	>50
Nitempyran	Neonicotinoid	4A	>50
Denotefuran	Neonicotinoid	4A	>50
Indoxacarb	Semicarbazone	22A	>400
Metaflumizone	Semicarbazone	22B	>50
Ivermectin	Macrocyclic lactone	6	>50
Selamectin	Macrocyclic lactone	6	40.1
Chlorpyrifos	Organophosphate	1B	12.5
Spinosad	Spinosyns	5	38.54
Amitraz	*N*-Methyl-formamide	19	<0.195[a)]

a) Stimulatory only, not a lethal effect.

lactone selamectin and spinosad have some weak effect. As expected, the neonicotinoids and semicarbazones had no effect at the levels tested.

The tick contact assay is a proven and validated test to determine the efficacy of novel compounds. Throughput is low, however, at about 50–100 compounds/week in SP testing.

Tick Ingestion Assay

A tick ingestion assay similar to the Georgi flea assay has been developed [2]. Another tick ingestion assay uses a capillary tube containing a solution of the compound to be tested, which is inserted into the mouth parts of a tick that is adhered to a glass platform by tape. Both of these assays have very low throughput; however, they may be the only approach to testing promising compounds *ex vivo* that are not active topically on the tick, short of conducting animal trials. Both feeding systems and associated assays are described further in Chapter 9.

The feeding apparatus has been used to test fipronil, permethrin, and ivermectin [2]. Fipronil at 10 ng/cm^2 area of membrane resulted in a 70% kill rate, while a concentration of 1 µg/cm^2 was 100% effective. Permethrin was 30% effective at 1 ng/cm^2 and 100% effective at 100 µg/cm^2. Ivermectin was effective at levels of ≥0.1 µg/ml of blood.

Mite Assays

Mites are very important infesting agents of all animals with *Sarcoptes*, *Demodex*, *Cheyletiella* (see Figure 8.7), and *Otodectes* infesting companion animals and *Sarcoptes*, *Dermanyssus*, *Chorioptes*, and *Psoroptes* infesting livestock.

Macrocyclic lactones such as ivermectin have been very effective against most mite infestations of companion animals and livestock. Mites are not a primary target for a discovery program in companion animals, but this activity of a new flea/tick compound is an important additional target for commercialization.

Ex vivo assays against mites have been reported in the literature [3–5]. A robust 72-h primary filter-paper-based *Psoroptes ovis ex vivo* screen [5] was used to test several essential oils, desiccants, and detergents. Furthermore, compounds of

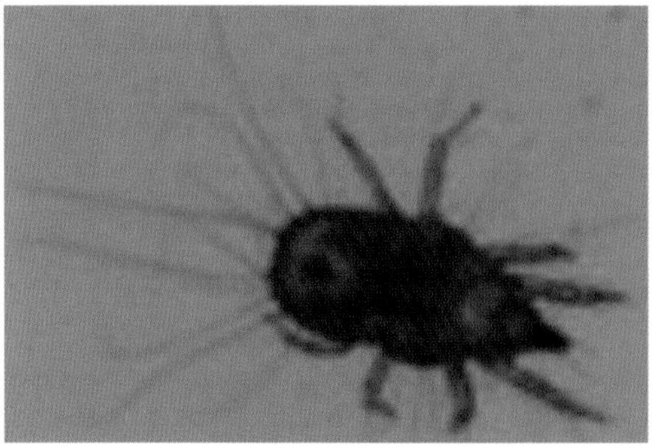

Figure 8.7 A *Cheyletiella* spp. mite.

specific mode of actions in mites such as growth regulators, lipid synthesis inhibitors, and nerve action and energy disruptors have also been tested. Many of these compound classes were effective against nymph or adult stages including fenpyroximate, spinosad, tolfenpyrad, and chlorantraniliprole as nerve action and energy disruptors.

Stable Fly Contact Assay

Stable flies are a ubiquitous and serious nuisance for livestock and can cause considerable loss of productivity (decreased weight gain and milk production) through their painful biting and harassment activity. The stable fly, *Stomoxys calcitrans,* also acts as a mechanical vector for *Trypanosoma* spp. protozoal disease in cattle and horses, which can cause very severe productivity losses and death. Stable flies can also act as an intermediate host for *Habronema* spp. larvae, the cause of summer sores in horses.

A contact assay was developed at Mycosynthetix to assess the activity of ectoparasiticide standards, unknown compounds, and natural product extracts and pure compounds against the adult stable fly. Pupae can be obtained from New Mexico State University (Dr. Ronnie Byford), and these develop into adult flies in a hatching cage (Bugdorm®, Figure 8.8a). Sugar water solution is provided as a nutrient source (Figure 8.8b). The closable sleeve on the left-hand side of the cage in Figure 8.8a allows access to the inside of the cage without losing flies to the laboratory. After the flies are hatched, the dish containing the spent or unhatched pupae is removed.

Compounds are prepared using a diluent solution containing acetone, Triton X 100, DMSO, and water so that 5, 1.25, 0.312, 0.078, and 0.0195 μg/cm^2 are applied to a 47-mm piece of filter paper contained in a 60-mm Petri dish as shown is Figure 8.8b. Manipulation of the flies during the assay is done by the use

(a) (b)

Figure 8.8 (a) Adult flies in Bugdorm® cage. (b) Flies in the Petri dish assay.

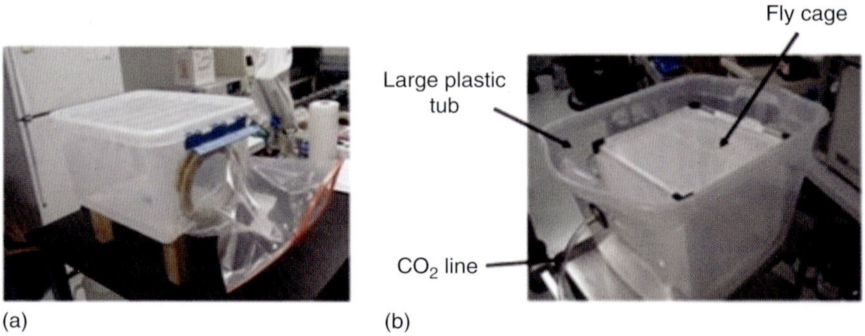

Figure 8.9 (a) Fly capture system. (b) Setup to anesthetize the flies.

of a large covered plastic tub (Figure 8.9a) into which the entire fly cage is placed (Figure 8.9b, top removed). One end of the tube from this tub (Figure 8.9b) is connected through a port and plastic tubing to a CO_2 source which is used to temporarily anesthetize the flies during the transfer procedures.

After the flies are immobilized, they are swept from the bottom of the hatching cage with a fine brush, and the cage is removed from the tub. A 2-inch hole in the bottom of the tub allows for the transfer of 10 anesthetized flies to each treated Petri dish. The flies recover from the CO_2 in the Petri dish after a few minutes, and thereafter the effect of the compounds on the live flies is noted at 1, 5, and 24 h post infestation. The flies are evaluated for death, gait abnormalities, twitching, righting reflex defects, and lateral recumbency and are scored as dead, affected, or live. Petri dishes where the filter paper is just treated with the solvent are used as a negative control to account for lethality during all the manipulation procedures not due to the test compounds.

Several ectoparasiticide standards can be used as positive controls, using compounds with different mode and speed of action. Activity of several standards is shown in Table 8.5.

Fipronil, permethrin, and chlorpyrifos are extremely fast acting, with systemic effects and death being seen within 5 h. Total kill is seen with all three compounds at 24 h with an EC_{100} as noted. The assay is low throughput, at about 100 compounds/week in an SP testing, so that only compounds of high ectoparasiticide promise based on other ectoparasiticide assays (see Figure 8.1b) should be considered for testing. The stable fly assay is very reproducible, as long as particular care is taken in the continued nourishment of the fly in the dish and during the CO_2 procedures. The typical survival rate of the solvent-treated control flies is ≥80% for valid assays. A similar assay for horn fly (*Haemotobia irritans*) was also developed at Mycosynthetix, but this fly requires blood as a nutrient source.

Other Ectoparasites of Animals

Other ectoparasites important in animal health, particularly in livestock, are not addressed by specific assays described here. These include many other fly genus/

Table 8.5 Activity of ectoparasiticide standards in the stable fly contact assay.

Standard	Chemical class	IRAC classification	EC100 (ug/cm^2)
Fipronil	Fiprole	2B	0.312
Chlorfenapyr	METI	13	1.35
Carbaryl	Carbamate	1A	>5
Carbofuran	Carbamate	1A	ND
Permethrin	Pyrethroid	3A	0.078
Imidacloprid	Neonicotinoid	4A	4.81
Nitempyran	Neonicotinoid	4A	>5
Indoxacarb	Semicarbazone	22A	ND
Metaflumizone	Semicarbazone	22B	ND
Eprinomectin	Macrocyclic lactone	6	>5
Ivermectin	Macrocyclic lactone	6	>5
Selamectin	Macrocyclic lactone	6	1.3
Chlorpyrifos	Organophosphate	1B	1.25
Spinosad	Spinosyns	5	ND
Amitraz	*N*-Methyl-formamide	19	>5

species such as face fly (*Musca autumnalis*), blow fly (*Lucilia* spp.), *Cuterebra, Hypoderma,* and *Culicoides.* Many of these are disease vectors and can cause considerable productivity losses. There is an *ex vivo* assay for *Lucilia cuprina* and animal models for others. Also not discussed here are the biting (Mallophaga) and sucking (Anoplura) lice which infest all livestock species as well as dogs and cats. Testing of compounds and botanicals in *ex vivo* assays has been reported for lice infesting water buffaloes (*Haematopinus tuberculatus*) [6], sheep (*Bovicola ovis* Schrank) [7, 8], and humans (*Pediculus humanus capitis*) [9, 10]. During the course of development of animal health ectoparasiticides, promising candidates would generally be tested against these fly and lice targets. However, these would not solely constitute a driving force for a new product, but would be valuable add-ons to the spectrum of the lead ectoparasiticide.

Relevance of Agricultural Chemical Assays to Animal Health

Agricultural chemical (agchem) companies have been discovering and developing insecticides and acaricides for crop protection targets for many decades. While the specific target species of agricultural pests may be different, they are nevertheless similar in many respects to those ectoparasites infesting animals. Many of the major chemical and mode-of-action classes of ectoparasiticides used in animal health were originally discovered for agricultural use and then adapted directly, altered chemically, or prepared in a different formulation or delivery system to be utilized in animals. These include organophosphates (chlorpyrifos), carbamates (carbaryl), fiproles (fipronil), neonicotinoids (imidacloprid), spinosyns (spinosad), semicarbazones (indoxacarb), pyrethroids (permethrin, cypermethrin), macrocyclic lactones (abamectin), and many others. In the course of his career, one of us (JNC) has identified several potential animal health ectoparasiticides by pursuing the new compounds being developed by agchem companies for their insecticide and acaricide targets. A number of companies have both agchem and animal health groups which mutually benefit from that association.

Conclusion

Whole-organism "*ex vivo*" assays have been used for many years in conjunction with target-based ("*in vitro*") and host animal ("*in vivo*") ectoparasite assays to discover and develop new compounds effective against parasitic and/or nuisance arthropods such as mosquitoes, fleas, ticks, flies, mites, lice, and many other related species. The whole-organism assays supply a valuable bridge between the relatively newly derived target-based screens and the pest on the target animal. This chapter describes the discovery process and the many whole-organism assays that have been developed for the laboratory. The assays vary in throughput and ease of manipulation. Ectoparasites are available from many suppliers, and the laboratory setup costs are modest unless very sophisticated phenotypic evaluation or liquid handling equipment is utilized. For most assays, a microscopic or naked eye observation is the usual procedure to determine efficacy, which is most often based on motility. Whole-organism *ex vivo* assays have been instrumental in the discovery of the initial novel chemistry, the selection of the eventual clinical candidate and the potency, spectrum, resistance expectations, phenotypic effects, and the onset of effect of most of the ectoparasiticides on the market today.

References

1. Georgi, J.R. (1992) Artificial system and method for breeding fleas. US Patent 5,133.289, July 28, 1992.
2. Krober, T. and Guerin, P.M. (2007) An *in vitro* feeding assay to test acaricides for control of hard ticks. *Pest Manag. Sci.*, **63**, 17–22.
3. Deloach, J.R. (1984) *In vitro* feeding of *Psoroptes ovis* (Acari: Psoroptidae). *Vet. Parasitol.*, **16**, 117–125.
4. Thind, B.B. and Ford, H.L. (2007) Maintaining *Psoroptes ovis* (Acari: Psorotidae) off-host in the laboratory. *Exp. Appl. Acarol.*, **43**, 205–211.
5. Dunn, J.A., Prickett, J.C., Collins, D.A., and Weaver, R.J. (2016) Primary screen for potential sheep scab control agents. *Vet. Parasitol.*, **224**, 68–76.
6. Khater, H.F., Ramadan, M.Y., and El-Madawy, R.S. (2009) Lousicidal, ovicidal and repellant efficacy of some essential oils against lice and flies infesting water buffaloes. *Vet. Parasitol.*, **14**, 257–266.
7. James, P.J. and Callander, J.T. (2012) Bioactivity of tea tree oil from *Melaleuca alternifolia* against sheep lice (*Bovicola ovis* Schrank) *in vitro*. *Vet. Parasitol.*, **187**, 498–504.

8. James, P.J., Hook, S.E., and Pepper, P.M. (2010) *In vitro* infection of sheep lice (*Bovicola ovis* Shrank) by steinernematid and heterorhabditid nematodes. *Vet. Parasitol.*, **174**, 85–91.
9. Asenov, A., Oliveira, F.A., Speare, R., Liesenfeld, O., Hengge, U.R., and Heukelbach, J. (2010) Efficacy of chemical and botanical over-the-counter pediculicides available in Brazil, and off-label treatments, against head lice *ex vivo*. *Int. J. Dermatol.*, **49**, 324–330.
10. Gallardo, A., Mougabure-Cueto, G., Vassena, C., Picollo, M.I., and Toloza, A.C. (2012) Comparative efficacy of new commercial pediculicides against adults and eggs of *Pediculus humanus capitis* (head lice). *Parasitol. Res.*, **110**, 1601–1606.

9
In vitro Feeding Methods for Hematophagous Arthropods and Their Application in Drug Discovery

Ard M. Nijhof[*] *and Katharine R. Tyson*

Abstract

In ectoparasiticide research, the use of artificial feeding systems (AFSs) to evaluate the *in vitro* efficacy of test compounds against hematophagous arthropods offers many potential advantages: only small amounts of test compounds are usually required, compound effects can be tested in a controlled laboratory setting with limited variables, and their use leads to a reduction in experimental animal use. However, initial assay development can be challenging and not all hematophagous arthropods can be maintained under laboratory conditions using AFSs. In this chapter, general requirements for the main components of each AFS, including the membrane, blood meal, and temperature control system are discussed and an overview of currently established *in vitro* feeding methods for major hematophagous arthropod groups is presented. AFSs have found successful application in drug discovery projects targeting fleas, which may serve as an example for the further development of improved and more consistent assays employing AFSs.

Introduction

Laboratory feeding of hematophagous arthropods usually requires the use of experimental animals. However, a long-standing interest exists in developing effective *in vitro* feeding systems of hematophagous arthropods for (i) the colonization and maintenance of some species, (ii) the mass rearing of insects for use in control programs (e.g., the sterile insect release method), and (iii) the facilitation of entomological studies, including studies on vector–pathogen interactions and drug discovery and efficacy. An additional factor that has stimulated further development of artificial feeding systems (AFSs) is the public demand for the use of alternatives

[*]Corresponding author.

Ectoparasites: Drug Discovery Against Moving Targets, First Edition.
Edited by Charles Q. Meng and Ann E. Sluder

to animal experimentation as enunciated in the 3R concept (*R*eplacement, *R*eduction, and *R*efinement of animal experiments) [1].

The earliest references to the artificial feeding of arthropods date back to the beginning of the twentieth century, with the publication of reports on the artificial feeding of tsetse flies and soft ticks to study *Trypanosoma* transmission and the function of Haller's organ, respectively [2, 3]. Both publications describe a feeding system similar to using an anticoagulated blood pool covered by rat skin or a rat diaphragm through which the arthropods can pierce their proboscis and imbibe a blood meal. Many improvements to these first AFSs have since been made, particularly by the replacement of animal skin or tissue with artificial membranes such as Parafilm or silicone. However, the basic principle remains the same, and most AFSs for hematophagous ectoparasites still consist of the following four components: (i) a confined space containing the arthropods; (ii) an artificial membrane or animal skin through which the arthropods can pierce and obtain a blood meal; (iii) the actual vertebrate blood meal, which is for most arthropods heated to the host body temperature; and (iv) a temperature control system. Optionally, attachment or foraging stimuli such as elevated CO_2 concentrations; exfoliated host skin, animal hair, and animal hair extracts; conspecific excreta or (synthetic) pheromones; and blood additives such as glucose, adenosine triphosphate (ATP), or reduced glutathione (GSH) and antibiotics can also be used to increase feeding success [4–7].

Artificial feeding methods are now regularly used for the rearing of some hematophagous insect laboratory cultures including the triatomine bug *Rhodnius prolixus* [8], the bed bug *Cimex lectularis* [9, 10], the soft tick *Ornithodoros erraticus* [11], and several mosquito species [12–14]. The technique has become indispensable for the mass rearing of certain species such as tsetse flies and *Aedes aegypti* mosquitoes that are used in release programs based on the sterile insect technique (SIT) [15, 16]. *In vitro* feeding methods are also used in many research applications including basic physiology studies of ectoparasites, vector–pathogen interactions, and drug discovery efforts using both *in vitro* feeding and surface contact efficacy tests. Similar systems also find use in research on non-hematophagous arthropods, for instance, to study plant pathogen transmission using artificial feeding media (reviewed in Ref. [17]).

The aim of this review is to provide an overview of currently established artificial feeding techniques for arthropods of veterinary relevance and to discuss how these methods were or could be implemented in drug discovery projects. Due to their relevance for the animal health industry, the primary focus lies on techniques developed for fleas and ticks, but AFSs for other ectoparasites, including lice, mosquitoes, flies, mites, and triatomine- and bed bugs are also presented and discussed.

AFS-Components: The Membrane

The use of animal skin membranes in AFSs most closely mimics natural feeding and has therefore been used extensively. However, since animal skin is prone to contamination and decay, especially when kept at warmer temperatures during *in vitro* feeding, may not be easily obtainable, and can raise bioethical issues, alternative (partial) synthetic membranes that mimic host skin have been developed. Of all the synthetic membranes tested, Parafilm is the easiest to handle and obtain and is now preferentially used to feed most hematophagous arthropods. Examples of other frequently reported alternatives include Baudruche membrane (or goldbeater's skin), which is prepared from the serosa of the intestine from cattle or sheep, and collagen sausage casings [18, 19]. For the feeding of ixodid ticks, silicone membranes find frequent use, as they do not decay and are, in contrast to Parafilm, self-sealing. This prevents blood from leaking into the feeding chamber when ticks withdraw their mouthparts [20, 21]. Silicone membranes reinforced with netting are also routinely used for feeding tsetse flies [22]. Synthetic membranes may require additional attachment stimuli such as animal odor extracts and hair to mimic animal skin and make them more attractive to the parasite, in particular for ixodid ticks.

AFS Components: The Blood Meal

The animal source of the blood meal impacts the *in vitro* feeding efficacy and fecundity of hematophagous arthropods. In rearing some tsetse fly species (*Glossina* spp.) and *R. prolixus*, pig blood was found to be superior to cow blood [8, 23]. However, for the cat flea, *Ctenocephalides felis*, no significant differences were observed in the survival and fecundity when fed cattle or dog blood *in vitro* [24]. In addition, *Ixodes ricinus* nymphs successfully fed on either chicken or sheep blood *in vitro* [25]. *Lutzomyia ovallesi* sandflies fed with chicken blood showed higher fecundity and longevity compared to sandflies fed with blood from six other animals. Unlike *L. ovallesi*, *Phlebotomus papatasi* sandflies displayed similar survival when fed with blood from eight different animals including chickens [26, 27]. Based on the cumulative results of these studies, it seems that for some hematophagous arthropods, survival and fecundity are dependent on the host source of blood. This specificity may be attributed to host immune factors present in the host blood that potentially adversely affect the physiology of the feeding arthropod.

Large amounts of blood for *in vitro* feeding systems can be collected at abattoirs during the exsanguination of animals. A disadvantage of this method is that the blood is not sterile and may require further treatment before use by gamma irradiation [28] or through the addition of (broad-spectrum) antibiotics [20]. Both treatment methods have limitations. Gamma irradiation requires a radiation source and results in accelerated hemolysis [28], while the addition of antibiotics to the blood affects the microbiome of the feeding arthropod, which potentially results in reduced fitness, fecundity, and vectorial capacity [29, 30]. Aside from sterility limitations, animal blood collected from slaughterhouses might also be infected with blood-borne pathogens or contain veterinary drug residues if the appropriate drug withdrawal times were not observed.

An alternative to the use of blood from a slaughterhouse is the use of aseptically collected blood from commercially obtained donor animals. The use of human blood from blood banks has also been reported. However, an increased age of the blood showed negative effects on the feeding rates and fecundity of mosquitoes and bed bugs [9, 19, 31]. The prolonged storage of blood at 4°C was also reported to decrease feeding success in AFSs for mosquitoes and tsetse flies [32, 33]. As an alternative to refrigeration, blood conservation techniques such as freezing [33], freeze-drying [34] or oven-drying [35] have in some instances proved to be useful for the long-term storage of blood meals, potentially reducing the cost of regular blood collections. Following collection, blood is defibrinated or treated with anticoagulants to prevent clotting. The selection of an appropriate anticoagulation method is important because some methods may affect feeding efficacy. For example, *Rhipicephalus appendiculatus* nymphs did not feed on acid citrate dextrose (ACD) or ethylenediaminetetraacetic acid (EDTA)-treated blood but did feed to repletion on defibrinated or heparinized blood, with the highest feeding success on heparinized blood [36]. Similar findings were reported for artificially fed *Amblyomma variegatum* ticks [37]. *Triatoma infestans* bugs fed on heparinized blood laid more viable eggs in comparison to bugs fed on sodium oxalate-, sodium citrate-, or sodium fluoride-treated blood [38]. Heparinized blood was also superior to defibrinated blood when fed to the bed bug, *C. lectularis* [10], but not when high heparin (10%) concentrations were used [9]. Besides affecting the feeding efficacy of hematophagous arthropods, specific anticoagulants may also affect downstream research applications. When *Anopheles albimanus* mosquitoes were artificially infected with blood containing *Plasmodium vivax* from patients, EDTA present in the blood inhibited the development of *P. vivax* in the mosquito [39].

The use of artificial diets to feed hematophagous arthropods could avoid many of the disadvantages associated with the use of animal or human blood. Artificial diets can be manufactured under controlled conditions ensuring a consistent and sterile composition, are pathogen-free, and do not rely on the direct use of (experimental) animals. Kogan *et al.* were able to rear *A. aegypti* for eight generations using a substitute blood meal consisting of proteins, salts, and the phagostimulant ATP, suggesting the promise of artificial diets [40]. A substitute diet developed and used to rear five generations of *Glossina palpalis* was also used to feed *Glossina morsitans, Stomoxys calcitrans,* and *Tabanus nigrovittatus* [41, 42]. Artificial diets such as these may however not be feasible for all arthropods, and the effects of the long-term use of an artificial diet on colony rearing are rarely reported.

AFS Components: Temperature Control Systems

Heating of the blood to a temperature near the physiological body temperature of the arthropod's natural host is a critical factor in each AFS. For instance, *A. variegatum* ticks did not attach to membranes at blood temperatures below 35°C or above 39°C [37], bed bugs did not attach at blood temperatures below 35°C [10], and the optimum temperature to induce probing in the black fly *Simulium venustum* was found to be 37°C [43]. The highest *in vitro* feeding rates for the poultry red mite (*Dermanyssus gallinae*) were obtained at blood temperatures of 40–41°C, within the range of the normal body temperature of chickens [44]. Examples of frequently used temperature control systems for heating the blood meal in AFSs include (circulating) water baths, microscope slide warmers, laboratory heating plates, and heated operating or aquarium mats [10, 20, 43, 45].

Artificial Feeding Methods and Applications

Fleas

The best known *in vitro* feeding method for cat fleas (*C. felis*) is the "artificial dog," which was sold commercially by its inventor Jay Georgi (FleaData, NY, USA). This system comprises a two-compartment Plexiglas box that normally contains 25 individual feeding units. Each feeding unit consists of a blood container and a flea cage. The blood container holds approximately 10 ml of blood and is sealed with stretched Parafilm at the bottom. The flea cage is made up of two rings that can be assembled to form a cylindrical cup. A feeding screen with 300 μm openings is cemented to the lower edge of the upper ring, whereas the bottom ring has a bottom screen with 30 μm openings to allow gas exchange while keeping flea eggs and faeces inside. Each flea cage accommodates approximately 25 to 30 fleas. The blood container is put in place on top of the feeding screen, and the feeding unit is placed into a hole in the Plexiglas bottom of the upper compartment. The top compartment contains a heater that warms the air and the blood through conduction to 38°C. Fleas can pierce the Parafilm membrane and imbibe the heated blood directly above them [46].

A modified version of the "artificial dog," the "Greyhound", was later developed for the simultaneous testing of insecticidal compounds against *C. felis* [47] (Figure 9.1). Instead of 25 feeding units with a 5-cm inner diameter, the "Greyhound" accommodates up to 104 cages with a diameter of 2.5 cm. This system has been used to identify and test several classes of ectoparasiticides, as described in more detail later in this chapter and in the chapter by Clark and Pearce in this volume.

An alternative AFS currently in use is the Rutledge feeder [48]. The feeder is attached to a laboratory support stand and then flea cages are brought into contact with a Parafilm membrane using a laboratory jack (Figures 9.2 and 9.3).

Besides their application in drug screening, flea AFSs have also been used to infect fleas *in vitro* with pathogens. When blood containing *R. felis*-infected ISE6 tick cells was fed to *C. felis* fleas, a persistent *R. felis* infection was detected in the fleas using quantitative real-time polymerase chain reaction (PCR). Vertical transmission to the F_1 progeny was not observed [49]. The AFS was also used to infect cat fleas with different *Bartonella* species, such as *Bartonella henselae* and *Bartonella quintata*, and then study their kinetics in the flea and the flea faeces [50, 51].

Ticks

Ticks are divided into two families, argasid (soft) ticks and ixodid (hard) ticks. A third family, the *Nuttallielidae*, is monotypic and will not be considered further

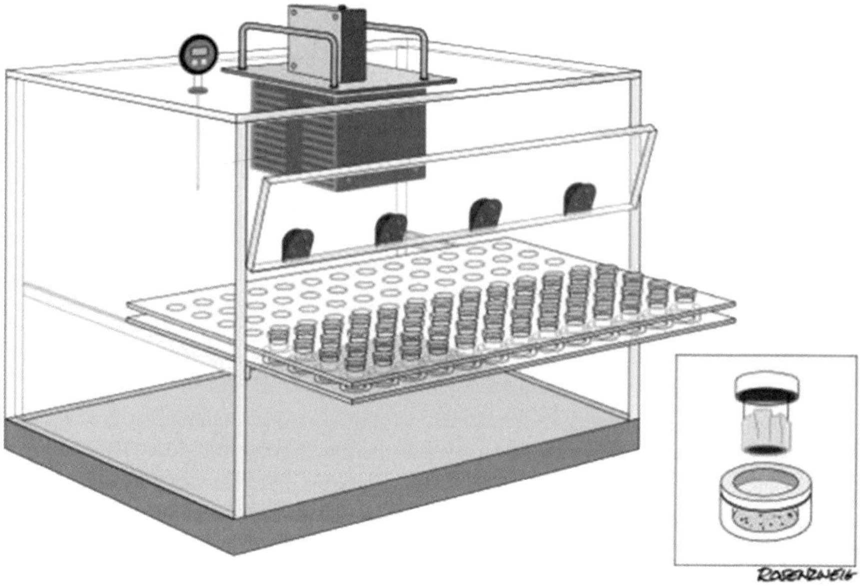

Figure 9.1 The "Greyhound" artificial feeding system. (Reproduced with permission from Zakson-Aiken *et al.* [47]. Copyright 2001, Oxford University Press.)

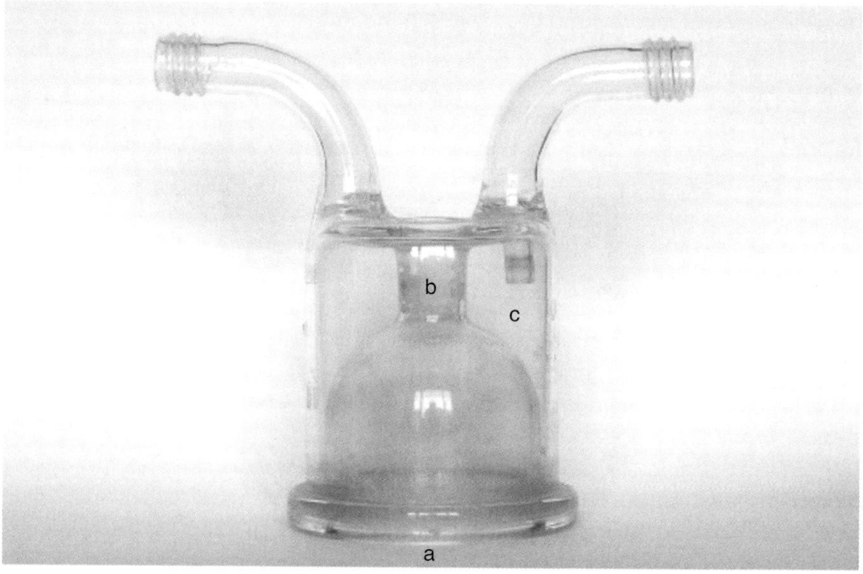

Figure 9.2 Rutledge-type feeder. The Rutledge feeding system consists of a jacketed hollow cone, the base of which can be covered by an artificial membrane or animal skin (a). Blood is introduced into the cone through a tube (b) that extends from its vertex. Using a circulating water bath, heated water is pumped through the cylindrical water jacket that surrounds the cone and tube (c), thereby warming the blood meal. (Courtesy of K. Seidl, Freie Universität Berlin.)

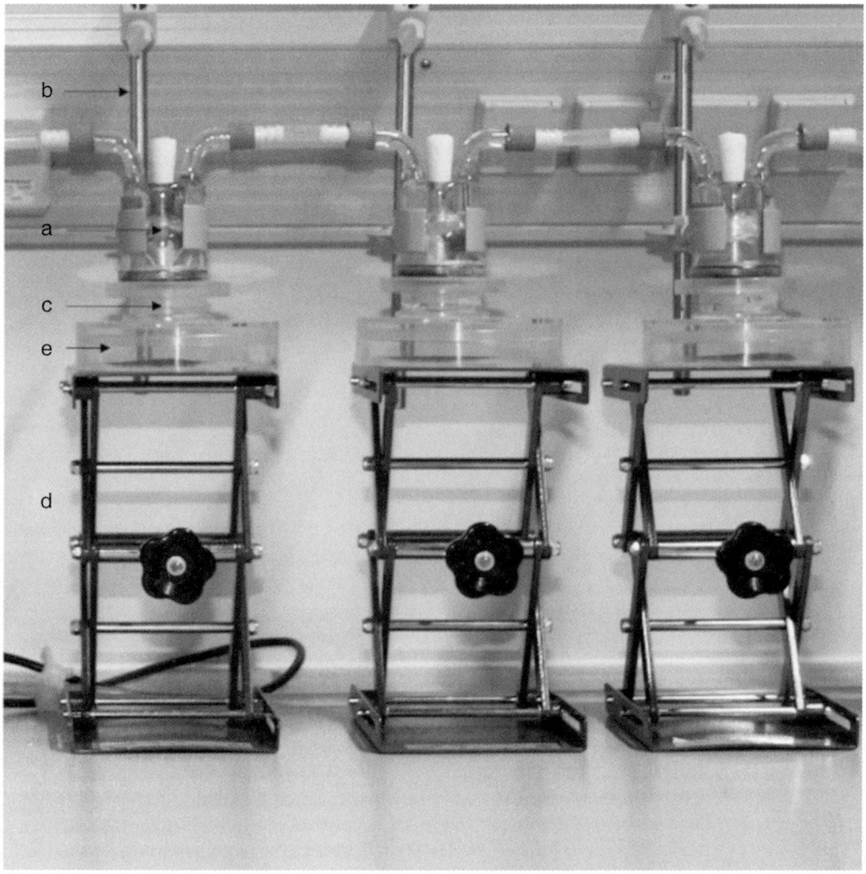

Figure 9.3 Example of an *in vitro* flea feeding system, showing three Rutledge-type feeders (a) attached to laboratory support stands (b) and connected in series to tubing attached to a circulating water bath. The continuous flow of warm water heats the blood meal inside the Rutledge feeders to approx. 37°C. Flea cages (c) are brought into contact with the Parafilm membrane covering the blood meal using a laboratory jack (d). Each flea cage has a feeding screen which directly touches the Parafilm membrane and a bottom screen which allows flea eggs and feces to fall through in a dish containing culture media (e), from where the eggs can be harvested. (Courtesy of K. Seidl, Freie Universität Berlin.)

here. The development of AFSs for ixodid ticks is particularly challenging, mainly due to the long duration of the feeding process and complex feeding behaviors. All life stages of ixodid ticks typically feed for several days to weeks [52], while nymphal and adult argasid ticks feed for relatively short time periods (15–60 min).

Researchers began experimenting with AFSs for ixodid ticks in the early 1900s [4]. Through the past century, advances in the development of AFSs for several species of ixodid ticks have been made, resulting in modified *in vitro* feeding systems that use various artificial membranes and complex combinations of tick attachment factors. Kuhnert *et al.* developed one of the first successful ixodid

AFSs using a silicone membrane rather than Baudruche [20]. In order to induce attachment of *Amblyomma hebraeum* to the silicone membrane, combinations of host hair and hair extracts, tick faeces, and a synthetic-attachment pheromone mixture were used. While all life stages of the tick were able to attach to the membrane and feed, adult females displayed prolonged pre-oviposition periods and reduced hatch rates. Kuhnert later investigated if other tick species were amenable to the same AFS [4]. Again, while *Rhipicephalus microplus*, *A. variegatum*, and *A. hebraeum* larvae, nymphs, and adults were able to attach to the silicone membranes and feed, adult female reproductive abilities were impaired. While maintaining ixodid tick colonies through AFSs remains a challenge, AFSs have successfully been used to test the efficacy of acaricides [4, 20, 53]. In more recent years, using a modification of Kuhnert's methods, Krober and Guerin fed *I. ricinus* various concentrations of two acaricides, fipronil and ivermectin. The improved method included the use of lens-cleaning paper-reinforced silicone membranes, acrylic glass tubing or polystyrene feeding units, glass fiber mosquito netting glued to the feeding unit, and a plastic cross and cow hair extract on the membrane (Figure 9.4). With this improved method, dose effects and efficacy differences were apparent in artificially fed *I. ricinus* ticks. This method has since been modified and used to feed and infect *Ixodes scapularis* with multiple pathogenic, tick-borne bacteria [54, 55]. A Rutledge-type feeder with animal skin membranes was used for studying artificial infection of *I. ricinus* with *Babesia divergens* [56].

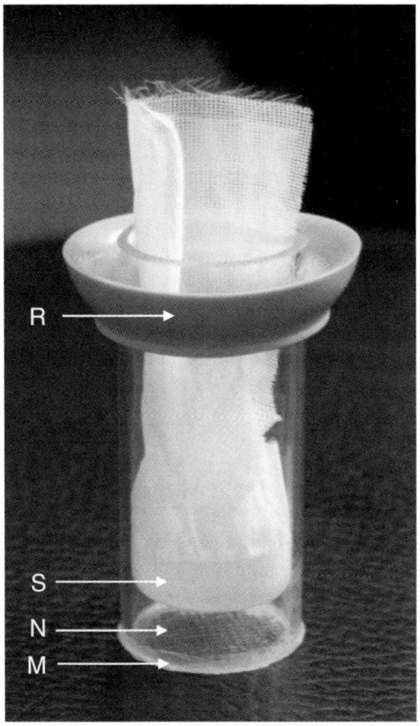

Figure 9.4 An *in vitro* feeding unit modified after Ref. [53]. It consists of glass tubing and mosquito netting (N) glued to a silicone membrane (M) which closes the unit on one side. Additional attachment stimuli such as hair or hair extracts can be placed on the membrane. A movable rubber ring (R) around the unit keeps the blood meal below the membrane when the feeding unit is placed in a blood container such as a glass beaker or the well of tissue culture plate. A perforated plastic stopper wrapped with muslin (S) is inserted in the unit to confine the ticks during feeding. (Courtesy of Bettina Böhme, Freie Universität Berlin.)

Because argasid ticks feed for shorter time periods and do not require complex mixtures of multiple components for attachment, the development of AFSs for argasid ticks has proved to be less challenging than the development of AFSs for ixodid ticks. In fact, argasid ticks appear to have increased reproductive abilities when fed by artificial means in comparison to natural host feeding. Specifically, Osborne and Mellor showed that when *Ornithodoros moubata* adults fed through either animal or silicone membranes in an artificial feeding apparatus, they became fully engorged faster and laid significantly more eggs than adult ticks fed naturally on mice [57]. Later, Schwan *et al.* further developed a Parafilm-based AFS using *O. moubata* with a capacity to feed up to 2000 nymphs or 200 adults at one time [58]. Using this feeding technique, adult females displayed no deleterious reproductive effects when fed different types of blood samples.

An alternative to membrane feeding is the capillary feeding of ticks [59]. In this method, ticks are adhered to a glass slide, and then pulled glass capillaries pre-filled with blood are placed over the ticks' mouthparts. A variation of this technique can be used to collect pilocarpine-induced tick saliva [60]. The capillary feeding of ticks has been used to artificially infect ticks with tick-borne pathogens (e.g., [59, 61–63]), for feeding double-stranded RNA (dsRNA) to ticks to induce RNA interference (RNAi) [64], and for feeding antibodies raised against potential tick vaccine antigens to evaluate their potential as anti-tick vaccine candidates [65]. The main disadvantage of this technique is that only partially fed ticks will imbibe larger amounts of blood, so the use of experimental animals to pre-feed ticks prior to the capillary feeding is usually required. This also makes the capillary feeding technique less suitable for the evaluation of systemic acaricidal compounds.

Mosquitoes

Several membrane feeding systems for mosquitoes have been developed and differ by their methods of temperature regulation, the nature of the membranes, and the blood meal compositions [19]. As a result of personal experience, preferences, and inherited methods, membrane feeding techniques vary between laboratories. The Rutledge feeder and the Hemotek system are probably the most frequently used feeding systems. A landmark in the development of AFSs for arthropods was the membrane feeder designed by Louis Rutledge and colleagues [66]. Initially used for feeding mosquitoes, this device has been adapted by many other laboratories for the feeding of other ectoparasites, including fleas and ticks [24, 56]. The Rutledge feeding system consists of a jacketed hollow cone, the base of which can be covered by an artificial membrane or animal skin. Blood is introduced into the cone through a tube that extends from its vertex. Using a circulating water bath, heated water is pumped through the cylindrical water jacket that surrounds the cone and lower part of the tube, thereby warming the blood meal [67] (Figure 9.2).

Another feeding system originally developed for the artificial feeding of *A. aegypti* is the Hemotek system, which is now sold commercially (Hemotek Limited, Lancashire, UK). The Hemotek system uses a low-voltage power supply to control the temperature of the blood meal inside an aluminum blood meal reservoir [68]. Multiple feeders can be connected to one power unit, and collagen membranes are provided with the system. A modified Hemotek system that provided a larger feeding surface to accommodate more mosquitoes was recently developed [69].

Lice

Until the 1940s, colonies of the human body louse (*Pediculus humanus*) and head louse (*Pediculus capitis*) could only be maintained on humans. This changed with the colonization of human lice that could be fed on rabbits [70]. A breakthrough in the *in vitro* feeding of lice using an artificial membrane was reported by Haddon in 1956. He was able to maintain the human body louse for two generations using membranes made of stretched Gutta-percha, a natural latex, and may have been the first to mention the successful use of Parafilm in AFSs for hematophagous arthropods [71, 72]. Other lice which have been artificially fed using membranes include the squirrel louse *Neohaematopinus sciuropteri* and the hog louse *Haematopinus suis* [73, 74]. To reduce human labor associated with AFSs, an automated system for the feeding of *P. capitis* was developed, in which the blood meal is pumped into a feeding reservoir, alternated with saline rinses. Fluid release into and drainage of the feeding reservoir were controlled by programmable timers [75].

Hematophagous Hemiptera

A laboratory colony of *C. lectularis* could be maintained for over 2 years by weekly *in vitro* feeding using a Rutledge-type feeding system [10]. Unfortunately, adaptation of field-collected bed bugs to an *in vitro* feeding system may be difficult [76]. Other cost-effective feeding systems include the use of heated blood bags made out of Parafilm [9] and an artificial host made of three stacked compact discs (CDs) [45]. In the latter system, the diameter of the middle CD was reduced to create a small void, which can be filled up with blood and covered with silicone membrane; an aquarium heat mat was used to heat the artificial host. An automated feeding apparatus designed for the human head louse has also been successfully used for the feeding of bed bugs [75].

Flies

As previously mentioned, AFSs using silicone membranes are widely used for the mass rearing of tsetse flies (*Glossina* spp.) for SIT; standard operating procedures are available from the International Atomic Energy Agency (IAEA) . Colonies of other hematophagous flies, such as the stable fly *S. calcitrans*, can also be reared using AFSs with silicone membranes, or by contact with blood-drenched cloth [77, 78]. Other telmophages (pool feeders), such as some horseflies (*Tabanus* spp.), can also be fed using the latter method [79, 80].

Mites

Various AFSs have been described for feeding the poultry red mite *D. gallinae* and the northern fowl mite *Ornithonyssus sylvarium in vitro* [44, 81–84]. The most frequently used membranes are day-old chicken skins, although some success with Parafilm and Parafilm-like products treated with skin extracts have been reported [85, 86].

Artificial Feeding Systems in Ectoparasiticide Drug Discovery

In order to discover effective insecticides and acaricides, animal health drug discovery programs have employed the use of AFSs. Successful integration of an AFS into a discovery program requires that certain factors be met. First, the objective of the screening program must be well defined in order to select the appropriate arthropod species for testing [87] and the desired phenotypic response to be measured (i.e., paralysis, mortality, feeding inhibition, etc.). If possible, the target species will be used for testing. However, if the target species is not amenable to laboratory conditions, an alternative species may be used. The desired activity on the alternative species must translate to activity on the target species for successful drug discovery. In addition, if the goal of the program is to discover a broad-spectrum parasiticide, multiple species representing a range of sensitivity to the drug should be used for testing.

Second, once the screening program objective is defined and the organism to be used and the phenotypes to be measured are determined, the assay design must be carefully considered. Assays need to be sensitive, reproducible, stable, economically feasible, and generate comparable, homogenous, and reliable data [88, 89]. To do this, assay variability should be reduced as much as possible, which is difficult with whole-organism screens due to inherent biological differences between individual insects that may result in different outcomes when the insects are exposed to experimental drugs. In addition, measuring whole-organism phenotypes in the presence of a drug typically requires laborious observations from highly trained individuals, which generates qualitative, subjective, and ambiguous data. Increasing the number of animals tested as well as using robotics and automated systems for assay setup and end point measurements help reduce variability [88]. Once an assay is established, various statistical parameters can be used for assay validation, specifically evaluating the performance and sensitivity of the assay [90, 91].

In the context of drug discovery, some AFSs have produced sensitive, reliable, and homogeneous data regarding the efficacy of multiple classes of ectoparasiticides. Specifically, the "Greyhound" AFS has been extensively used as a primary assay to evaluate the efficacy of multiple ectoparasiticides because of its ease of use, reduced compound testing needs in comparison to animal models, sensitivity, and reproducibility. In one study, the "Greyhound" was used to test the relative potencies of three different concentrations (1, 10, and 20 μg/ml) of avermectin analogs in heparinized bovine blood fed to fleas [47]. Flea mortality was scored after 48 h of drug exposure. The "Greyhound" feeding method was sensitive enough to differentiate potencies between different compounds, and one compound had superior LC_{90} and LC_{50} values when compared to ivermectin. When that same compound was administered subcutaneously in a dog at >100 times the commercial ivermectin dosage, it showed zero efficacy against fleas. These *in vitro* and *in vivo* results led to the conclusion that the low systemic activity against fleas of ivermectin is likely to be a class-wide phenomenon.

The "Greyhound" system has also been used to test the activity of several other ectoparasiticides, including nodulisporic acid derivatives [92], pyriproxifen [93], and the most recently discovered novel class of ectoparasiticides, the isoxazolines [94–96]. Gassel *et al.* determined that the isoxazoline, fluralaner, displayed superior insecticidal activity to fipronil and imidacloprid when all compounds were fed to dieldrin-resistant *C. felis* [94]. Shoop *et al.* used the "Greyhound" to determine the optimal blood concentration of afoxolaner [95]. In this study, fleas were fed *in vitro* on a sixfold serial dilution of afoxolaner in citrated bovine blood (0.32, 0.16, 0.08, 0.04, 0.02, and 0.01 μg/ml), and flea mortality was assessed after 24, 48, and 72 h. The concentration of 0.16 μg/ml killed all fleas within 24 h in the *in vitro* system, and this concentration was subsequently chosen as a target blood concentration for *in vivo* studies in dogs. The afoxolaner plasma concentration estimated to provide 90% efficacy (EC_{90}) against fleas in dogs was later found to be 0.023 μg/ml [97], different from the *in vitro* data.

The observed difference between the *in vivo* and *in vitro* results was also found in a similar study using the fungal metabolite nodulisporic acid. This difference could be partially explained by the absence of host behavioral responses, such as grooming, in the *in vitro* system, which would otherwise increase the mortality rate of ataxic fleas [98]. A reduced uptake of blood in the *in vitro* system compared to *in vivo* feeding [99] and possibly a prolonged acclimatization to the *in vitro* system, which may result in a delayed start of feeding compared to feeding *in vivo*, may be other contributing factors. While effective concentrations of ectoparasiticides have differed between *in vitro* and *in vivo* flea feeding experiments, the "Greyhound" is a sensitive and reliable method to identify and differentiate active compounds for further progression in a drug discovery program.

AFSs for ticks have also been used to determine the efficacy of acaricides. In a recent study, fluralaner concentrations as low as 0.001 μg/ml resulted in 83% mortality when fed to *O. moubata* nymphs using a membrane feeding system similar to the methods described earlier in this chapter [96]. Because ixodid tick AFSs are laborious to set up, often get contaminated, and are typically low throughput, they have not been used as extensively as "the Greyhound" and argasid tick AFSs in drug discovery programs as a primary screening tool. Alternatively, various types of tick contact assays, which are not described here, are frequently used to determine the acaricidal activity of ectoparasiticides on ixodid ticks; see the chapter by Clark and Pearce in this volume for descriptions of these assays.

Conclusions

Over the past century, several advances have been achieved in the development of AFSs for hematophagous arthropods. These advances have allowed for the laboratory maintenance of some species of blood feeders, the infection of hematophagous arthropods with various microbial pathogens, RNAi and vaccine studies, and drug discovery efforts. The main advantages of AFSs in ectoparasiticide research are that small amounts of test compounds are usually required, compound effects can be tested in controlled systems with limited variables, and experimental animal use is reduced. However, initial assay development can be challenging. Some hematophagous arthropods cannot be maintained under laboratory conditions. Many hematophagous arthropods fed by artificial means, including triatomine bugs, ixodid ticks, fleas, and lice [20, 24, 75, 100], have reduced weights and egg masses when compared to naturally fed arthropods. This phenomenon may be caused by several factors: (i) the uptake of smaller blood meals *in vitro* compared to arthropods which fed on mammals due to a lack of natural host factors that attract the arthropods and signal them to initiate feeding; (ii) a disruptive effect of the addition of antibiotics to the blood meal on the microbiota and symbionts of arthropods that may in turn affect the fecundity; (iii) decreased mating activity of *in vitro* fed arthropods; (iv) decreased nutritive quality of stored blood; (v) bacterial or fungal contamination; or (vi) the use of anticoagulants [4, 9, 101, 102]. Elucidation of these factors is essential to obtain optimal arthropod growth and reproduction *in vitro* and improve existing AFSs for drug discovery.

Another limitation of AFSs for drug discovery is assay variability and reduced sensitivity due to the use of whole organisms. Robotics and automation serve to help reduce assay variability and increase performance. While several automated systems are available for use with arthropod screening programs, they have multiple limitations and typically cannot be used to assay larger life stages of arthropods (i.e., nymphs and adults) [90]. In order to reduce variability and improve assay performance, future assay improvements need to be aimed at developing more rapid, consistent, and automated methods for setting up assays and measuring a variety of desired phenotypes using the appropriate target organisms and life stages.

References

1. Russel, W.M.S. and Burch, R.L. (1959) The Principles of Humane Experimental Technique, Methuen, London.
2. Hindle, E. and Merriman, G. (1912) The sensory perception of *Argas persicus* (Oken). *Parasitology*, **5** (3), 203–216.
3. Rodhain, J., Pons, C., Vandenbranden, C., and Bequaert, J. (1912) Contribution au mécanisme de la transmission des trypanosomes par les glossines. *Arch. f. Schiffs u. Tropen Hyg.*, **16**, 732.
4. Kuhnert, F. (1996) Feeding of hard ticks *in vitro:* new perspectives for rearing and for the identification of systemic acaricides. *Altex*, **13** (2), 76–87.
5. Galun, R. and Kindler, S.H. (1965) Glutathione as an inducer of feeding in ticks. *Science*, **147** (3654), 166–167.
6. de Moura, S.T., da Fonseca, A.H., Fernandes, C.G., and Butler, J.F. (1997) Artificial feeding of *Amblyomma cajennense* (Fabricius, 1787) (Acari:Ixodidae) through silicone membrane. *Mem. Inst. Oswaldo Cruz*, **92** (4), 545–548.
7. Hosoi, T. (1959) Identification of blood components which induce gorging of the mosquito. *J. Insect Physiol.*, **3** (3), 191–218.

8. Langley, P.A. and Pimley, R.W. (1978) Rearing triatomine bugs in the absence of a live host and some effects of diet on reproduction in *Rhodnius prolixus* Stål (Hemiptera: Reduviidae). *Bull. Entomol. Res.*, **68** (02), 243–250.
9. Aak, A. and Rukke, B.A. (2014) Bed bugs, their blood sources and life history parameters: a comparison of artificial and natural feeding. *Med. Vet. Entomol.*, **28** (1), 50–59.
10. Montes, C., Cuadrillero, C., and Vilella, D. (2002) Maintenance of a laboratory colony of *Cimex lectularius* (Hemiptera: Cimicidae) using an artificial feeding technique. *J. Med. Entomol.*, **39** (4), 675–679.
11. Basto, A.P., Nix, R.J., Boinas, F., Mendes, S. *et al.* (2006) Kinetics of African swine fever virus infection in *Ornithodoros erraticus* ticks. *J. Gen. Virol.*, **87** (Pt 7), 1863–1871.
12. Phasomkusolsil, S., Tawong, J., Monkanna, N., Pantuwatana, K. *et al.* (2013) Maintenance of mosquito vectors: effects of blood source on feeding, survival, fecundity, and egg hatching rates. *J. Vector Ecol.*, **38** (1), 38–45.
13. Benzon, G.L. and Apperson, C.S. (1987) An electrically heated membrane blood-feeding device for mosquito colony maintenance. *J. Am. Mosq. Control Assoc.*, **3** (2), 322–324.
14. Deng, L., Koou, S.Y., Png, A.B., Ng, L.C. *et al.* (2012) A novel mosquito feeding system for routine blood-feeding of *Aedes aegypti* and *Aedes albopictus*. *Trop. Biomed.*, **29** (1), 169–174.
15. Feldmann, U. (1994) Guidelines for the rearing of tsetse flies using the membrane feeding technique, in Techniques of Insect Rearing for the Development of Integrated Pest and Vector Management Strategies (ed. J.P.R. Ochieng'-Odero), ICIPE Science Press, Nairobi, pp. 449–471.
16. Carvalho, D.O., Nimmo, D., Naish, N., McKemey, A.R. *et al.* (2014) Mass production of genetically modified *Aedes aegypti* for field releases in Brazil. *J. Vis. Exp.*, **83**, e3579.
17. Bosco, D. and Tedeschi, R. (2013) Insect vector transmission assays, in Phytoplasma: Methods and Protocols (eds M. Dickinson and J. Hodgetts), Humana Press, Totowa, NJ, pp. 73–85.
18. Waladde, S.M., Kemp, D.H., and Rice, M.J. (1979) Feeding electrograms and fluid uptake measurements of cattle tick *Boophilus microplus* attached on artificial membranes. *Int. J. Parasitol.*, **9** (2), 89–95.
19. Wirtz, R.A. and Rutledge, L.C. (1980) Reconstituted collagen sausage casings for the blood feeding of mosquitos. *Mosq. News*, **40** (2), 287–288.
20. Kuhnert, F., Diehl, P.A., and Guerin, P.M. (1995) The life-cycle of the bont tick *Amblyomma hebraeum in vitro*. *Int. J. Parasitol.*, **25** (8), 887–896.
21. Habedank, B. and Hiepe, T. (1994) Untersuchungen zur *In-Vitro*-Fütterung von zecken - Argasidasae und Ixodidae. *Mitt. Osterr. Ges. Tropenmed. Parasitol.*, **16**, 107–114.
22. Bauer, B. and Wetzel, H. (1976) A new membrane for feeding *Glossina morsitans* Westw (Diptera, Glossinidae). *Bull. Entomol. Res.*, **65** (4), 563–565.
23. Mews, A.R., Langley, P.A., Pimley, R.W., and Flood, M.E.T. (1977) Large-scale rearing of Tsetse flies (*Glossina* spp.) in absence of a living host. *Bull. Entomol. Res.*, **67** (1), 119–128.
24. Wade, S.E. and Georgi, J.R. (1988) Survival and reproduction of artificially fed cat fleas, *Ctenocephalides felis* Bouche (Siphonaptera: Pulicidae). *J. Med. Entomol.*, **25** (3), 186–190.
25. Liu, X.Y., Cote, M., Paul, R.E., and Bonnet, S.I. (2014) Impact of feeding system and infection status of the blood meal on *Ixodes ricinus* feeding. *Ticks Tick Borne Dis.*, **5** (3), 323–328.
26. Noguera, P., Rondón, M., and Nieves, E. (2006) Effect of blood source on the survival and fecundity of the sandfly *Lutzomyia ovallesi* Ortiz (Diptera: Psychodidae), vector of *Leishmania*. *Biomédica*, **26**, 57–63.
27. Harre, J.G., Dorsey, K.M., Armstrong, K.L., Burge, J.R. *et al.* (2001) Comparative fecundity and survival rates of *Phlebotomus papatasi* sandflies membrane fed on blood from eight mammal species. *Med. Vet. Entomol.*, **15** (2), 189–196.

28. Bauer, B., Iwannek, K.H., Hamann, H.J., and Adamsky, G. (1979) Use of gamma-irradiated blood for feeding tsetse flies. Symposium on Isotope and Radiation Research on Animal Diseases and Their Vectors. International Atomic Energy Agency, Vienna, pp. 319–325.

29. Gendrin, M., Rodgers, F.H., Yerbanga, R.S., Ouedraogo, J.B. *et al.* (2015) Antibiotics in ingested human blood affect the mosquito microbiota and capacity to transmit malaria. *Nat. Commun.*, **6**, 5921.

30. Narasimhan, S. and Fikrig, E. (2015) Tick microbiome: the force within. *Trends Parasitol.*, **31** (7), 315–323.

31. Pothikasikorn, J., Boonplueang, R., Suebsaeng, C., Khaengraeng, R. *et al.* (2010) Feeding response of *Aedes aegypti* and *Anopheles dirus* (Diptera: Culicidae) using out-of-date human blood in a membrane feeding apparatus. *J. Vector Ecol.*, **35** (1), 149–155.

32. Luo, Y.P. (2014) A novel multiple membrane blood-feeding system for investigating and maintaining *Aedes aegypti* and *Aedes albopictus* mosquitoes. *J. Vector Ecol.*, **39** (2), 271–277.

33. Wetzel, H. and Luger, D. (1978) The use of deep frozen, stored bovine blood for *in vitro* feeding of tsetse flies. *Z. Parasitenkd.*, **57** (2), 163–168.

34. Wetzel, H. (1980) The use of freeze-dried blood in the membrane feeding of tsetse flies (*Glossina p. palpalis*, Diptera: Glossinidae). *Tropenmed. Parasitol.*, **31** (3), 259–274.

35. Kabayo, J.P., Taher, M., and Barnor, H. (1988) Use of oven-dried blood for *in vitro* feeding of tsetse flies. *Experientia*, **44** (9), 802–803.

36. Waladde, S.M., Young, A.S., Ochieng, S.A., Mwaura, S.N. *et al.* (1993) Transmission of *Theileria parva* to cattle by *Rhipicephalus appendiculatus* adults fed as nymphae *in vitro* on infected blood through an artificial membrane. *Parasitology*, **107** (Pt 3), 249–256.

37. Voigt, W.P., Young, A.S., Mwaura, S.N., Nyaga, S.G. *et al.* (1993) *In vitro* feeding of instars of the ixodid tick *Amblyomma variegatum* on skin membranes and its application to the transmission of *Theileria mutans* and *Cowdria ruminatium*. *Parasitology*, **107** (Pt 3), 257–263.

38. Niel, L.L., Ubiergo, G.O., Combin, O., and Giraldez, E. (1980) [Effects of the addition of anticoagulants in the artificial feeding of *Triatoma infestans*]. *Medicina (Buenos Aires)*, **40** (Suppl 1), 259.

39. Solarte, Y., Manzano Mdel, R., Rocha, L., Castillo, Z. *et al.* (2007) Effects of anticoagulants on *Plasmodium vivax* oocyst development in *Anopheles albimanus* mosquitoes. *Am. J. Trop. Med. Hyg.*, **77** (2), 242–245.

40. Kogan, P.H. (1990) Substitute blood meal for investigating and maintaining *Aedes aegypti* (Diptera: Culicidae). *J. Med. Entomol.*, **27** (4), 709–712.

41. Kabayo, J.P., Taher, M., and Van Der Vloedt, A.M. (1985) Development of a synthetic diet for Glossina (Diptera: Glossinidae). *Bull. Entomol. Res.*, **75** (4), 635–640.

42. Stoffolano, J.G. and Majer, J. (1997) Comparison between tsetse fly synthetic diet and bovine blood on reproduction and survivorship in *Tabanus nigrovittatus* (Diptera : Tabanidae). *J. Med. Entomol.*, **34** (6), 738–740.

43. Sutcliffe, J.F. and McIver, S.B. (1975) Artificial feeding of simuliids (*Simulium venustum*): factors associated with probing and gorging. *Experientia*, **31** (6), 694–695.

44. Kirkwood, A.C. (1971) *In vitro* feeding of *Dermanyssus gallinae*. *Exp. Parasitol.*, **29** (1), 1–6.

45. Naylor, R. (2012) Ecology and Dispersal of the Bedbug, The University of Sheffield.

46. Georgi, J.R. and Georgi, M.E. (1991) Mass production of cat fleas with an artificial dog, in First Symposium on Ectoparasites of Pets (ed. F.W. Knapp), University of Kentucky, Lexington, KY, pp. 84–86.

47. Zakson-Aiken, M., Gregory, L.M., Meinke, P.T., and Shoop, W.L. (2001) Systemic activity of the avermectins against the cat flea (Siphonaptera: Pulicidae). *J. Med. Entomol.*, **38** (4), 576–580.

48. Wedincamp, J. Jr. and Foil, L.D. (2002) Vertical transmission of *Rickettsia felis* in the cat flea (*Ctenocephalides felis* Bouche). *J. Vector Ecol.*, **27** (1), 96–101.

49. Reif, K.E., Kearney, M.T., Foil, L.D., and Macaluso, K.R. (2011) Acquisition of

Rickettsia felis by cat fleas during feeding. *Vector Borne Zoonotic Dis.*, **11** (7), 963–968.

50. Bouhsira, E., Franc, M., Boulouis, H.J., Jacquiet, P. *et al.* (2013) Assessment of persistence of *Bartonella henselae* in *Ctenocephalides felis*. *Appl. Environ. Microbiol.*, **79** (23), 7439–7444.

51. Kernif, T., Leulmi, H., Socolovschi, C., Berenger, J.M. *et al.* (2014) Acquisition and excretion of *Bartonella quintana* by the cat flea, *Ctenocephalides felis felis*. *Mol. Ecol.*, **23** (5), 1204–1212.

52. Waladde, S.M. and Rice, M.J. (1982) The sensory basis of tick feeding behaviour, in Physiology of Ticks (eds F.D. Obenchain and R. Galun), Pergamon Press, Oxford, pp. 71–118.

53. Krober, T. and Guerin, P.M. (2007) An *in vitro* feeding assay to test acaricides for control of hard ticks. *Pest Manag. Sci.*, **63** (1), 17–22.

54. Andrade, J.J., Xu, G., and Rich, S.M. (2014) A silicone membrane for *in vitro* feeding of *Ixodes scapularis* (Ixodida: Ixodidae). *J. Med. Entomol.*, **51** (4), 878–879.

55. Oliver, J.D., Lynn, G.E., Burkhardt, N.Y., Price, L.D. *et al.* (2016) Infection of immature *Ixodes scapularis* (Acari: Ixodidae) by membrane feeding. *J. Med. Entomol.*, **53** (2), 409–415.

56. Bonnet, S., Jouglin, M., Malandrin, L., Becker, C. *et al.* (2007) Transstadial and transovarial persistence of *Babesia divergens* DNA in *Ixodes ricinus* ticks fed on infected blood in a new skin-feeding technique. *Parasitology*, **134** (Pt 2), 197–207.

57. Osborne, R.W. and Mellor, P.S. (1985) Use of a silicone membrane feeding technique in the laboratory maintenance of a colony of *Ornithodoros moubata*. *Trop. Anim. Health Prod.*, **17** (1), 31–38.

58. Schwan, E.V., Hutton, D., Shields, K.J., and Townson, S. (1991) Artificial feeding and successful reproduction in *Ornithodoros moubata moubata* (Murray, 1877) (Acarina: Argasidae). *Exp. Appl. Acarol.*, **13** (2), 107–115.

59. Chabaud, A.G. (1950) Sur la nutrition artificielle des tiques. *Ann. Parasitol. Hum. Comp.*, **25** (1–2), 42–47.

60. Patton, T.G., Dietrich, G., Brandt, K., Dolan, M.C. *et al.* (2012) Saliva, salivary gland, and hemolymph collection from *Ixodes scapularis* ticks. *J. Vis. Exp.*, (60), pii: 3894.

61. Broadwater, A.H., Sonenshine, D.E., Hynes, W.L., Ceraul, S. *et al.* (2002) Glass capillary tube feeding: a method for infecting nymphal *Ixodes scapularis* (Acari: Ixodidae) with the lyme disease spirochete *Borrelia burgdorferi*. *J. Med. Entomol.*, **39** (2), 285–292.

62. Inokuma, H. and Kemp, D.H. (1998) Establishment of *Boophilus microplus* infected with *Babesia bigemina* by using *in vitro* tube feeding technique. *J. Vet. Med. Sci.*, **60** (4), 509–512.

63. Purnell, R.E. (1970) Infection of the tick *Rhipicephalus appendiculatus* with *Theileria parva* using an artificial feeding technique. *Res. Vet. Sci.*, **11** (4), 403–405.

64. Soares, C.A., Lima, C.M., Dolan, M.C., Piesman, J. *et al.* (2005) Capillary feeding of specific dsRNA induces silencing of the isac gene in nymphal *Ixodes scapularis* ticks. *Insect Mol. Biol.*, **14** (4), 443–452.

65. Antunes, S., Merino, O., Mosqueda, J., Moreno-Cid, J.A. *et al.* (2014) Tick capillary feeding for the study of proteins involved in tick-pathogen interactions as potential antigens for the control of tick infestation and pathogen infection. *Parasit. Vectors*, **7**, 42.

66. Rutledge, L.C., Ward, R.A., and Gould, D.J. (1964) Studies on the feeding response of mosquitoes to nutritive solutions in a new membrane feeder. *Mosq. News*, **24** (4), 407–419.

67. Das, S., Garver, L., and Dimopoulos, G. (2007) Protocol for mosquito rearing (*A. gambiae*). *J. Vis. Exp.*, (5), 221.

68. Cosgrove, J.B., Wood, R.J., Petric, D., Evans, D.T. *et al.* (1994) A convenient mosquito membrane feeding system. *J. Am. Mosq. Control Assoc.*, **10** (3), 434–436.

69. Damiens, D., Soliban, S.M., Balestrino, F., Alsir, R. *et al.* (2013) Different blood and sugar feeding regimes affect the productivity of *Anopheles arabiensis* colonies (Diptera: Culicidae). *J. Med. Entomol.*, **50** (2), 336–343.

70. Culpepper, G.H. (1948) Rearing and maintaining a laboratory colony of body lice on rabbits. *Am. J. Trop. Med. Hyg.*, **28** (3), 499–504.

71. Haddon, W. (1956) An artificial membrane and apparatus for the feeding of the human body louse *Pediculus-Humanus-Corporis*. *Am. J. Trop. Med. Hyg.*, **5** (2), 315–325.

72. Haddon, W. (1956) The maintenance of the human body louse *Pediculus-Humanus-Corporis* through complete cycles of growth by serial feeding through artificial membranes. *Am. J. Trop. Med. Hyg.*, **5** (2), 326–330.

73. Hafner, P. and Ludwig, H.W. (1969) A method for membrane-feeding of hog louse *Haematopinus suis*. *Zeitschrift für Parasitenkunde*, **33** (2), 177.

74. Lauer, D.M. and Sonenshine, D.E. (1978) Adaptations of membrane feeding techniques for feeding the squirrel flea, *Orchopeas howardi*, and the squirrel louse, *Neohaematopinus sciuropteri*, with notes on the feeding of the human body louse, *Pediculus humanus var. corporis*. *J. Med. Entomol.*, **14** (5), 595–596.

75. Takano-Lee, M., Velten, R.K., Edman, J.D., Mullens, B.A. *et al.* (2003) An automated feeding apparatus for *in vitro* maintenance of the human head louse, *Pediculus capitis* (Anoplura: Pediculidae). *J. Med. Entomol.*, **40** (6), 795–799.

76. Kilpinen, O., Kristensen, M., and Jensen, K.M. (2011) Resistance differences between chlorpyrifos and synthetic pyrethroids in *Cimex lectularius* population from Denmark. *Parasitol. Res.*, **109** (5), 1461–1464.

77. Salem, A., Franc, M., Jacquiet, P., Bouhsira, E. *et al.* (2012) Feeding and breeding aspects of *Stomoxys calcitrans* (Diptera: Muscidae) under laboratory conditions. *Parasite*, **19** (4), 309–317.

78. Venkatesh, K. and Morrison, P.E. (1980) Some aspects of oogenesis in the stable fly *Stomoxys calcitrans* (Diptera, Muscidae). *J. Insect Physiol.*, **26** (10), 711–715.

79. Magnarelli, L.A. (1985) Blood feeding and oviposition by tabanids (Diptera) in the laboratory. *J. Med. Entomol.*, **22** (6), 600–603.

80. Stoffolano, J.G. (1979) Comparison of methods for feeding blood to *Tabanus nigrovittatus* in the laboratory (Diptera, Tabanidae). *Ann. Entomol. Soc. Am.*, **72** (2), 216–217.

81. Crystal, M.M. (1986) Artificial feeding of northern fowl mites, *Ornithonyssus sylviarum* (Canestrini and Fanzago) (Acari: Macronyssidae), through membranes. *J. Parasitol.*, **72** (4), 550–554.

82. Harrington, D., Din, H.M., Guy, J., Robinson, K. *et al.* (2009) Characterization of the immune response of domestic fowl following immunization with proteins extracted from *Dermanyssus gallinae*. *Vet. Parasitol.*, **160** (3–4), 285–294.

83. Makert, G.R., Vorbruggen, S., Krautwald-Junghanns, M.E., Voss, M. *et al.* (2016) A method to identify protein antigens of *Dermanyssus gallinae* for the protection of birds from poultry mites. *Parasitol. Res.*, **115** (7), 2705–2713.

84. Wright, H.W., Bartley, K., Nisbet, A.J., McDevitt, R.M. *et al.* (2009) The testing of antibodies raised against poultry red mite antigens in an *in vitro* feeding assay; preliminary screen for vaccine candidates. *Exp. Appl. Acarol.*, **48** (1–2), 81–91.

85. Harrington, D.W., Guy, J.H., Robinson, K., and Sparagano, O.A. (2010) Comparison of synthetic membranes in the development of an *in vitro* feeding system for *Dermanyssus gallinae*. *Bull. Entomol. Res.*, **100** (2), 127–132.

86. Zeman, P. (1988) Surface skin lipids of birds – a proper host kairomone and feeding inducer in the poultry red mite, *Dermanyssus gallinae*. *Exp. Appl. Acarol.*, **5** (1–2), 163–173.

87. Rutledge, L.C., Collister, D.M., Meixsell, V.E., and Eisenberg, G.H. (1983) Comparative sensitivity of representative mosquitoes (Diptera: Culicidae) to repellents. *J. Med. Entomol.*, **20** (5), 506–510.

88. Giacomotto, J. and Segalat, L. (2010) High-throughput screening and small animal models, where are we? *Br. J. Pharmacol.*, **160** (2), 204–216.

89. Inglese, J., Johnson, R.L., Simeonov, A., Xia, M. *et al.* (2007) High-throughput screening assays for the identification of chemical probes. *Nat. Chem. Biol.*, **3** (8), 466–479.

90. Iversen, P.W., Eastwood, B.J., Sittampalam, G.S., and Cox, K.L. (2006) A comparison of assay performance measures in screening assays: signal window, Z' factor, and assay variability ratio. *J. Biomol. Screening*, **11** (3), 247–252.

91. Zhang, J.H., Chung, T.D.Y., and Oldenburg, K.R. (1999) A simple statistical parameter for use in evaluation and validation of high throughput screening assays. *J. Biomol. Screening*, **4** (2), 67–73.

92. Meinke, P.T., Colletti, S.L., Fisher, M.H., Wyvratt, M.J. *et al.* (2009) Discovery of the development candidate *N*-tert-butyl nodulisporamide: a safe and efficacious once monthly oral agent for the control of fleas and ticks on companion animals. *J. Med. Chem.*, **52** (11), 3505–3515.

93. Meola, R., Meier, K., Dean, S., and Bhaskaran, G. (2000) Effect of pyriproxyfen in the blood diet of cat fleas on adult survival, egg viability, and larval development. *J. Med. Entomol.*, **37** (4), 503–506.

94. Gassel, M., Wolf, C., Noack, S., Williams, H. *et al.* (2014) The novel isoxazoline ectoparasiticide fluralaner: selective inhibition of arthropod gamma-aminobutyric acid- and l-glutamate-gated chloride channels and insecticidal/acaricidal activity. *Insect Biochem. Mol. Biol.*, **45**, 111–124.

95. Shoop, W.L., Hartline, E.J., Gould, B.R., Waddell, M.E. *et al.* (2014) Discovery and mode of action of afoxolaner, a new isoxazoline parasiticide for dogs. *Vet. Parasitol.*, **201** (3–4), 179–189.

96. Williams, H., Zoller, H., Roepke, R.K., Zschiesche, E. *et al.* (2015) Fluralaner activity against life stages of ticks using *Rhipicephalus sanguineus* and *Ornithodoros moubata* IN *in vitro* contact and feeding assays. *Parasit. Vectors*, **8**, 90.

97. Letendre, L., Huang, R., Kvaternick, V., Harriman, J. *et al.* (2014) The intravenous and oral pharmacokinetics of afoxolaner used as a monthly chewable antiparasitic for dogs. *Vet. Parasitol.*, **201** (3–4), 190–197.

98. Shoop, W.L., Gregory, L.M., Zakson-Aiken, M., Michael, B.F. *et al.* (2001) Systemic efficacy of nodulisporic acid against fleas on dogs. *J. Parasitol.*, **87** (2), 419–423.

99. Koehler, P.G. and Hinkle, N.C. (1991) Methods of laboratory rearing of fleas, in First Symposium on Ectoparasites of Pets (ed. F.W. Knapp), University of Kentucky, Lexington, Kentucky, pp. 78–81.

100. de Isola, E.D., Sanchez, D., and Katzin, V. (1980) [*Triatoma infestans*: effect of artificial feeding on its life cycle]. *Medicina*, **40** (Suppl 1), 207–212.

101. Zug, R. and Hammerstein, P. (2015) Bad guys turned nice? A critical assessment of *Wolbachia* mutualisms in arthropod hosts. *Biol. Rev. Camb. Philos. Soc.*, **90** (1), 89–111.

102. Dean, S.R. and Meola, R.W. (2002) Factors influencing sperm transfer and insemination in cat fleas (Siphonaptera: Pulicidae) fed on an artificial membrane system. *J. Med. Entomol.*, **39** (3), 475–479.

10
Testing in Laboratory Animal Models for Ectoparasiticide Discovery and Development

Sandra Schorderet-Weber and Ronald Kaminsky*

Abstract

Animal models are commonly used in drug discovery for the identification of novel ectoparasiticides but few are disclosed in detail. They can be important selection tools in early phases of chemistry optimization, when compound availability and information on toxicity are very limited. Most models have been designed using ticks and fleas, mainly due to their direct medical and veterinary importance but also as good general representatives of acari and insects. This review summarizes and describes laboratory rodent and rabbit models currently available for compound evaluation. Some models developed for other purposes are also mentioned because they could be easily adapted to drug discovery. Important parameters to take into account when establishing models with ectoparasites are also presented, such as choice of surrogate host, parasite confinement and recovery, and group size.

Introduction

The search for novel ectoparasiticides in the agrochemical, pharma, or animal health industry is pursued today through the screening of hundreds of thousands of compounds from synthetic or natural origin [1]. In the first selection steps, high-throughput screening (HTS) tests, often extensively automated, are performed either on validated target receptors or directly on the whole parasites, monitoring their survival over time in phenotypic assays (for more details, see Chapter 8). As ectoparasites are, by definition, living at the interface between the environment and their host, assays mimicking the real world can be reproduced in standardized laboratory conditions rather accurately. This is particularly true for agrochemical science since the final product is usually intended to be deposited on plants for protection against various pests. Compounds to screen can be

*Corresponding author.

Ectoparasites: Drug Discovery Against Moving Targets, First Edition.
Edited by Charles Q. Meng and Ann E. Sluder

sprayed or poured on leaves or fragments of them, in Petri dishes or multi-well systems in a scenario that mimics the final use. Parasites will be soaked in the compound formulation, or walk on a treated surface, similar to the actual field situation [2]. Ectoparasiticides developed for animal and human health use are often identified and selected similarly in the first steps of the discovery process [3]. However, as they are eventually intended to be administered to companion/ farm animals, or humans, the chemistry optimization process and selection tools should preferentially integrate a mammalian host as early as possible. *In vitro*, or more precisely off-host testing, might be sufficient to discover the intrinsic potency of a molecule, but will be unable to predict how the molecule may behave in, or on a warm-blooded vertebrate (e.g., dog, cattle, or human) with respect to bioavailability, duration, and so on. [4].

Efficacy studies that target companion or farm animals such as dogs or cattle are time- and resource demanding, requiring the synthesis of significant amounts of active ingredient. They may also raise animal welfare concerns given the need to confine individual test animals for long periods of time inside the testing facility in order to avoid cross-contamination and interference by wild ectoparasites. For those reasons, studies with target hosts are usually limited to the final phase of the profiling process, when selection has been narrowed down to a handful of compounds from which the final drug development candidate will be identified.

Chemistry optimization to improve ectoparasiticidal activity and reduce toxicity in the mammalian host relies on the establishment and understanding of a reliable structure–activity relationship (SAR) [5]. *In silico* selection for the best physicochemical properties in combination with off-host efficacy testing against the parasite very often results in insufficient selectivity, with too many candidate molecules showing equivalent potencies and properties. There is, therefore, a gap or a missing selection step that would help narrowing down to fewer potential candidate compounds, which could be synthetized in gram amounts for further evaluation in target hosts. This gap in the drug discovery process can be successfully filled with appropriate laboratory animal models.

Parasite models using rodents are widely used in research against endoparasites, in particular helminths, for drug discovery [6], vaccine research [7], and immunology studies (e.g., [8]). For example, the chemistry optimization leading to the discovery of the new anthelmintic monepantel was mainly achieved through the use of a gastrointestinal nematode model in gerbils (*Meriones unguiculatus*) [9]. In a similar way, models using surrogate hosts like rodents or rabbits are strong tools for chemistry optimization of novel ectoparasticide classes. They require limited amounts of compound, very often within the tens of milligrams range, provide short testing cycle times, allow the parallel comparison of several analogs, and can be adapted to different modes of application.

The smaller physical size of those animals is well suited to laboratories, and even small facilities can successfully house rodent units that are far less constraining and resource demanding than companion or farm animal facilities. Most pharmaceutical companies do use such models and have often developed them internally, tailoring them to their own research workflow. These types of efficient models can be considered as providing a competitive advantage and, for this reason, are rarely published, except in patent applications without many experimental details [10–13]. The present review summarizes information available in the public domain and discusses the advantages, limitations, and difficulties in developing such models for ectoparasiticidal drug discovery.

Developing Laboratory Animal Models of Ectoparasite Infestation

In this section we discuss a few parameters we consider to be important when establishing new laboratory animal models with ectoparasites based on our own experience at developing, validating, and running rodent models for compound screening.

As for any parasite model, finding the right surrogate host can be challenging. Ectoparasites (some ticks, mites, fleas, or lice) can be rather specific to their natural host. Other parasites, while perhaps less restrictive, may still have their host preferences. Therefore, an extensive knowledge of the parasite life cycle and behavior is a prerequisite for the development of a good representative model. Rats and mice are often chosen as surrogate hosts because they are relatively easy and affordable to purchase and maintain. However, some ectoparasites may attach faster and feed better on non-murine rodents or non-rodent hosts. Therefore, other common laboratory species such as gerbils (*M. unguiculatus,* Picture 10.1) [13], hamsters (*Mesocricetus auratus*), guinea pigs (*Cavia porcellus*) [14], or even birds such as chickens [15], should be considered during model design.

Parasite confinement is another difficulty, particularly for flying or jumping arthropods. It is not always possible simply to release the parasites on a treated

Picture 10.1 Non-murine rodents like gerbils can be used successfully for evaluating ectoparasiticides. To our experience, they are easy to handle and are very good hosts for ticks and cat fleas.

rodent. The main end point in those models, like in target species, is normally the collection and counting of parasites, dead or alive, engorged, attached or not, in order to calculate a percentage of efficacy in comparison with numbers of parasites collected from control animals [16]. In order to get solid and reproducible data, parasites should be collected in sufficient numbers from each individual. It is, therefore, important to find the best confinement technique, and prevent the host from removing, killing, or eating the ectoparasites during the study. Neck collars are often used for that purpose [17]. We also observed that collars are very good at preventing licking and grooming, which is important if the rodent host is treated on its entire body surface. Feeding chambers can be used as confinement tools for preventing ticks from escaping into the environment [18]. They can be constructed of different materials, generally plastic, silicone, or even metal [19]. The top is closed with a lid that can be opened to insert the parasites. The lid generally holds a mesh that allows both air circulation and visual inspection without opening the chamber. The chambers are glued onto a shaved area on the animal's body. The choice of the glue is crucial for avoiding parasite escapes. It has to be strong enough to hold the device in place for several days and also be supple enough to fit closely with the folds of the skin. It should not provoke any inflammation that could interfere with the parasite attachment or cause animal welfare issues. The number of parasites fitting in such confinement devices will depend on the interior surface available and the stage and feeding behavior of the parasite chosen for the model. For example, one should remember that ticks significantly increase their volume during feeding, and that nymphs take more blood than larvae. Putting too many parasites in a confinement device would result in poor and irregular engorgement, and excessive mortality in controls. If feeding will be interrupted after a defined period of time for purposes of scoring results, numbers can often be increased accordingly.

When parasites are released freely on the host, good confinement should be imposed at the cage level. In some cases, cages may have to be adapted to the arthropod behavior. For example, cages holding rodents infested with fleas may have to be elevated to avoid escapes into the environment and loss of data [20]; ticks dropping off their host after engorgement may try to escape the cage to find a quiet place to molt. Placing the cages over or into a container of water to catch the dropping ticks could prevent contamination of the environment and facilitate easy collection of ticks [17].

The recent interest in long-lasting systemic ectoparasiticides [3, 21–23] has introduced an additional complexity in the design of good compound selection tools, both off-host and in laboratory animal models. Efficacy data generated via oral or injection application in rodents should be used with care, keeping in mind that drug metabolic fate is species-specific and could be very different among models and even more varied when compared to target host species.

Regular, proper evaluation of drug candidate pharmacokinetics in the final host, whether or not combined with efficacy, is certainly needed to monitor drug persistence and to complement the potency information provided by either a laboratory animal model or by off-host testing [24, 25]. Despite this limitation, with appropriate caution, drug SAR can still be established using laboratory models and used to guide chemistry optimization. Testing for systemic activity requires the parasite to be in contact with the drug upon feeding. In this situation, parasites are often released on the host before or in parallel to treatment [24, 26].

Finding the right treatment group size for generating solid data on which to base optimizational decisions is another difficult challenge for the model design. Scarcity of the compounds to be tested and requests for parallel evaluations of multiple analogs may restrict group size to a level not allowing statistical comparison between test groups. Individual variations should be reduced to their minimum, to allow clear differences between controls and treatment groups. Our own experience shows that efficacy ranking can be established with small group sizes ($n = 2$–4) through successive dose down-titrations. Group size may be increased at a later stage if statistical analysis is requested at important decision points, by repeating the test with as many or more animals and pooling all data generated at the dose of interest. Testing with only one animal per group could be tempting to save active ingredient. However, we would not recommend this option as there is high risk that one could end up with no data at all, resulting in a loss of time and the need to repeat the full test with a potentially inactive compound.

Finally, validation with known ectoparasiticides (positive standards) is strongly advised before a model is integrated into a discovery workflow [4, 18]. Additional compounds known to be poorly or not active against the arthropod of interest could be also added as "negative standards" to strengthen the validation process. As a prerequisite, positive and negative standards should behave similarly in the model compared to the target host species, preferentially but not necessarily at comparable dosages. Ideally, activity ranking among standards should be maintained in the same order in the model. However, exceptions are often seen, and should not lead to the complete withdrawal of the model if restricted to one or two standards. We often observed that model correlation with the target species is chemistry class-specific, and extra validation with the class of interest should be performed with early leads to guide the correct interpretation of the data generated with the model. We could note here that the choice of the route of administration (topical, oral, injection) in the model for drug selection is independent of the mode of action of the chemistry class (contact or systemic activity). In general, overinterpretation of results might lead to an incorrect understanding of the SAR, resulting in mistakes during chemistry optimization.

Available Ectoparasite Models

The animal models most commonly used in the animal health industry are certainly those for testing compounds against ticks and fleas and they will be the main focus of this chapter. However, models established with other ectoparasites do exist and are also described here as interesting alternatives (Table 10.1).

Ticks are primary targets, impacting farm animal growth performance, causing skin damage and secondary bacterial infections. Some species like *Ixodes*

Table 10.1 Principal features of laboratory models used for drug screening and characterization.

Arthropod	Laboratory host	Parasite confinement device used	Treatment	Constraints			References
				Anesthesia	Neck collar	Immobilization w/o anesthesia	
Ticks	Rabbit	Yes	Topical restricted to confined area	No	Yes	Yes	[10, 27, 28]
	Mouse	Yes	Topical restricted to confined area	Yes	No	No	[12]
	Rat	Yes	Topical restricted to confined area, and oral	Yes	No	No	[5, 18]
	Gerbil	No	Topical whole body, oral	No	Yes	No	[13, 26]
Fleas	Mouse	No	Topical whole body	No	NIA[a]	NIA	[29]
		No	Oral	Yes	No	No	[4]
	Rat	No	Topical whole body	No	NIA	No	[30]
	Gerbil	No	Topical whole body	No	Yes	No	[31]
Bed bugs	Mouse	No	Oral	No	No	Yes	[32, 33]
Lice	Mouse	No	Topical whole body	NIA	No	Yes	[26]
	Chicken	No	Topical whole body	No	No	Yes	[15]
Mosquitoes	Mouse	No	Topical whole body	Yes	No	No	[34]
Stable flies	Mouse	No	Topical whole body	Yes	No	No	[35]

[a] NIA, no information available.

holocyclus can cause lethal paralysis in dogs. Ticks are also vectors of a broad variety of pathogens of medical and veterinary importance (for more details, see Chapter 3). Fleas are nuisance parasites and their bite can induce severe allergic disorders in companion animals. Both ticks and fleas can be utilized as good representatives of other acari and insects, respectively, for which models do not exist, are difficult to establish, or are of lower medical or economic importance. Rodents and rabbits are natural hosts for immature stages (or even adults) of most ixodid tick species. Setting up models for testing against ticks should therefore be rather straightforward.

Tick Models

Rabbits

Rabbits have been used for a very long time in ectoparasitology for tick rearing in the laboratory. Almost all ixodid species of veterinary importance have been raised on this host, including the one-host cattle tick *Rhipicephalus (Boophilus) microplus* [36] and the brown dog tick *Rhipicephalus sanguineus* [19]. Rabbits have been widely used for studying tick–host immune interactions [37, 38], and are often the preferred surrogate host for evaluating novel anti-tick vaccines [39, 40]. Rabbits are also of a reasonable size that allows the parallel feeding of up to 2–3 tens of adult ticks, which is a real advantage compared to the smaller rodents. Ticks of any stage can be released in a confinement device glued on the shaved back of an animal restrained from grooming by a neck collar. This setup was used for comparing tick repellents for activity against *R. sanguineus* [27]. Treatments were applied on skin only inside the confinement devices, at different incremental dosages, and left to dry for 24 h before the ticks were released. Tick attachment was monitored daily. The length of time needed to reach 50% tick attachment was used to define the level of efficacy for each dosage of repellent tested. Based on the percentage of ticks that did not attach or engorge, repellency was qualified as slight (50–75%), moderate (76–90%), or strong (>90%).

Alternatively, and probably more useful for compound screening, ticks can be confined into nylon or cotton bags wrapped around each rabbit ear. In this setup, the treatment is limited to one or both ears, and delivered as a spray. This procedure limits the amount of active ingredient needed for the testing. A neck collar is also required to prevent the rabbit from licking or removing the ear bags. If cross-contamination between ears is effectively prevented, two different compounds could be evaluated on the same animal, one on each ear, reducing the number of animals needed per compound tested. As ear bags can be opened at any time during the experiment, the percentage of mortality, attachment, and engorged ticks can be recorded daily or even on an hourly base, making the model very close to the procedure used in cats and dogs [41]. Such a system was described for testing repellents [28], and was similarly used for evaluating acaricides [10]. Rabbits were treated by spray on their ears, and infested one day later with adult *R. sanguineus* ticks. Efficacy was measured 24, 48, and 72 h after infestation by counting live attached ticks, in comparison with a placebo-treated group [16]. After 72 h, remaining ticks were removed and the treated animals were challenged with repeated infestations at weekly intervals until efficacy dropped.

Rodents

Rabbit maintenance is nowadays more space demanding than in the past due to a significant improvement in animal welfare considerations for this species. Consequently, many companies have developed screening models in smaller laboratory animals, primarily in rats [11, 18, 42], but also in mice [12] and gerbils [13]. A rat model specifically designed for the evaluation and optimization of chemical entities using nymphs of *Amblyomma americanum* has been described in detail (Figure 10.1) [18].

Containment units were used for confining the ticks on the animal. Each unit was made of the upper portion of a 15-ml screw-cap conical tube inserted into a commercial orthodontic nipple cut across at the top. The tube lid was drilled at the center and closed with a self-adhesive mesh allowing air circulation, tick insertion, and subsequent engorgement monitoring. An additional rubber washer encircling the base of each nipple helped stabilize the unit and improve fastening to the skin. Containment units were attached to rats on a hair-clipped area within the dorsothoracic region using an instant adhesive. Anesthesia was necessary for the installation. Compounds to be evaluated were dissolved in absolute ethanol and poured inside the units. After the ethanol had completely evaporated, nymphs were released and left to attach and feed. Control animals received ethanol only. Concentrations of the active ingredient were calculated in milligram per square centimeter and due to the small surface of application, only few tens of milligrams were needed to run the test. Five animals per group and 10 nymphs per unit were used in the standard study design. The number of surviving ticks was assessed after 4 days' infestation, in comparison with the controls to calculate a percentage

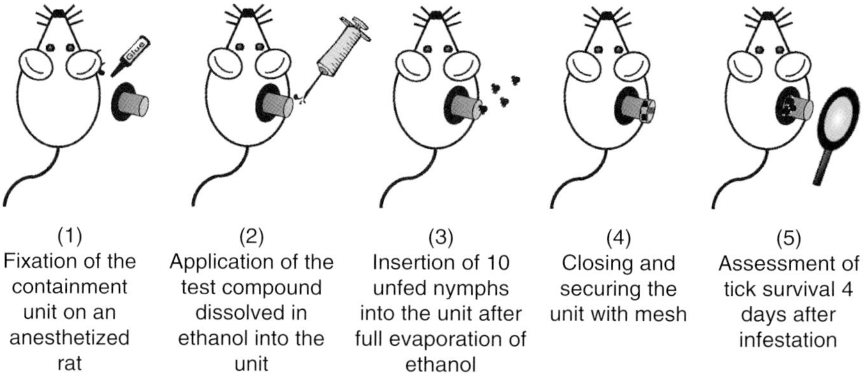

Figure 10.1 Tick rat model, diagrammed based on description by Gutierrez *et al.*, 2006 [18].

of tick mortality [16]. Nymphs collected were weighed to obtain additional information on their engorgement status. Serial dilutions were used to describe dose-dependent responses, evaluated via linear and nonlinear regression analysis. Four commercial acaricides (fipronil, ivermectin, chlorpyrifos, and permethrin) were tested to validate the model. All compounds showed activity in a dose-dependent manner, allowing 50% and 90% effective concentrations (EC_{50} and EC_{90}) to be calculated. Compared to the recommended doses of commercial products (Table 10.2), effective surface calculated dosages were in a similar range in the model. Ivermectin blocked feeding at much lower dosages than was required to kill ticks, confirming the usefulness of tick weight as an additional end point in this model. The EC values of permethrin measured in rats against nymphal stages of *A. americanum* were also very close to those obtained in cattle against adult *A. americanum* (EC_{50} 0.035 mg/cm^2 in rats, vs 0.028 mg/cm^2 in cattle). This model was later used successfully to characterize a compound of the benzimidazole-carbamate class as overcoming tick resistance to other acaricides [42]. A very similar model with *Dermacentor variabilis* nymph ticks was used for the optimization and patenting of a novel ectoparasitic control agent [11]. In this case, tick mortality was assessed after 48 h only.

Mouse models have also been developed for compound screening. A protocol using the same type of tick containment units as used for rats was run on mice with *Haemaphysalis longicornis* larvae [12]. Tick mortality was monitored after 72 h feeding. Other mouse-tick combinations have been described for studying tick–host immune interactions [44–46]. In all cases, parasite confinement was achieved via containment devices. Larvae have been the preferred stage for infestation, and were usually left to full engorgement, although nymphs have also been used [44]. Based on these data, one could suppose that the screening model described for *H. longicornis* could easily be adapted to many other tick species, depending on tick availability and veterinary importance in a specific country.

A model implemented in gerbils differed from the others as *R. sanguineus* nymph ticks were freely released on their host [13]. Test compounds diluted in spray or spot-on formulations were applied and left to dry for 24–48 h before infestation. Ticks were allowed to feed to repletion, then collected and kept under controlled conditions for molting to the next stage. Molting success was scored as an additional metric of tick survival. Compounds that disrupt the tick life cycle and/or are slow acting can thus be identified. Efficacy was expressed in comparison with controls [16]. An advantage of this model was to avoid the use of anesthesia and tick containment device, and thus could more closely mimic the situation found on target host species. This model might however have been more compound demanding in some situations, as the total body surface of the gerbils was treated. Nevertheless, since gerbil weight normally does not exceed 100 g, compound requirements should still remain within a range of a few tens of milligrams.

Recently, a novel class of acaricides with long-lasting systemic efficacy in dogs was brought to the animal health market [3, 21–23]. Different subclasses of isoxazolines have been described and in some cases results obtained using rodent tick

Table 10.2 Efficacious doses for known ectoparasiticides in the corresponding laboratory models in comparison with dosages used in the marketed product.

Model	Reference	Compound	Effective dose (EC_{90}) (mg/cm^2)	Marketed product (cattle)	Marketed dose (mg/kg)	Application mode	Surface calculated dose (mg/cm^{2a})
Rat-Tick	[18]	Fipronil	0.00084	Ectonil®	1	Pour-on	0.0095
		Ivermectin	0.0577	Ivomec®	0.5	Pour-on	0.0048
		Chlorpyrifos	0.11485	Warrior®	100	Ear-tag	0.95
		Permethrin	0.1767	Durvet Permethrin 1% Pour-On®	3.3–33	Pour-on	0.03–0.3

Model	References	Compound	Oral effective dose (mg/kg)	Marketed product (dog)	Marketed dose (mg/kg)	Application mode
Flea-Mouse	[4]	Nitenpyram	<1 mg/kg	Capstar®	1	Oral
	[4]	Selamectin	10 mg/kg	Stronghold®	6	Spot-on
	[4]	Fipronil	10 mg/kg	Frontline TopSpot®	6.5	Spot-on
	[4]	Cythioate	10 mg/kg	Proban®	3.3	Oral
	[4, 33]	Nodulosporic acid	10 mg/kg	–	10	Oral
	[4]	Ivermectin	>30 mg/kg	Ivomec®	0.3	Oral
	[4]	Lufenuron	>30 mg/kg	Program®	30	Oral

[a] Based on a cattle average weight of 400 kg: Body surface area (cm^2) = 1470 × Body weight (kg) $^{0.56}$ [43].

models were used to guide the chemistry optimization [5, 24]. The rat model with *D. variabilis* nymphs [18] was adapted for evaluation of systemic activity by putting the ticks into the units 24 h before the oral treatment. Compound efficacy was evaluated 48 h later. The gerbil model with *R. sanguineus* nymphs was used similarly. Ticks were released on the animals 24 h before oral treatment [26] and left to engorge until repletion. In most cases, the best candidate compounds selected with models were further evaluated for persistence in dogs through pharmacokinetic and/or efficacy studies [5, 11, 24].

Flea Models

Animal models for testing efficacy against fleas are less numerous. Flea confinement can be a real challenge and cat fleas, *Ctenocephalides felis*, although rather permissive in terms of host selection, are not frequently reported to infest rodents in the field. Few epidemiological studies mention *C. felis* collected from wild rodents [47]. The main flea species is generally the oriental rat flea *Xenopsylla cheopis*, and *C. felis* is only found at low percentages [48]. Attempts to improve feeding success by selecting a flea population physiologically better adapted to rodents have been made by rearing cat fleas on mice in the laboratory [20]. Mice were maintained on a grid, allowing flea eggs to fall through for collection and rearing. Successful breeding was achieved as fleas were able to survive on mice for more than 40 days. A model for testing compounds was established using this approach, and the general methods have been partially disclosed [29]. Mice were treated topically and adult unfed fleas were freely released on the animals one day later. Flea survival was assessed after 24 h. Another attempt to establish a flea model on mice has been described more precisely [4]. To allow fleas to feed freely without any particular confinement, mice were anesthetized to prevent grooming and active removal and killing of the insects. Different variations were evaluated during model development, including putting fleas on nude or shaved mice, but this did not provide any experimental advantage. Feeding failed when only the bare tail was available to the fleas, although the mice did not have to be anesthetized in this case. Body temperature was shown to be an important parameter for feeding success, as there was a clear correlation between dosages of sedative, the subsequent drop in mouse body temperature, and the percentage of fleas successfully feeding. The model was validated with known systemic insecticides applied orally at three dosages (nitenpyram, selamectin, fipronil, cythioate, nodulisporic acid, ivermectin, and the insect growth regulator lufenuron). About 1 h after treatment, mice ($n = 3$) were sedated, housed individually, and each infested with 30 fleas for a 2-h period. Fleas were removed by combing and vacuum aspiration and incubated for 24 h to monitor survival. The treatment efficacy was expressed by calculating flea mortality: for each individual mouse, the number of live fleas recovered after 24-h incubation was divided by the total number of fleas collected from the animal. For each treatment group, individual data were averaged and multiplied by 100, resulting in a percentage of flea survival. To calculate mortality/

efficacy, those percentages were subtracted from 100. Control animals ($n = 14$) treated with the vehicle (10% dimethyl sulfoxide (DMSO)) were handled similarly; but results were used only to compare feeding success, and were not integrated in the drug efficacy calculation. The percentage of mortality/efficacy was qualified as poorly (<50%), moderately (50–70%), very active (70–90%), and highly active (>90%),

The model efficiently discriminated and ranked the seven tested insecticides and could be used for novel drug profiling. Nitenpyram was the best compound, being active at the lowest dosage, followed by selamectin. Fipronil, cythioate, and nodulisporic acid were less potent than the two former molecules. Ivermectin and lufenuron were not active at any dosage tested. As lufenuron is an inhibitor of the larval development, no efficacy on adult fleas was expected. These results were consistent with dosages used in the marketed formulations, even if some active ingredients of topical products were administered orally in the model (Table 10.2). Ivermectin could be considered an exception, but Ivomec® is not specifically indicated for flea control in dogs, and this molecule was previously shown to be poorly active against cat fleas [30].

Besides mice, some published patent applications provide examples of flea models in rats [31] or gerbils [32]. Rats were infested with fleas and treated topically 24 h later with the test compounds. Placebo-treated animals served as controls, and a positive group treated with a known standard insecticide (not disclosed) was also included in each study. Fleas were left to feed for 48 h and were collected by combing. Geometric means were calculated and efficacy expressed in comparison with placebo [16]. Gerbils were conversely treated first by spray and infested with fleas 24 h later [32]. Efficacy was recorded 24 and 48 h after infestation by collecting and counting live fleas from treated animals, in comparison with live flea numbers from a placebo-treated control group [16]. Infestations could be repeated at weekly intervals until efficacy decreased to gain additional information about the potential long-term persistence of the drug. This model was also used for evaluating compounds with systemic efficacy [26]. The protocol was slightly changed in order to make sure that the parasites would be in contact with the compound during the peak of absorption. Gerbils were therefore infested immediately after the oral treatment. Efficacy was assessed only after 48 h of infestation, in the same way as previously described.

Other Ectoparasite Models

Bedbugs

Bedbugs (*Cimex lectularius*) have been successfully used in a mouse model to identify systemic activity of novel natural products [33, 49]. Test compounds were applied orally by gavage or through medicated feed for 5 days. Mice were restrained in plastic containers and their tail inserted into a tube containing unfed fourth instar bedbugs. The parasites were left to feed for 10–20 min, and then incubated until molting to the next stage. Activity criteria were mortality, paralysis, or molting delay. Fifty percent effective dosage (ED_{50}) [49] or chi-square comparison of results from treated versus control groups were used to express efficacy [33]. The model served also as preliminary filter for mammalian toxicity [49].

Lice

Lice are parasites of human and veterinary importance, and resistance to standard insecticides is widespread [50]. Novel insecticide classes should be screened against hemimetabolous insects such as lice as they may respond differently from species undergoing complete metamorphosis. Lice are very host specific and do not allow the development of animal models with human, sheep, or cattle lice. A mouse model with *Polyplax serrata*, the mouse sucking louse, has been established for insecticide selection and optimization [26]. Mice naturally infested with lice were treated topically by line-on application (one line on the back from the neck to the base of the tail). Live lice (all instars) were counted before, and at 4 and 14 days after treatment. The efficacy of each treatment was calculated using the Henderson and Tilton formula [51], which considers the numbers of live lice found on each individual mouse before and after treatment, weighed with the lice numbers found at the same time-points on untreated control mice held under similar conditions. Killing efficacy and life cycle disruption could be assessed with this two-time-point assessment.

For testing insecticides and infestation deterrents against human head lice, a model has been developed in chickens naturally infested with the chewing lice *Menopon* spp. and *Menacanthus* spp. [15, 35]. Chickens were treated topically by spray, as would be done to treat human hair. Besides the treated group and an infested untreated "seeder" control group, an additional lice-free group was added to monitor the speed of reinfection over time, as might happen among children in a school setting. The three groups were mixed for 24 h, and lice counts performed after 8, 12, and 24 h. Arithmetic and geometric means of pretreatment and posttreatment counts were statistically compared for efficacy evaluation. This model could probably also be used for selecting compounds for control of lice on sheep or cattle.

Flying Insects

Very few small-animal models have been described for testing compounds against flying insects. Adequate confinement and prevention of premature killing by the host are more challenging with flying insects than with fleas and ticks. A mouse model with adult stable flies was used for testing spinosad and could be representative of the methodology required for evaluating drugs against flying insects [34]. Mice were treated topically, then anesthetized and left for 40–60 min in fly boxes containing unfed adult stable flies (*Stomoxys calcitrans*). Efficacy was expressed by counting dead flies in the box after exposure to the treated animal, in comparison with controls. In a similar design, an insecticide preparation was tested topically on mice as a line-on application against *Aedes albopictus* mosquitoes [52]. Interestingly, the volume of administration was calculated on the basis of the dog dose to obtain a similar surface area dose on mice. Rodents were anesthetized and exposed to female mosquitoes for 1 h on days 1, 7, 14, 21, and 28 post treatment. Insect survival and engorgement status were evaluated immediately after exposure. Live mosquitoes were kept for 24 h and observed for additional mortality. Both end points were used to calculate efficacy in comparison with untreated controls. Alternatively, repellents or deterrents potentially active against mosquitoes or biting midges [53, 54] have been tested directly on human arms, circumventing the use of animal models. This procedure must however be limited to compounds of known toxicity, and certainly is not suited to optimization of efficacy in novel chemistry classes.

Mange Mites

Like lice, mites are specific to their host. Mites of veterinary importance are phylogenetically very diverse, and sometimes occupy only specific body niches, such as ears. It is a challenge to find mite species that parasitize laboratory species and also qualify as a good surrogate representative for a different target species. For that reason, tick animal models are often used for selecting compounds that are later tested directly against mites on target hosts. However, possibilities may exist to develop models related to these particular ectoparasites. *Myobia musculi* and *Myocoptes musculinus* are pests often found on mice in animal facilities [55], and both species are related to mites of higher veterinary importance: *Myobia* to *Demodex* spp., and *Myocoptes* to *Psoroptes* and *Sarcoptes* spp. [56]. Using mice naturally infected with one or both species could be a valid option for evaluating compounds potentially active against mange mites. In a similar approach, *Psoroptes cuniculi* commonly infects the ears of rabbits and could be a possible laboratory model for compound screening. The *Psoroptidae* family includes *Psoroptes, Otodectes* spp. (dog and cat ear mite) and *Chorioptes* spp. (cattle ear mite) and is also related to *Sarcoptes* mites [56]. The methodology established for testing compounds against ticks could probably be adapted to this parasite. However, the severity of the disease in rabbits [57] should not be underestimated in the development of such a model and should be kept to a minimum.

Conclusions

The successful development of animal models relies on identification of the best surrogate host for the parasite of interest, and on knowledge of the ectoparasite's requirements on the model and on the target host in order to mimic the field situation as closely as possible. Compromises will have to be made to meet the needs of drug discovery research (e.g., small group sizes, fast turnaround times, parasite confinement, simplified setup, and use of another related parasite species). Despite these limitations, models have proved their usefulness in drug discovery and certainly help in filling the gap between off-host high-throughput screening and efficacy studies on target species. They are valid components in often complex workflows leading to the discovery of novel classes of ectoparasiticides.

Acknowledgments

The authors thank Dr. Léonore Lovis and Dr. François Pautrat for their support in the literature search, and Dr. Jean-Marc Weber, Dr. Christian Epe, Dr. Ann E. Sluder, Dr. Charles Meng, and Dr. Stephen Peterson for the critical review of the manuscript.

References

1. Sparks, T.C. (2013) Insecticide discovery: an evaluation and analysis. *Pestic. Biochem. Physiol.*, **107**, 8–17.
2. Loso, M.R., Nugent, B.M., Huang, J.X., Rogers, R.B., *et al.* (2011) Insecticidal composition comprising N-substituted (6-haloalkylpyridin-3-yl)alkyl sulfoximines. European patent: EP 2351740.
3. Gassel, M., Wolf, C., Noak, S., Williams, H. *et al.* (2014) The novel isoxazoline ectoparasiticide fluralaner: selective inhibition of arthropod γ-aminobutiric acid- and l-glutamate-gated chloride channels and insecticidal/acaricidal activity. *Insect Biochem. Mol. Biol.*, **45**, 111–124.
4. Santora, K.A., Zakson-Aiken, M., Rasa, C., and Shoop, W. (2002) Development of a mouse model to determine the systemic activity of potential flea-control compounds. *Vet. Parasitol.*, **104**, 257–264.
5. Zhang, Y.K., Plattner, J.J., Easom, E.E., Akama, T. *et al.* (2016) Optimization of isoxazoline amide benzoxaboroles for identification of a development candidate as an oral long acting animal ectoparasiticide. *Bioorg. Med. Chem. Lett.*, **26**, 3182–3186.
6. Fankhauser, R., Cozzie, L.R., Nare, B., Powell, K. *et al.* (2012) Use of rodent models in the discovery of novel anthelmintics, in Parasitic Helminths: Targets, Screens, Drugs and Vaccines, 1st e edn (ed. C.R. Caffrey), Wiley-VCH Verlag GmbH & Co KCaA, pp. 181–199.
7. Morris, C.P., Evans, H., Larsen, S.E., and Mitrea, E. (2013) A comprehensive, model-based review of vaccine and repeat infection trials for filariasis. *Clin. Microbiol. Rev.*, **26**, 381–421.
8. Hurst, R.J.M. and Else, K.J. (2013) *Trichuris muris* research revisited: a journey through time. *Parasitology*, **140**, 1325–1339.
9. Schorderet-Weber, S., Ducray, P., Gauvry, N., Goebel, T., *et al.* (2009) Use of a rodent model for the optimization and characterization of a novel class of anthelmintics. Abstract Volume. World Association for the Advancement of Veterinary Parasitology Conference, August 9–13, 2009, Calgary, Canada.
10. Früchtel, J., Gauvry, N., Schorderet-Weber, S., and Bouvier, J. (2007) N- (Hetero) aryl indole derivatives as pesticides. World patent: WO2007051619.
11. Iwasa, T. and Hirata, K. (2014) Ectoparasite control agent. European Patent Application: EP2965627.
12. Nosomu, N., Yoshimasa, F, Shigeki, K., and Ikuya, O. (2014) Method for producing pest controlling agent. European Patent Application: EP2749555.
13. Gauvry, N., Pautrat, F., Bouvier, J., Früchtel, J., *et al.* (2008) Pyrimidine derivatives and their use as pesticides. World Patent: WO2008009691.
14. Jones, L.D., Davies, C.R., Steele, G.M., and Nuttall, P.A. (1988) The rearing and maintenance of ixodid and argasid ticks in the laboratory. *Anim. Technol.*, **39**, 99–106.
15. Ketzis, J.K., Clements, K., and Honraet, K. (2014) Use of a poultry model to assess the transfer inhibition effect of head lice (*Pediculus humanus capitis*) products. *Parasitol. Res.*, **113**, 1943–1948.
16. Abbott, W.S. (1987) A method of computing the effectiveness on an insecticide. *J. Am. Mosq. Control Assoc.*, **3**, 302–303.
17. Immler, R.M. (1973) Untersuchungen zur Biologie und Oekologie der Zecke *Dermacentor reticulatus* (Fabricius 1794) (*Ixodidae*) in einem endemischen Vorkommensgebiet. *Mitt. Schweiz. Entomol. Ges.*, **46**, 1–70.

18. Gutierrez, J.A., Zhao, X., Kemper, C.J., Plummer, P.R. *et al.* (2006) An *in vivo* rodent model for identifying and characterizing acaricides. *J. Med. Entomol.*, **43**, 526–532.
19. Srivastava, S.C. and Varma, M.G.R. (1964) The culture of the tick *Rhipicephalus sanguineus* (Latreille) (Ixodidae) in the laboratory. *J. Med. Entomol.*, **1**, 154–157.
20. Sembo, S. (2002) A mass-rearing method for the cat flea, *Ctenocephalides felis* (Siphonaptera: Pulicidae) on mice. *Jpn. J. Appl. Entomol. Zool.*, **46**, 61–65.
21. Weber, T. and Selzer, P.M. (2016) Isoxazolines: a novel chemotype highly effective on ectoparasites. *ChemMedChem*, **11**, 270–276.
22. Shoop, W.L., Hartline, E.J., Gould, B.R., Waddell, M.E. *et al.* (2014) Discovery and mode of action of afoxolaner, a new isoxazoline parasiticide for dogs. *Vet. Parasitol.*, **201**, 179–189.
23. McTier, T.L., Chubb, N., Curtis, M.P., Hedges, L. *et al.* (2016) Discovery of sarolaner: a novel, orally administered, broad-spectrum, isoxazoline ectoparasiticide for dogs. *Vet. Parasitol.*, **222**, 3–11.
24. Zhang, Y.K., Plattner, J.J., Easom, E.E., Zhou, Y. *et al.* (2015) Discovery of an orally bioavailable isoxazoline benzoxaborole (AN8030) as a long acting animal ectoparasiticide. *Bioorg. Med. Chem. Lett.*, **25**, 5589–5593.
25. Curtis, M.P., Vaillancourt, V., Goodwin, R.M., Chubb, N.A. *et al.* (2016) Design and synthesis of sarolaner, a novel, once-a-month, oral isoxazoline for the control of fleas and ticks on dogs. *Bioorg. Med. Chem. Lett.*, **26**, 1831–1835.
26. Dupont, E., Gauvry, N., Nanchen, S., Ogawa, C., *et al.* (2014) (Hetero) Arylacrylamides for the control of ectoparasites. World Patent: WO2014096381.
27. Kumar, S., Prakash, S., and Kaushik, M.P. (1992) Comparative activity of three repellents against the ticks *Rhipicephalus sanguineus* and *Argas persicus*. *Med. Vet. Entomol.*, **6**, 47–50.
28. Mwangi, E.N., Hassanali, A., Essuman, S., Myandat, E. *et al.* (1995) Repellents and acaricidal properties of *Ocimum suave* against *Rhipicephalus appendiculatus* ticks. *Exp. Appl. Acarol.*, **19**, 11–18.
29. Naono, N., Kishun, I., and Kaori, I. (2015) Composition for controlling animal ectoparasite. Japanese Patent: JP 2015151363.
30. Zakson-Aiken, M., Gregory, L.M., Meinke, P.T., and Shoop, W.L. (2001) Systemic activity of the avermectins against the cat flea (*Siphonaptera: Pulicidae*). *J. Med. Entomol.*, **38**, 576–580.
31. Soll, M.D. and Meng, C.Q. (2011) Insecticidal 4-amio-thieno [2, 3-D]-pyrimidine compounds and methods of their use. World Patent: WO 2011014660.
32. Früchtel, J., Dupont, E., Goebel, T., Schorderet-Weber, S., *et al.* (2009) Organic compounds. World Patent: WO 2009071500.
33. Ostlind, D.A., Cifelli, S., Conroy, J.A., Mickle, G.W. *et al.* (2001) A novel *Cimex lectularius* rodent assay for the detection of systemic ectoparasiticide activity. *Southwest. Entomol.*, **26**, 181–185.
34. Kirst, H.A., Creemer, L.C., Naylor, S.A., Pugh, P.T. *et al.* (2002) Evaluation and development of spinosyns to control ectoparasites on cattle and sheep. *Curr. Top. Med. Chem.*, **2**, 675–699.
35. Yoon, K.S., Ketzis, J.K., Andrewes, S.W., Wu, C.S. *et al.* (2015) *In vitro* and *in vivo* evaluation of infestation deterrents against lice. *J. Med. Entomol.*, **52**, 970–978.
36. Zacarias do Amaral, M.A., de Azevedo Prata, M.C., Daemon, E., and Furlong, J. (2002) Biological parameters of cattle ticks fed on rabbits. *Rev. Bras. Parasitol. Vet.*, **21**, 22–27.
37. Willadsen, P. (1980) Immunity to ticks. *Adv. Parasitol.*, **18**, 293–313.
38. Schorderet, S. and Brossard, M. (1993) Changes in immunity to *Ixodes ricinus* by rabbits infested at different levels. *Med. Vet. Entomol.*, **7**, 186–192.
39. Coumou, J., Wagemakers, A., Trentelman, J.J., Nijhof, A.N. *et al.* (2015) Vaccination against Bm86 homologues in rabbits does not impair *Ixodes ricinus* feeding or oviposition. *PLoS One*, **10** (4), e0123495. doi: 10.1371/journal.pone.0123495
40. Contreras, M. and de la Fuente, J. (2016) Control of *Ixodes ricinus* and *Dermacentor reticulatus* tick infestations in rabbits

vaccinated with the Q38 Subolesin/Akirin chimera. *Vaccine*, **34**, 3010–3013.

41. Marchiondo, A.A., Holdsworth, P.A., Fourie, L.J., Rugg, D. *et al.* (2013) World Association for the Advancement of Veterinary Parasitology (W.A.A.V.P.) second edition: guidelines for evaluating the efficacy of parasiticides for the treatment, prevention and control of flea and tick infestations on dogs and cats. *Vet. Parasitol.*, **194**, 84–97.
42. White, W.H., Bauer, S.M., Zhao, X., Gutierrez, J.A. *et al.* (2005) Comparison of *in vitro* and *in vivo* ectoparasiticide activity of experimental Benzimidazole-Carbamate with Permethrin and Amitraz. *J. Med. Entomol.*, **42**, 207–211.
43. Elting, E. (1926) A formula for calculating surface area of dairy cattle. *J. Agric. Res., Washington DC*, **33**, 269–279.
44. Mbow, M.L., Christe, M., Rutti, B., and Brossard, M. (1994) Absence of acquired resistance to nymphal *Ixodes ricinus* ticks in BALB/c mice developing cutaneous reactions. *J. Parasitol.*, **80**, 81–87.
45. DenHollander, N. and Allen, J.R. (1985) *Dermacentor variabilis*: acquired resistance to ticks in BALB/c mice. *Exp. Parasitol.*, **59**, 118–129.
46. Matsuda, H., Fukui, K., Kiso, Y., and Kitamura, Y. (1985) Inability of genetically mast cell-deficient W/W^{v} mice to acquire resistance against larval *Haemaphysalis longicornis* ticks. *J. Parasitol.*, **71**, 443–448.
47. Loftis, A.D., Reeves, W.K., Szumlas, D.E., Abbassy, M.M. *et al.* (2006) Surveillance of Egyptian fleas for agents of public health significance: *Anaplasma*, *Bartonella*, *Coxiella*, *Ehrlichia*, *Rickettsia*, and *Yersinia pestis*. *Am. J. Trop. Med. Hyg.*, **75**, 41–48.
48. Bakr, M.E., Morsy, T.A., Nassef, N.E., and El Meligi, M.A. (1996) Flea ectoparasites of commensal rodents in Shebin El Kom, Menoufia Governorate, Egypt. *J. Egypt. Soc. Parasitol.*, **26**, 39–52.
49. Meinke, P.T., Colletti, S.L., Fisher, M.H., Wyvratt, M.J. *et al.* (2009) Discovery of the development candidate N-tert-butyl nodulisporamide: a safe and efficacious once monthly oral agent for the control of fleas and ticks on companion animals. *J. Med. Chem.*, **52**, 3505–3515.
50. Levot, G. (2000) Resistance and the control of lice on humans and production animals. *Int. J. Parasitol.*, **30**, 291–297.
51. Holdsworth, P.A., Vercruysse, J., Rehbein, S., Peter, R.J. *et al.* (2006) World Association for the Advancement of Veterinary Parasitology (WAAVP) guidelines for evaluating the efficacy of ectoparasiticides against biting lice, sucking lice and sheep keds on ruminants. *Vet. Parasitol.*, **136**, 45–54.
52. Tahir, J., Davoust, B., Almeras, L., Bérenger, J.-M., *et al.* (2016) Insecticidal repellency by a topical administration of dinetofuran-pyriproxifen-permethrin spot-on (Vectra® 3D) on mice against *Aedes albopictus* mosquitoes. Proceedings of the 61st meeting of the American Association of Veterinary Parasitologists, August 6–9, 2016, San Antonio, TX, USA.
53. Chou, J.T., Rossignol, P.A., and Ayres, J.W. (1997) Evaluation of insect repellents on human skin against *Aedes aegypti* (Diptera: Culicidae). *J. Med. Entomol.*, **34**, 624–630.
54. Stuart, A.E., Brooks, C.J.W., Prescott, R.J., and Blackwell, A. (2000) Repellent and antifeedant activity of salicylic acid and related compounds against biting midges, *Culicoides impunctatus* (Diptera: Ceratopogonidae). *J. Med. Entomol.*, **37**, 223–227.
55. Carty, A.J. (2008) Opportunistic infections in mice and rats: Jacoby and Lindsey revisited. *ILAR J.*, **49**, 272–276.
56. Roskov, Y., Abucay, L., Orrell, T., Nicolson, D., *et al.* (eds) (2016) Species 2000 & ITIS Catalogue of Life, 27 June 2016. Digital resource at www.catalogueoflife.org/col. Species 2000: Naturalis, Leiden, the Netherlands. ISSN 2405-8858.
57. Shang, X., Wang, D., Miao, X., Wang, X. *et al.* (2014) The oxidative status and inflammatory level of the peripheral blood of rabbits infested with *Psoroptes cuniculi*. *Parasit. Vectors*, **7**, 124–129.

11
Testing in Target Hosts for Ectoparasiticide Discovery and Development

Jeffrey N. Clark[*]

Abstract

Animal models of ectoparasite infestations form a pivotal bridge in the discovery and development process, guiding the advancement of active ectoparasiticidal compounds identified by laboratory studies against key molecular targets and whole organisms to new product candidates ready for field or clinical testing. Rodent infestation models have the advantage of requiring less compound for testing than do target animal studies, while still providing valuable *in vivo* information on efficacy, safety, metabolism, bioavailability, and the pharmacokinetic (PK) properties of a new compound, thereby identifying possible issues before the compound is evaluated in larger target animals such as dogs, cats, and livestock. Rodent models are often used as a screening tool to identify the best lead candidate prior to efficacy, safety, and PK studies in the target animal. Recommendations for the procedures for controlled testing of ectoparasiticides against most arthropods infesting target animals have been published. This chapter describes the reasons for using animal models, an outline of the screening paradigm to identify lead and clinical candidates and the test methods used. The discussion focuses primarily on controlled laboratory efficacy studies in target animals, but also briefly discusses other types of studies. Several new ectoparasiticide compound classes that have been evaluated in both rodent and target animals using a variety of routes of administration are described. Extension of the utility of these animal models to the development of human health ectoparasiticides is also briefly discussed.

Role of Animal Models in Ectoparasiticide Discovery and Development

Controlled laboratory infestation models for arthropod external parasite pests are used to evaluate novel ectoparasiticide candidates that have previously been identified during the discovery process using *in vitro* target-based assays and/or

[*]Corresponding author.

Ectoparasites: Drug Discovery Against Moving Targets, First Edition.
Edited by Charles Q. Meng and Ann E. Sluder

ex vivo whole-organism assays (see Chapters 7 and 8). Such animal models of infestation can be utilized to address several questions in early product development prior to field testing, including compound pharmacokinetics (PKs) and bioavailability, product safety, duration of efficacy, ectoparasite spectrum, efficacy against different life cycle stages, route of application, development and testing of formulations, and product aesthetics. The target pests are established early in a discovery campaign, based on the desired market: fleas and ticks are most important for dog and cats, and flies, ticks, lice and, mites are most important for livestock. Throughout the process, the initial lead compound is always scrutinized to determine if it is the best candidate to move forward into late-stage development. In some cases, a backup compound will be selected as a better product candidate for various reasons, a decision in which animal model studies can play a pivotal role. The discussions that follow are designed to introduce the reader to the basic concepts of the use of animal models in this workflow, using primarily the dog as the example model.

The flow charts in Figure 11.1 present a prototypical, albeit simplified, workflow for an ectoparasiticide discovery/development program in animal health. Animal models are employed once a lead compound (or a lead series) has been selected on the basis of the discovery screening in target-based and/or whole-organism assays. Throughput of compounds in animal model ectoparasite assays is very low, and trials are expensive in both time and amount of compound required. Thus, in the earlier stages these studies may utilize rodent models of infestation, while later stages shift to target hosts amenable to controlled laboratory use (e.g., dogs). Still, even in rodent assays, no more than 100–200 compounds/week can be evaluated at a single concentration depending on the duration of the assay and the manpower available. Dog studies can only evaluate a few compounds/week (<20 in single point) and can require up to several grams of sample.

Historically, leads were selected directly in the whole organism screens to be used in animal models, and target-based screening has been a more recent addition to the workflow. A compound, or compound series, selected as a screening lead, ideally will exhibit satisfactory potency against the selected *in vitro* target, lack apparent off-target activity, display good potency, spectrum, speed of kill in the whole-organism *ex vivo* screens, and lack obvious mammalian cell toxicity in cell-based assays. Then the compound is advanced to the rodent ectoparasite model. Evaluation in a rodent ectoparasite infestation model can then be used to identify a specific lead candidate from a series of molecules in a lead optimization program that evaluates efficacy, safety, and PKs.

If the ultimate product is intended for oral or injectable administration, based on a response to the need in the market, the lead will be evaluated in mouse PK models to determine oral or systemic bioavailability, PK parameters such as $t_{1/2}$, T_{max}, C_{max}, and area under the curve (AUC) and also initial metabolism and any toxicity issues. If these studies show the compound remains promising, it advances to the rodent ectoparasite efficacy model, if available; thereafter moving on to dog PK studies and finally dog ectoparasite efficacy trials. If no rodent

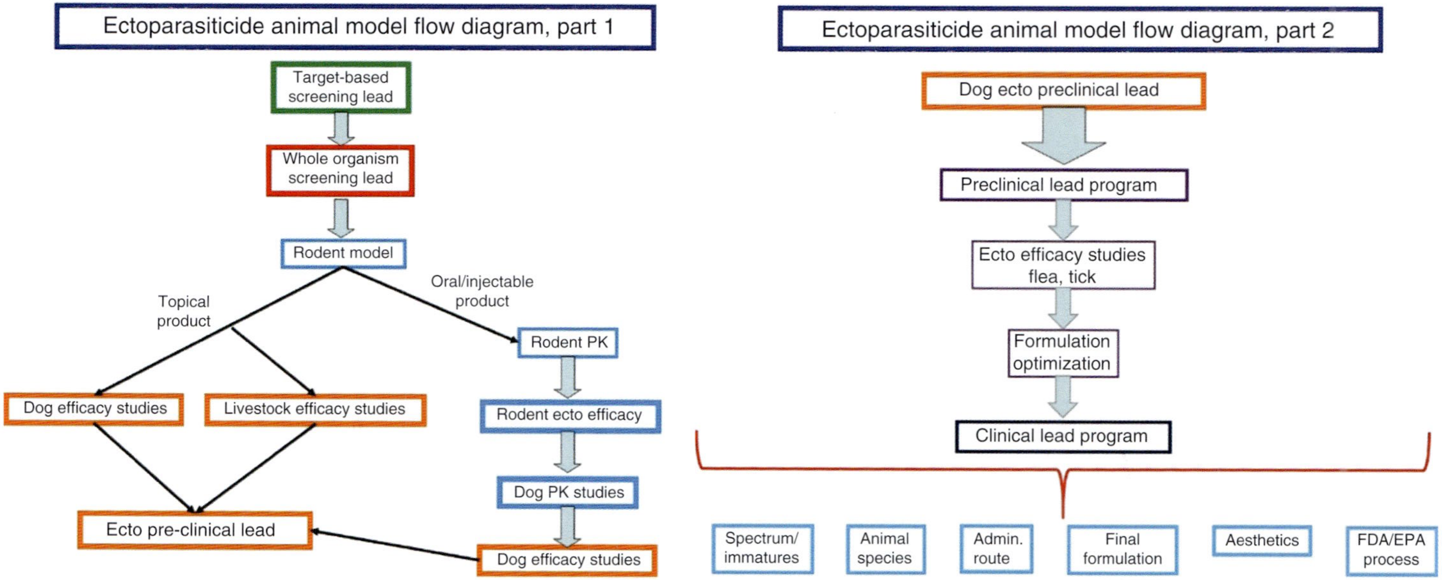

Figure 11.1 Prototypical workflow and decision tree for ectoparasiticide development.

ectoparasite model is available, the compound can advance directly to the dog PK evaluation prior to dog efficacy studies. In the past, PK monitoring was not done extensively in animal health drug research, but now it is an essential part of any discovery/development program.

If the primary goal of the program is to develop a topical product, following *in vitro* and whole-organism screening the compound can go directly into the rodent efficacy model (if available) and then proceed on to the dog efficacy model studies. The goal at this point is to determine if the compound is active against the primary ectoparasite targets, fleas, and ticks, on dogs. Preliminary information on potency, duration, and spectrum of activity and any safety issues are determined at this time. Safety issues include gastrointestinal disturbances, neurological changes (shaking, seizures, mydriasis, etc.), skin irritation, injection or administration site issues, and other sensitivities. At this time, the biological data generated in rodents and dogs and in previous *ex vivo* studies are scrutinized to determine if the compound might also hold promise for ectoparasite targets in livestock. In recent times, as the livestock animal sector holds less value (in terms of profitability) than the flea/tick market in companion animals, livestock ectoparasite targets may be an extension to the flea/tick market, unless the whole-organism screening indicates that the principal ectoparasite targets are only in this livestock sector, and there is a large and viable market for a new product.

If there are no outstanding issues, then the compound proceeds into preclinical development studies (Figure 11.1, part 2) in the target animal. At this time in the process, different product candidates will progress through the flowchart in a similar manner, regardless of route of administration, target pest, or target animal species.

During a preclinical program in dogs, the compound is further evaluated in more advanced efficacy and safety studies to determine if it continues to exhibit the desired attributes of efficacy, spectrum, duration, and safety. Also, heretofore, a discovery formulation (often dimethyl sulfoxide (DMSO) or similar) will have been used; but in the preclinical program, a series of more suitable formulations for an actual product are evaluated to identify the likely lead formulation. Often this exercise is initially based on previously used and acceptable formulations for similar molecules. During this process, any effects on hair coat quality can be ascertained, although this aspect will be more thoroughly evaluated when a more advanced formulation is available.

If all tests continue to show the promise of the compound, a clinical lead program is initiated using the lead compound and the final formulation. During this program, the compound undergoes several efficacy and safety evaluations in larger trials while looking at parasite spectrum, including secondary targets such as the mite, other animal species such as the cat, the best route of administration and, for topical products, any aesthetic issues with the hair coat following treatment.

The final product proceeds through the U.S. Food and Drug Administration (FDA; for oral or injectable products) or Environmental Protection Agency (EPA; for topical products) regulatory requirements. Both agencies require a long list of efficacy and safety studies as well as many additional studies in the case of food production livestock to evaluate residues and effects on reproduction as a measure of potential safety issues for humans. Protocols for all these studies are well established but require the target animal and the compound in its final formulation. It is beyond the scope of this chapter to discuss regulatory requirements and the associated study designs in full detail. A good reference series for more detailed information on various ectoparasite targets in various companion and livestock animals can be found in Refs [1–6].

Alignment of Animal Model Assays with Desired Product Characteristics

Animal models are used to evaluate the efficacy of novel compounds against several arthropod species in both companion animals and livestock. While scoring for efficacy against ectoparasites is very simple in animal models, as lethality is the desired effect, design of specific animal model assays must anticipate the combination of properties desired in the final ectoparasitide product. These models can be used to determine speed of kill, duration of activity, ectoparasite spectrum, repellency effects, spreading of the formulated active compound(s), effects on insect or acarine immature stages, cosmetic effects, and so on, for both single and multiple active formulations administered by a variety of routes (oral, spot-on, pour-on, and injectable). Efficacy evaluation at various early time points after infestation can be used to determine speed of kill. Also, fleas from different life cycle propagation suppliers can be used to determine if there are any differences between flea lines. Furthermore, efficacy against adult and immature fleas and ticks can be evaluated in comparison to untreated control or to another product or active (positive control). Some life stages (egg, larvae, nymph) may occur off the host, so inclusion of off-animal end points for product activity may be important. For example, when testing for insect growth regulator (IGR) effects in flea studies, eggs produced by fleas on the treated animal are collected and then placed in an incubator and scored a few days later for egg hatching and larval effects compared to control. Effects on later stages of the life cycle such as pupae and adults (for fleas) can also be assessed by providing the appropriate growth medium to the larvae.

The desired route of exposure of the ectoparasite to the compound is another consideration. For companion animals, a topical product that has good contact activity against fleas and ticks and kills these pests before they can draw blood is generally preferred over a systemic product that the ectoparasite encounters only via a blood meal, as the latter does not prevent transfer of blood-borne pathogens or immunological reactions such as flea bite dermatitis. In contrast, systemic product activity may be required for full control of some fly species in livestock, where immature stages can attack internally. Thus, animal model assay designs often incorporate steps to assess the impact of compounds on ectoparasite feeding, such as recording tick attachment and engorgement or testing fleas collected post-exposure for evidence of a blood meal.

Exposure to the ectoparasiticide topically can also be through the process of translocation in which the compound moves from a single spot-on, small-volume formulation over the skin and hair of the animal through the hair follicles and sebaceous glands. In this way, the pest becomes exposed to the compound anywhere on the body of the host, and the duration of efficacy can be extended. Translocation is a well-known characteristic for fipronil, as noted in the Frontline® products [7]. Translocation studies can be conducted using radiolabeled active [7], in this case [^{14}C]-fipronil, which is administered once in a spot-on formulation, and then, hair snippets, skin swabs, and full-thickness skin biopsies are taken weekly for several weeks. From this procedure, one can determine the amount/concentration and duration of active on the surface of the animal at each time period, and the appearance and skin penetration of the radiolabel by autoradiography of the skin biopsies. In this study, fipronil was found to penetrate only the upper layers of the skin and the pilo-sebaceous structures and did not penetrate into the dermal or hypodermal layers. Radiolabel was found in the skin/hair samples for up to 56 days post application.

To guide selection of compounds with the desired potency, animal model studies are conducted using controls in parallel with the test compound. Negative or solvent controls establish the baselines of infestation levels against which the efficacy of a compound in reducing ectoparasite burdens will be evaluated. Often, studies also use as positive control a marketed compound in the same or different class but having the same or similar spectrum of activity and duration of efficacy. In these cases, the proposed drug is expected to be at least as good as the marketed product but preferably lasts longer, has better spectrum, acts faster, has fewer side effects, and so on, to be in a position to compete against existing products. Animal model studies can be designed to answer all these objectives (assuming that they can be met). For example, if 1 month of efficacy against fleas and ticks is required, then the study will be conducted for at least 4 weeks, preferably at least 6 weeks or until the efficacy is below 90% to make certain that there is a margin of error. Such duration studies are best conducted in the target animal host rather than in rodent models. Studies using a combination of products present several challenges, as the efficacy of each active in the combination must be evaluated by itself (to get a baseline) and then in the presence of one of more actives in separate study groups. The objective of these studies is to determine whether other actives in the same formulation interfere with the activity of any other active (see Ref. [1]).

Animal models are also used to evaluate different solvents and excipients that may be used in formulations, especially for topical use, to aid in exposure of the compound to the host and the target pest, extending duration of effect, minimizing effects of photodegradation and limiting exposure to off-target hosts such as humans, the environment, and beneficial organisms. Aesthetic studies are conducted to determine if the formulation leaves any visible residue (white powder, crystals, stickiness, etc.) that might affect the acceptability and possible human exposure to the product. Formulations for systemic use are more limited, as there is a finite list of solvents and excipients that can be injected or given orally without causing adverse reactions in the animal host. In all these studies, a negative control (no active compound) or an approved positive control with known characteristics will have to be included in any efficacy study to verify that there is no contribution to the activity against the parasite by the formulation alone.

Additional examples of the types of animal model studies that are conducted are included in following sections.

Specific Animal Model Ectoparasite Assays

Laboratory Animal Models

Rodent and rabbit ectoparasite models have been used for many years to determine the efficacy of various compounds. Summarized here are a few specific examples of use of rodent model assays in the identification of active compounds that became products or were very promising leads, to illustrate use of these models in ectoparasiticide development. Rodent models of ectoparasite infection are discussed further in Chapter 10.

A *Cimex lectularius* (bedbug) assay in mice was used for many years at Merck Research Laboratories [8] and was instrumental in the discovery, optimization, and development of macrocylic lactones and nodulisporamides and many other actives [9]. In this model, mice are treated orally with the test compound, and 24 h later are restrained in plastic containers which allow their tail to be exposed. The tail is then inserted into a vial containing fourth instar bedbug larvae. The subsequent engorged larvae are observed daily until ≥80% have molted to the fifth instar and then scored for death, paralysis, and molt delay relative to controls. Several nodulisporamides were found to have activity of ≤1 mg/kg body weight in this assay, and this test along with the flea ingestion assay (discussed in Chapter 8), directed the chemical optimization program to find the eventual lead compound, *N-tert*-butyl nodulisporamide [9].

Adult fleas were also evaluated in a rat model at Merck Research Laboratories to find successors to ivermectin and abamectin as ectoparasiticides, and also to find a novel macrocyclic lactone for use in a topical flea treatment product. Rats were treated topically and infested with fleas. Efficacy was determined on a weekly basis by counting live and dead fleas. A mouse model was also used to evaluate the systemic activity of several existing and experimental compounds of many different chemical classes against fleas [10]. The mouse was sedated with acepromazine, and ectoparasiticides were given orally at 1, 10, and 30 mg/kg body weight. An early evaluation (2 h) indicated that 77% of the fleas fed. Efficacy was measured as the percent reduction of live fleas on treated animals relative to untreated controls. At 24 h, nitempyran was found to be active (94%) at 1 mg/kg, selamectin active (86%) at 10 mg/kg, fipronil active (83%) at 30 mg/kg, cythioate active (64%) at 10 mg/kg, and nodulisporic acid active (55%) at 10 mg/kg. Lufenuron and ivermectin were not active at 30 mg/kg. This model thus provided a very good measure of the flea adulticide activity of known compounds, and was subsequently used to help prioritize experimental compound leads.

Target Host Animal Models

The dog has been the primary animal model for flea and tick studies, and many chemical series have been tested in this model using various routes of administration. Several meetings have been held among flea/tick investigators to develop guidelines for the evaluation of ectoparasiticides in companion animals. Such a meeting recently took place at the World Association for the Advancement of Veterinary Parasitology (W.A.A.V.P.) conference, which resulted in a key publication [1] that outlines recommended procedures for dog and cat model assays. Key aspects of these recommendations include use of controls, standardization of animal handling and housing during a study, recommended numbers of parasites per animal, definition of quantitative quality control thresholds for assay performance in a given study, and standardized formulae for calculating and reporting compound efficacy. In addition to this publication on flea and tick companion animal model assays, W.A.A.V.P. members have also published guidelines for conducting ectoparasiticide studies in several other host animal species. These include guidelines for mange mites on livestock [2], flies on livestock [3], myiasis on ruminants [4], lice and sheep keds on ruminants [5], and ticks on ruminants [6]. A thorough familiarity with these published best practices is recommended for the practitioner preparing to implement animal model assays in an ectoparasiticide development program.

Specific Examples of Animal Model Uses in the Development of Ectoparasiticide Products

In this section I review the use of animal models in studies conducted for the registration of several insecticide/acaricide products, with the goal of further illustrating the principles presented in the preceding sections.

Ectoparasite Control Products for Topical Administration on Companion Animals

Early compounds to control flea infestations systemically in dogs include the orally administered organothiophosphate cythioate (Proban®) and the topically administered organophosphate fenthion (ProSpot®). These were developed using a dog adult flea model where the product was given once either orally or topically, respectively, followed by weekly infestations with fleas and determination of adulticide activity.

These products were followed by the pyrethroids, which included a topical 60% solution of permethrin delivered in a spot-on formulation (ExSpot®), whereby a small volume of the product is placed between the shoulder blades of the dog to achieve efficacy against fleas and ticks as well as repellency of these arthropods. A recent paper [11] reviewed the efficacy of several permethrin-containing products against the tick *Dermacentor reticulatus* in induced infestation trials using dogs. Inherent in the study design was an evaluation of efficacy against ticks located in various parts of the body, thereby giving information on the movement of permethrin across the dog. Groups of six dogs were treated topically with a recommended dose of the product, while control dogs were untreated, and thereafter all dogs were infested with ticks on days 1, 14, and 28. Efficacy was determined 6 h after tick infestation to evaluate speed of kill and any repellent effects. All the products were found to be very effective within 6 h of infestation through the 28 days of the study. Permethrin was found to be widely distributed within 24 h of treatment. It was speculated that this rapid action of permethrin could help prevent the transmission of rickettsial, bacterial, and protozoal diseases to the dog and to man.

The evaluation of a two-way combination of the insecticide fipronil and the IGR methoprene was conducted [12] against adult and immature stages of the cat flea (*Ctenocephalides felis*) using an induced infestation model in dogs. Test groups of eight animals each included an untreated control group and groups treated with fipronil, methoprene, and the fipronil/methoprene combination. Dogs were treated topically with a spot-on formulation on day 0, and thereafter were infested with ~200 fleas weekly. Flea eggs were also collected for 24 h at 3 days after each infestation to determine effects of the combination on immature flea stages, for example, egg production, larval hatch, as a measure of ovicidal effects, and adult flea emergence from subsequent pupae. The fipronil-only group provided excellent control of adult fleas for 5 weeks. However, the combination of fipronil and methoprene provided an additional effect (>90%) on ovicidal activity and inhibition (91.4%) of adult flea emergence for 12 weeks. A synergistic effect of the combination of the two compounds was seen on ovicidal activity and adult flea emergence. In this development program, the dog flea model proved very useful for demonstrating high efficacy of fipronil and of a fipronil/methoprene combination against the adult stage of the target parasite, high efficacy against various stages of the target parasite (the combination), and a synergistic combination of two actives where high efficacy could be maintained for longer than that observed with either active alone.

Another series of two-way topical spot-on combination studies evaluated the effect of two adulticides, fipronil and permethrin, using induced flea and fly infestations in dogs [13–15]. This combination has been developed to control fleas, ticks, mosquitoes, sandflies, and biting flies on dogs, and, therefore, different animal-arthropod models were used for each of these indications. For fleas, the dogs were treated on day 0 followed by weekly infestations and determination of efficacy at 1 and 6 h for rapid adulticide activity and repellency effects and then at 24 h after each infestation. Excellent activity was seen for fast-acting, 24-h infestation efficacy and repellent effects throughout the 28-day study. These effects were observed even after shampooing and washing of the dog on several days during the study [14]. Dogs were also exposed to stable flies weekly following treatment on day 0 with this two-way active formulation. After 1 h, the flies were removed and crushed to determine feeding status. Insecticidal and repellent effects observed in this dog fly model showed the product to have excellent effects (≥98.3% and 88.7%, respectively) on these two parameters throughout the 35-day trial.

A two-way spot-on formulation containing the insecticides metaflumizone and amitraz (Promeris Duo®) was tested against fleas and ticks on dogs [16] in an induced infestation model. Dogs were infested with both fleas (started on day 1) and ticks (started on day 2) followed by treatment on day 0 and thereafter infested weekly. Counts were made on day 1 to estimate knockdown activity and thereafter 48 h after infestation to determine standard efficacy. The six treatment groups consisted of a solvent control, three dosage groups of the metaflumizone/amitraz formulation to provide a dose titration (10, 20, and 40 mg/kg of both actives), and one group was treated with fipronil (Frontline®) to act as a positive control. This study was therefore designed to select a dosage for the new product and compare the efficacy at the preferred dosage to a positive control, fipronil. The 20 mg/kg dosage of the metaflumizone/amitraz combination exhibited excellent efficacy against both fleas and ticks for at least 35 days, which was similar to that of fipronil.

An induced infestation cat model was used to evaluate a four-way topical spot-on formulation containing fipronil, (S)-methoprene, eprinomectin, and praziquantel (e.g., Broadline®) against the cat flea [17] compared to untreated control. As susceptibility to ectoparasiticides can vary between flea colonies, fleas from different locations in the United States, South Africa, and Germany were used in the infestations. Cats were left untreated (control) or treated with Broadline® on day 0. The animals were infested with 100 fleas on day 2 and thereafter weekly through day 35. Live adult fleas were counted 24 h after infestations. In a related study in the same publication [17], gravid fleas previously fed on an untreated host were placed on treated animals. Excellent efficacy was achieved for the duration of the studies against adult fleas, and the product also had excellent activity in preventing adult flea emergence arising from eggs laid following treatment. Therefore, the animal model was used to demonstrate activity against both adult and immature flea stages.

Animal studies can be conducted to compare the efficacy against adult fleas of one active in comparison to another. In one study [18], cats received induced flea infestations on days 28 and 21 and thereafter were treated with either selamectin (Revolution®) or fipronil (Frontline®) monthly for 0, 30, 60, 90, and 120 days. Adult flea comb counts to measure efficacy were conducted every 2 weeks for the duration of the study (day 150). Cats were also re-infested starting on day 91 every 7 days until the end of the study. Excellent adulticide (97–100%) activity was observed with both products.

Ectoparasite Control Collars for Companion Animals

Flea and tick collars have been used for many years to control ectoparasite infestations in companion animals. Most of the early collars used carbamates or organophosphates as the insecticide/acaricide. With time, increased resistance to these actives plus occasional skin irritation and the arrival of the much more effective spot-on products, decreased the use of collars. Novel polymer chemistry to make superior slow-release products and the advent of much more effective insecticides/acaricides have resurrected the collar as an administration option.

One such improved product, a combination imidacloprid/flumethrin slow-release collar, (Seresto®), has recently been developed to combat ectoparasites, nematodes, and protozoa on or in dogs [19–23]. Field trials using natural infections [19, 20] in dogs and/or cats evaluated the efficacy of the collar versus untreated control against heavy flea and tick infestations. Some animals also had severe cases of flea allergic dermatitis (FAD) caused by flea bites. The collar was shown to have a dramatic effect on reducing the flea infestation by day 2 with reduction to zero by day 14. Tick numbers were dramatically dropped by day 7 and then to zero by day 14. Furthermore, the incidence of FAD dropped precipitously as well. These superior effects were observed until the end of the study (day 250). One study [23] also showed, in addition to excellent long-term efficacy against natural flea and tick infestations, that tick-borne pathogen transmission was also dramatically (91.6%) decreased by use of the collar. The combination collar was compared to a moxidectin/imidacloprid spot-on topical against the transmission of *Thelazia callipaeda* to dogs in natural infections [21]. The spot-on formulation proved superior in this situation. The collar was also tested for protection of dogs from infection by the protozoa *Leishmania infantum*, which is transmitted by the sandfly (*Phlebotomus* spp., *Sergentomyia* spp.). Again, natural infestation models were used in hyperendemic sandfly areas in Italy [22]. Two groups of dogs were used, one receiving the collar and one for the negative control. Laboratory tests for the presence of *L. infantum* were conducted over the course of 300 days. At the end of the study, the collar was able to decrease the presence of the parasite in the dog by 93.4%. Some of the dogs in the control group also showed cutaneous signs of *L. infantum* infection.

Ectoparasite Control Products for Oral Administration to Companion Animals

As mentioned, two products, Proban® (oral) and ProSpot® (topical), were developed in the 1980s following the decline in the usage of the collar, spray, shampoo, and powder organophosphate and carbamate products, as clients wanted more effective and easy-to-use products. Exspot® was subsequently a major topical spot-on flea/tick product, which was followed by many other chemistries in the 1990s including fiproles (Frontline®), imidacloprid (Advantage®), and macrocyclic lactones (selamectin). Most of the new products were topically administered as a small volume spot-on, and later products combined one or more actives to broaden the ectoparasiticide spectrum or to effect immature pest stage efficacy. The only other oral products during the 1990s were lufenuron (Program®), a chitin synthesis inhibitor, which controlled flea reproduction by its effects on immature stages but did not kill adult fleas, and later nitempyram (Capstar®), which provided rapid but very short-term adult flea control. The latter product was used in conjunction with lufenuron, as the lufenuron did not have rapid-onset flea adulticide efficacy but only provided long-term efficacy to control flea populations by controlling immature stages.

Therefore, for many years, oral products were not in favor for ectoparasite control. There were many reasons for this, including (i) the concern about toxicity to the dog with the active, (ii) the need for the ectoparasite to take a blood meal before being affected by the drug, thus allowing immunological issues (such as FAD and pathogen transfer) to take place prior to the control of the infestation, and (iii) the difficulties in discovering compounds with sufficient systemic bioavailability and rapid onset of effect to control the ectoparasite. One advantage to the oral route of administration was lower potential exposure to the owner or to the environment than the topical.

One such oral new entity compound was the natural product nodulisporic acid and its chemically synthesized derivative, *N-tert*-butyl nodulisporamide [9]. This series was discovered in a *Cimex lectularis* (bed bug) assay conducted in a mouse model as described earlier [8]. This mouse model was used along with a laboratory *ex vivo* flea feeding assay (see Chapters 8 and 9) to prioritize the chemical derivatization work to result in the clinical candidate. Compounds selected on the basis of the flea feeding ($LC_{50} \leq 1$ ppm) and *Cimex* mouse (ED50 ≤ 1.0 mg/kg) assays were then evaluated in dogs. The first two assays provided valuable data without the use of much compound, which was limited in supply. Probe-dog-induced infestation studies were conducted at 5 mg/kg body weight orally, which was reduced to 2.5 mg/kg as more potent derivatives were synthesized. PK studies were conducted at the same time to select compounds with superior bioavailability. Larger dog studies using fleas and ticks commenced using a small selection of the analogs and coupled with PK to help explain the relative efficacy between the analogs. The *N-tert*-butyl nodulisporamide analog at 10 mg/kg orally proved to have 100% flea control on dogs for 6 weeks, with partial losses coming at 7 and 8 weeks; and finally at week 9, efficacy was less than 50%. Dogs were also infested with *Rhipicephalis sanguineus* ticks during this study (in between the flea challenges), and an efficacy of 80%, 78%, and 41% was observed during days 12, 19, and 26, respectively. This same analog at 20 mg/kg body weight in induced flea infestation trials in cats showed 100% efficacy against fleas at week 3 and 97% and 94% efficacy at week 4 and 5, respectively. The difference in efficacy between the dog and cat was determined to be the shorter half-life and lower C_{max} of the compound in cats, as determined from the simultaneous PK studies. Control (100%) of two tick species at 30 mg/kg in the dog of the lead compound was shown for 14 days, but this subsequently decreased to 85–87% at 21 days and 84–92% at 28 days.

Spinosad is a natural product originally developed for the agricultural chemical insecticide market, which was also later developed as an oral flavored tablet for the flea market in dogs and cats [24–27]. During product development, several dog and cat models were used to establish the most effective dose, to determine if the compound provided a rapid knockdown and then a sustained effect against adult fleas and had any effect on environmental flea stages (eggs, larvae, pupae), and to determine whether adding another active (milbemycin oxime) could provide efficacy against internal parasite infections. In the adult flea knockdown studies [25, 26], flea counts were conducted at 0.5, 2, 4, 8, and 24 h post treatment. The product provided >90% efficacy at 2 h post dosing and 100% kill by 24 h. The cage of the dog was adapted (solid walls, carpeted floors) to provide an ideal habitat for flea reproduction. Treatment of animals in this model at days 0,

30, and 60 showed excellent activity to prevent infestations. Spinosad was shown in a related trial to have no direct effect on flea environmental stages, but showed dramatic effects on the flea population following heavy infestations [25]. Field trials with natural infections were also conducted at several sites which showed that the product worked very well [96%] against fleas on cats for at least 60 days [26]. A further trial [27] was conducted with spinosad/milbemycin oxime in dogs, which confirmed the dosage of the two actives and showed that the compounds did not interfere with each other in their respective activity against fleas (spinosad) or hookworms (milbemycin oxime).

A new chemistry, the isoxazolines, for control of flea and tick infestations by oral administration was recently launched by three separate companies [28–34].

Afoxolaner (NexGard®), formulated as a flavored chewable, was tested in induced dog flea and tick models [28, 29]. Against ticks, the product was effective (99%) against existing infestations and maintained efficacy following a single dose against weekly tick infestations of over 96% for up to 30 days [28]. Studies conducted to evaluate efficacy against adult fleas at 12 and 24 h post infestation showed excellent activity at both evaluation times through day 35. NexGard also reduced egg counts by 99% for all evaluation time points between days 7 and 35 [29].

Fluralaner (Bravecto®) is a similar isoxazoline compound presented in a flavored chew for fleas and ticks, which was evaluated in induced dog ectoparasite infestation models [30–32]. Given at a much higher dose (25 mg/kg) than NexGard (2.5 mg/kg) and with flea counting at 48 rather than 24 h after infestation, a single oral dose of Bravecto provided 100% efficacy against adult fleas after weekly infestations for 4 months and completely prevented egg production within 48 h during this time after each infestation [30]. A study [31] was conducted against *Ixodes ricinis* ticks in induced infestations in dogs at the 25 mg/kg dose and was shown to be highly effective in quick kill (4, 8, and 12 h post infestation) and at 24 h post infestation evaluation times. Efficacy of at least 98.1% was maintained at both 12 and 24 h post infestation for up to 12 weeks. PK studies in dogs and cats [32] showed that fluralaner is readily absorbed from the skin into adjacent dermal tissues and into the blood where the tick can become exposed upon biting the dog. Blood levels remain high for several weeks and are still measurable at 12 weeks post single topical dosing. Intravenous PK studies also support this long residence time in both cats and dogs. Several induced tick infestation studies in dogs showed efficacy of 91.1–100% within 48 h and 95.4–100% thereafter for 12 weeks.

Sarolaner (Simparica™) is the latest isoxazoline to be introduced [33, 34]. This molecule was the result of testing of several thousand compounds in a combination of whole-organism and animal model testing. These tests included an adult flea ingestion assay followed by tick ingestion models (see Chapters 8 and 9) using the soft tick, *Ornithodoros turicata*. Compounds of interest were then evaluated in a mouse safety model followed by further tests in dog models for safety, PK, and efficacy. Mice used for safety testing were orally treated with up to 30 mg/kg body weight of sarolaner in an aqueous micelle formulation and were observed at several time points for up to 6 h after administration. Observations included posture, activity (hypo or hyper), convulsions, ptosis, and grip strength. No adverse effects were found. Active compounds that were acceptable in this mouse safety assay were then evaluated in dog models. Beagles were used to evaluate the safety of sarolaner at 0, 2, 6, or 10 mg/kg body weight orally using capsules. Dogs were dosed three times with a 28-day dosing interval, followed by serum chemistry, hematology, urinalysis, and clinical and neurological evaluation. Finally, a complete necropsy was conducted after the last dose. A similar study was conducted in 8-week old puppies, with dosing at 4, 12, and 20 mg/kg body weight given two times at 28-day intervals. No adverse effects were found in these target animal safety studies. PK was conducted in dogs at 2 (intravenous) or 20 (oral) mg/kg body weight. From these PK studies the oral bioavailability was 85% and the $t_{1/2}$ was 11–12 days. Flea and tick efficacy studies were conducted in dogs per the standard guidelines [5]. These studies showed that sarolaner was 100% effective against fleas and ticks at 48 h post infestation and maintained 100% efficacy against fleas and *R. sanguineus* ticks and 98.0% against *D. reticulatus* ticks for 35 days at 2.5 mg/kg body weight given orally. It also provided nearly 100% efficacy against *Ixodes ricinus* ticks through day 57 at 2.5 and 5.0 mg/kg body weight given orally. No adverse reactions were seen in treated dogs.

Products for Control of Other Ectoparasites in Companion Animals

Macrocyclic lactones such as ivermectin and selamectin have been used to treat mite and fly infestations in companion animals [35]. Evaluation of treatments for these ectoparasites have been done using natural infestations, as induced models either do not exist or the prevalence of such infestations does not warrant a full development program. However, veterinarians can prescribe such products for these infestations and often publish the results to alert others. Dosing is taken from the approved levels for the more common ectoparasites such as fleas and ticks. Ear mites (*Otodectes cynotis*), Notoedric mange (*Notoedres cati*), nasal mites (*Pneumonyssoidses caninum*), walking dandruff (*Cheyletiella* spp.), and miasis (*Cordylobia anthropophaga*) in either cats or dogs have responded well to selamectin treatment at 6–24 mg/kg body weight. Dog models do exist for *Demodex* spp. and scabies (*Sarcoptes scabiei*) mite infestations. Although neither ivermectin nor selamectin is effective against *Demodex* spp. infection, ivermectin and other macrocyclic lactone derivatives are effective against scabies. Ivermectin and selamectin were also found to be effective against *Cheyletiella* spp. natural infestations in rabbits [36].

Products for Control of Livestock Ectoparasites

Induced infestation models have been implemented in cattle for a few ectoparasites such as flies (*Stomoxys calcitrans*, stable fly; and *Haemotobia irritans*, hornfly) and mites. However, many trials are conducted using naturally infested production animals, as induced models are very difficult to establish. For example, the fly models for cattle are difficult to conduct, few animals can be included in a study, and the fly life cycle maintenance is very labor intensive. Therefore, field trials with natural or acquired rather than induced infestations are generally used to demonstrate efficacy of the product.

Treatment of ectoparasites on livestock can occur with oral, systemic, or topical treatment including sustained release formulations. For example, ivermectin was found to eliminate scabies (*S. scabiei* var *suis*) in pigs when provided in an in-feed formulation [37]. Similarly, topically applied ivermectin using naturally acquired infestations was found to be highly effective against cattle grubs (*Hypoderma bovis* and *H. lineatum*) and lice (*Damalina bovis*) [38] and horn flies (*H. irritans*) [39].

A subcutaneously administered sustained release formulation of doramectin was shown to be highly effective against the lone star tick (*Amblyomma americanum*) and horn flies (*H. irritans*) on cattle [40]. In this study, animals were administered the formulation, and ticks were placed in a stockinet sleeve adhered to the side of the animal. Thereafter, the ticks were observed to determine the percentage of engorged females, weight of the engorged females, the weight of the egg mass produced, and the numbers of eggs to determine an index of fecundity. The effect of blood from treated animals on horn flies was determined using *ex vivo* whole-organism assays. Doramectin was shown to be highly effective against both ectoparasites.

Ivermectin in a sustained release bolus administered orally was shown to be effective against horn fly infestations in cattle [41]. When two boluses were used, resulting in serum levels of 31.2 ppb at 13 weeks post treatment, efficacy against feeding female horn flies was 96.2%. Horn fly larvae placed in dung pats from treated animals were completely prevented from development through week 19 posttreatment. In the development of the ivermectin bolus, stomach contents from a rumen fistulated cattle model was evaluated for ivermectin levels in addition to blood levels to help monitor the duration of efficacy based on PK.

Extension to Human Conditions

As mentioned in the chapter for whole-organism assays in this volume (Chapter 8), information gained from the testing of new compounds in animals also provides data useful for human infestations. Humans can become infested naturally with the same or similar ectoparasite organisms as in animals and can become infected with the zoonotic organisms carried by the ectoparasite. It is therefore important to develop novel ectoparasiticides for humans, particularly for tick and mite infestations. Studies conducted in animals which have reference to human infestations include efficacy information in regard to effective dosage; PK requirements for minimum blood levels and residence time; and safety information (toxic levels via different exposure routes, exposure to residues in animal tissues, etc.). These data derived from animal models are useful to determine the appropriate treatment of humans using off-label procedures or in actually developing a new drug specifically for humans. An example of the latter is that of ivermectin approval in humans for a topical head lice product Sklice®.

Humans can be bitten by the same or similar ectoparasites as animals and can be exposed to a number of highly debilitating or even lethal pathogens such as ricketisia, other bacteria, and protozoa. The zoonotic potential of ectoparasites either directly or by the pathogens that an ectoparasite carries as observed in animal models provide very valuable data regarding exposure time between the bite and the delivery of the pathogen to the human and potential lethal or serious debilitation. Experiments in animal models can be highly useful to identify a safe, effective therapy for human diseases.

Conclusion

This chapter has explored the use of rodent and target animal (dog, cat, livestock) ectoparasiticide models to determine the efficacy; potency; duration of effect; therapeutic margin; PK parameters; and reproductive, target animal, young animal, and food safety aspects (residues, safety margins) of new lead candidates for potential ectoparasiticide products. Rodent models are used to screen many compounds using little compound to determine safety, efficacy, and PK characteristics of the lead candidate. These findings are used to prioritize the compounds for studies in the target host animal that require far more compound, and are very expensive with low throughput and often long duration. Induced infestation target animal models have been established for both companion animals and livestock, although natural infestations are most often used in livestock. The use of PK has become much more important in the discovery process to prioritize compounds based on bioavailability and metabolism. Several examples of the use of animal models to find many different classes of ectoparasiticides which are delivered by a variety of routes of administration are provided. Recommended procedures for the use of animal models to approve new ectoparasiticides have been published and have been written by leading academic and industry scientists and provide a valuable blueprint. Finally, mention is made of the utility of animal models to discover new ectoparasiticides for use in human health or to provide efficacy and safety data suitable to allow off-label use of a veterinary product in humans.

References

1. Marchiondo, A.A., Holdsworth, P.A., Fourie, L.J., Roug, D., Hellman, K., Snyder, D.E., and Dryden, M.W. (2013) World Association for the Advancement of Parasitology (W.A.A.V.P.) second edition: guidelines for evaluating the efficacy of parasiticides for the treatment, prevention and control of flea and tick infestations on dogs and cats. *Vet. Parasitol.*, **194**, 84–97.
2. Vercruysse, J., Rehbein, S., Holdsworth, P.A., Letonja, T., and Peter, R.J. (2006) World Association for the Advancement of Veterinary Parasitology (W.A.A.V.P.) guidelines for evaluating the efficacy of acaricides against (mange and itch) mites on ruminants. *Vet. Parasitol.*, **136**, 55–66.
3. Holdsworth, P.A., Vercruysse, J., Rehbein, S., Peter, R.J., De Bruin, C.,

Letonja, T., and Green, P. (2006) World Association for the Advancement of Veterinary Parasitology (W.A.A.V.P.) guidelines for evaluating the efficacy of ectoparasiticides against biting and nuisance flies on ruminants. *Vet. Parasitol.*, **136**, 3–13.

4. Holdworth, P.A., Vercruysse, J., Rehbein, S., Peter, R.J., De Bruin, C., Letonja, T., and Green, P. (2006) World Association for the Advancement of Veterinary Parasitology (W.A.A.V.P.) guidelines for evaluating the efficacy of ectoparasiticides against myiasis causing parasites on ruminants. *Vet. Parasitol.*, **136**, 15–28.
5. Holdsworth, P.A., Vercruysse, J., Rehbein, S., Peter, R.J., Letonja, T., and Green, P. (2006) World Association for the Advancement of Veterinary Parasitology (W.A.A.V.P.) guidelines for evaluating the efficacy of ectoparasiticides against biting lice, sucking lice, and sheep keds on ruminants. *Vet. Parasitol.*, **136**, 45–54.
6. Holdsworth, P.A., Kemp, D., Green, P., Peter, R.J., De Bruin, C., Jonsson, N.N., Leltonja, T., Rehbein, S., and Vercruysse (2006) World Association for the Advancement of Veterinary Parasitology (W.A.A.V.P.) guidelines for evaluating the efficacy of acaricides against ticks (Ixodidae) on ruminants. *Vet. Parasitol.*, **136**, 29–42.
7. Cochet, P., Birckel, P., Bromet-Petit, M., Bromet, N., and Weil, A. (1997) Skin distribution of fipronil by microautoradiography following topical administration to the beagle dog. *Eur. J. Drug Metab. Pharmacokinet.*, **22**, 211–216.
8. Ostlind, D.A., Cifelli, S., Conroy, J.A., Mickle, W.G., Ewanciw, D.V., Andriuli, F.J., and Ho, P. (2001) A novel *Cimex lectularius* – rodent assay for the detection of systemic ectoparasiticide activity. *Southwest. Entomol.*, **26**, 181–186.
9. Menke, P.T., Colletti, S.L., Fisher, M.H., Wyvratt, M.J., Shih, T.L., Ayer, M.B., Li, C., Lim, J., Ok, D., Salva, S., Warmke, L.M., Zakson, M., Michael, B.F., deMontigny, P., Ostlind, D.A., Fink, D., Drag, M., Schmatz, D.M., and Shoop, W.L. (2009) Discovery of the development candidate *N-tert*-butyl nodulisporamide: a safe and efficacious once monthly oral agent for the control of fleas and ticks on companion animals. *J. Med. Chem.*, **52**, 3505–3515.
10. Santora, K.A., Zakson-Aiken, M., Rasa, C., and Shoop, W. (2002) Development of a mouse model to determine the systemic activity of potential flea-control compounds. *Vet. Parasitol.*, **104**, 257–264.
11. Lussenhop, J., Baumer, W., Kietzmann, M., Schneider, M., and Wolken, S. (2011) Dynamics of distribution and efficacy of different spot-on permethrin formulations in dogs artificially infested with *Dermacentor reticulatus*. *Parasit. Vectors*, **4**, 45–54.
12. Young, D.R., Jeannin, P.C., and Boeckh, A. (2004) Efficacy of fipronil/(S)-methoprene combination spot-on for dogs against shed eggs, emerging and existing adult cat fleas (*Ctenocephalides felis*, Bouche). *Vet. Parasitol.*, **125**, 397–407.
13. Beugnet, F., Soll, M.D., Bousira, E., and Franc, M. (2015) Sustained speed of kill and repellency of a novel combination of fipronil and permethrin against *Ctenocephalides canis* flea infestations in dogs. *Parasit. Vectors*, **8**, 52–58.
14. Fankhauser, B., Dumont, P., Halos, L., Hunter, J.S., Kunkle, B., Everett, W.R., Chester, S.T., Fourie, J.J., and Soll, M.D. (2015) Efficacy of a new combination of fipronil and permethrin against *Ctenocephalides felis* flea infestation in dogs. *Parasit. Vectors*, **8**, 62–68.
15. Fankhauser, B., Irwin, J.P., Stone, M.L., Chester, S.T., and Soll, M.D. (2015) Repellent and insecticidal efficacy of a new combination of fipronil and permethrin against stable flies (*Stomoxys calcitrans*). *Parasit. Vectors*, **8**, 61–67.
16. Rugg, D. and Hair, J.A. (2007) Dose determination of a novel formulation of metaflumizone plus amitraz for control of cat fleas (*Ctenocephalides felis felis*) and brown dog ticks (*Rhipicephalis sanguineus*) on dogs. *Vet. Parasitol.*, **150**, 203–208.

17. Baker, C., Tielemans, E., Prullage, J.B., Chester, S.T., Knaus, M., Rehbein, S., Fourie, J.J., Young, D.R., Everett, W.R., and Rosentel, J.K. (2011) Efficacy of a novel combination of fipronil, (S)-methoprene, eprinomectin and praziquantel against adult and immature stages of the cat flea (*Ctenocephalides felis*) on cats. *Vet. Parasitol.*, **179**, 324–329.
18. Ritzhaupt, L.K., Rowan, T.G., and Jones, R.L. (2000) Evaluation of efficacy of selamectin and fipronil against *Ctenocephalides felis* in cats. *J. Am. Vet. Med. Assoc.*, **217**, 166–168.
19. Brianti, E., Falsone, L., Napoli, E., Prudente, C., Gaglio, G., and Giannetto, S. (2013) Efficacy of a combination of 10% imidacloprid and 4.5% flumethrin (Seresto®) in slow release collars to control ticks and fleas in highly infested dog communities. *Parasit. Vectors*, **6**, 210–218.
20. Stanneck, D., Rass, J., Radeloff, I., Kruedewagen, E., Le Sueur, C., Hellman, K., and Krieger, K. (2012) Evaluation of the long term efficacy and safety of an imidacloprid 10%/flumethrin 4.5% polymer matrix collar (Seresto®) in dogs and cats naturally infested with fleas and/or ticks in multicenter clinical field studies in Europe. *Parasit. Vectors*, **5**, 66–77.
21. Lechat, C., Simeon, N., Pennant, O., Desquilbet, L., Chahory, S., Le Sueur, C., and Guillot, J. (2015) Comparative evaluation of the prophylactic activity of a slow release insecticide collar and a moxidectin spot-on formulation against *Thelazia callipaeda* infection in naturally exposed dogs in France. *Parasit. Vectors*, **8**, 93–98.
22. Brianti, E., Gaglio, G., Napoli, E., Falsone, L., Prudente, C., Basano, F.S., Latrofa, M.S., Tarallo, V.D., Dantas-Torres, F., Capelli, G., Stanneck, D., Giannetto, S., and Otranto, D. (2014) Efficacy of a slow-release imidacloprid (10%)/flumethrin (4.5%) collar for the prevention of canine leishmaniosis. *Parasit. Vectors*, **7**, 327–337.
23. Dantes-Torres, F., Capelli, G., Giannelli, A., Ramos, R.A.N., Lia, R.P., Cantacessi, C., de Caprariis, D., de Tommasi, A.S., Latrofa, M.S., Lacasella, V., Tarallo, V.D., Di Paola, G., Qurollo, B., Breitschwerdt, E., Stanneck, D., and Otranto, D. (2013) Efficacy of an imidacloprid/flumethrin collar against fleas, ticks and tick-borne pathogens in dogs. *Parasit. Vectors*, **6**, 245–253.
24. Snyder, D.E., Meyer, K.A., Wiseman, S., Trout, C.M., and Young, D.R. (2013) Speed of kill efficacy and efficacy of flavored spinosad tablets administered orally to cats in a simulated home environment for the treatment and prevention of cat flea (*Ctenocephalides felis*) infestations. *Vet. Parasitol.*, **196**, 492–496.
25. Blagburn, B.L., Young, D.R., Moran, C., Meyer, J.A., Leigh-Hoffman, A., Paarlberg, T., Zimmerman, A.G., Mowrey, D., Wiseman, S., and Synder, D.E. (2010) Effects of orally administered spinosad (Comfortis®) in dogs on adult and immature stages of the cat flea (*Ctenocephalides felis*). *Vet. Parasitol.*, **168**, 312–317.
26. Franc, M., Houhsira, E., Bohm, C., Wolken, S., Wolf, O., Lohlein, W., Wiseman, S., Hayes, B., Schnitzler, B., and Fisher, M. (2014) Evaluation of spinosad for the oral treatment and control of flea infestations on cats in Europe. *Vet. Rec. Open*, **1** (1), e000047.
27. Snyder, D.E. and Wiseman, S. (2012) Dose confirmation and non-interference evaluations of the oral efficacy of a combination of milbemycin oxime and spinosad against the dose limiting parasites, adult cat flea (*Ctenocephalides felis*) and hookworm (*Ancylostoma caninum*), in dogs. *Vet. Parasitol.*, **184**, 284–290.
28. Dumont, P., Blair, J., Fourie, J.J., Chester, T.S., and Larsen, D.L. (2014) Evaluation of the efficacy of afoxolaner against two European tick species: *Dermacentor reticulatus* and *Ixodes ricinus*. *Vet. Parasitol.*, **201**, 216–219.
29. Hunter, J.S., Dumont, P., Chester, T.S., Young, D.R., Fourie, J.J., and Larsen, D.L. (2014) Evaluation of the curative and preventive efficacy of a single oral administration of afoxolaner against cat

flea *Ctenocephalides felis* infestations in dogs. *Vet. Parasitol.*, **201**, 207–211.

30. Dryden, M.W., Smith, V., Bennett, T., Math, L., Kallman, J., Heaney, K., and Sun, F. (2015) Efficacy of fluralaner flavored chews (Bravecto®) administered to dogs against the adult cat flea, *Ctenocephalides felis felis* and egg production. *Parasit. Vectors*, **8**, 364–371.
31. Wengenmayer, C., Williams, H., Zschiesche, E., Moritz, A., Langenstein, J., Roepke, R.K.A., and Heckeroth, A.R. (2014) The speed of kill of fluralaner (Bravecto™) against *Ixodes ricinus* ticks on dogs. *Parasit. Vectors*, **7**, 525–530.
32. Kilp, S., Ramirez, D., Allan, M.J., and Roepke, R.K.A. (2016) Comparative pharmacokinetics of fluralaner in dogs and cats following single topical or intravenous administration. *Parasit. Vectors*, **9**, 296–303.
33. McTier, T.L., Chubb, N., Curtis, M.P., Hedges, L., Inskeep, G.A., Knauer, C.S., Menon, S., Mills, B., Pullins, A., Zinser, E., Woods, D.J., and Meeus, P. (2016) Discovery of sarolaner: a novel, orally administered, broad-spectrum, isoxazoline ectoparasiticide for dogs. *Vet. Parasitol.*, **222**, 3–11.
34. Geurden, R., Becskei, C., Grace, S., Strube, C., Doherty, P., Liebenberg, J., Mahabir, S.P., Slootmans, N., Lloyd, A., and Six, R.H. (2016) Efficacy of a novel oral formulation of sarolaner (Simparica™) against four common tick species infesting dogs in Europe. *Vet. Parasitol.*, **222**, 33–36.
35. Fisher, M.A. and Shanks, D.J. (2008) A review of the off-label use of selamectin (Stonghold®/Revolution®) in dogs and cats. *Acta Vet. Scand.*, **50**, 46–51.
36. Mellgren, M. and Bergvall, K. (2008) Treatment of rabbit cheyletiellosis with selamectin or ivermectin: a retrospective case study. *Acta Vet. Scand.*, **50**, 1–6.
37. Alva-Valdes, R., Wallace, D.H., Foster, A.G., Ericsson, G.F., and Wooden, J.W. (1989) Efficacy of an in-feed ivermectin formulation against gastrointestinal helminths, lungworms, and sarcoptic mites in swine. *Am. J. Vet. Res.*, **50**, 1392–1395.
38. Alva-Valdes, R., Wallace, D.H., Holste, J.E., Egerton, J.R., Cox, J.L., Wooden, J.W., and Barrick, R.A. (1986) Efficacy of ivermectin in a topical formulation against induced gastrointestinal and pulmonary nematode infections and naturally acquired grubs and lice in cattle. *Am. J. Vet. Res.*, **47**, 2389–2392.
39. Marley, S.E., Hall, R.D., and Corwin, R.M. (1993) Ivermectin cattle pour-on: duration of a single late spring treatment against horn flies, *Haematobia irritans* (L.) (Diptera: Muscidae) in Missouri, USA. *Vet. Parasitol.*, **51**, 167–172.
40. Lohmeyer, K.H., Miller, J.A., Pound, J.M., and Klavons, J.A. (2009) A sustained release gel formulation of doramectin for control of lone star ticks (Acari: Ixodidae) and horn flies (Diptera: Muscidae) on cattle. *Vet. Entomol.*, **102**, 804–808.
41. Miller, J.A., Davey, R.B., Oehler, D.D., Pound, J.M., and George, J.E. (2003) Efficacy of the ivomec SR bolus for control of horn flies (Diptera: Muscidae) on cattle in South Texas. *J. Econ. Entomol.*, **96**, 1608–1611.

Part Three
Isoxazolines

Ectoparasites: Drug Discovery Against Moving Targets, First Edition.
Edited by Charles Q. Meng and Ann E. Sluder

12
Isoxazolines: A Novel Chemotype Highly Effective on Ectoparasites

*Tina Weber and Paul M. Selzer**

Abstract

Efficient control of arthropod ectoparasite infestations has a long-standing history in the agriculture and veterinary sectors, aiming to decrease the parasite burden of affected crops and animals. Ligand-gated chloride channels (LGCCs) modulated by γ-aminobutyric acid (GABA) and glutamate have been identified as suitable molecular targets, and several classes of potent parasiticides have been devised. Due to the increase in cross-resistance and decreased development of new chemical entities, an urgent need for new parasiticides or prevention schemes has emerged. In the past decade, an innovative isoxazoline chemotype appeared to offer promise for inhibiting LGCCs with a new mode of action and distinct binding site from that of historical agents. Considerable efforts have focused on optimizing the antiparasitic activity of isoxazolines and may provide the potential for future human use.

Arthropod Ectoparasites: Burden to the Agricultural and Veterinary Sectors

Safe and efficient control of pests has become pivotal not only for agricultural productivity [1] but also in the veterinary sector, including companion animals and livestock [2]. In this regard, antiparasitic drugs strive to regulate, mitigate, and ideally prevent infestation by parasites such as arthropods, helminths, or protozoa and the consequences associated with them [3, 4]. From an animal health perspective, acarid and insect ectoparasites are particularly key target species, with ticks, fleas, and flies affecting cattle as well as cats and dogs severely. Prominent species are *Ctenocephalides felis* (flea) and *Dermacentor variablis* (tick) in cats and dogs, respectively, and *Rhipicephalus microplus* (tick) in cattle. Many of them are primarily blood feeders; however, the immediate and more

*Corresponding author.
Note: This article has been pre-published in: ChemMedChem 2016, 11, 270–276; DOI: 10.1002/cmdc.201500516

Ectoparasites: Drug Discovery Against Moving Targets, First Edition.
Edited by Charles Q. Meng and Ann E. Sluder

apparent pathology linked with the initial bite of the parasites should not deflect from their role as vectors for parasitic microorganisms and as a source of infections. Extensive research has consequently gone into the development of efficient and selective veterinary ectoparasiticides [5]. As humans are also heavily burdened by ectoparasites, they might also benefit from these developments.

Ligand-gated Chloride Channels as Suitable Targets for Ectoparasiticides

Considering the mode of action, a significant number of active compounds target ligand-gated ion channels [6, 7]. Three representatives of the major structural insecticide classes that act on those targets are depicted in Table 12.1.

Dieldrin belongs to the organochlorides, more specifically to the cyclodienes, which were extensively used in the agricultural sector for over three decades since the 1940s [8]. Long persistence in the environment, however, proved detrimental, as bioaccumulation of dieldrin gave rise to toxic effects in species not initially targeted, including humans, along with the development of resistance. A more recent alternative is fipronil, a phenylpyrazole with contact as well as systemic activity that was introduced in the 1990s [9, 10]. The broad-spectrum ectoparasiticides are distributed as the racemate and are used in plant protection schemes

Table 12.1 Ectoparasiticides acting on ligand-gated chloride channels.

Structure	Name	References
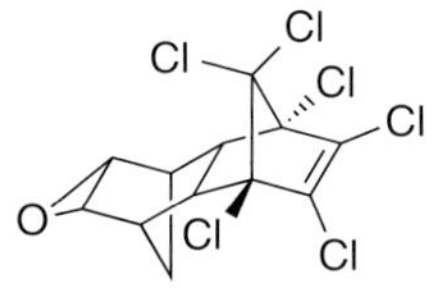	Dieldrin	[8]
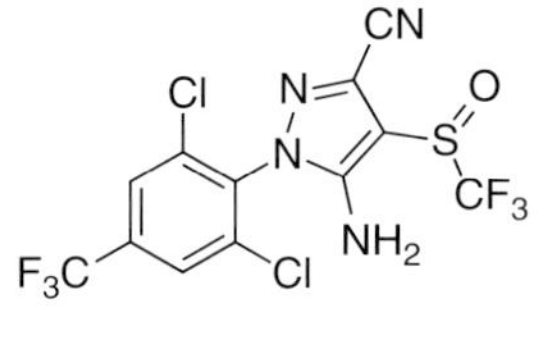	Fipronil	[9, 10]
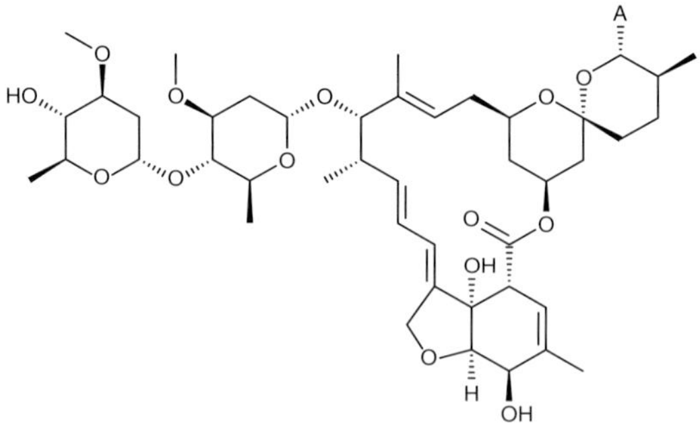	Ivermectin (A = iPr/sBu)	[11]

as well as in the veterinary sector. Structurally more complex are naturally occurring macrocyclic lactones and derivatives thereof, such as ivermectin [11–15]. The potent parasiticides exhibit better safety and selectivity profiles, with greater affinity for invertebrate receptors than mammalian analogs [16].

Mode of Action

The three parasiticide classes described herein act by disrupting the signaling ability of the synapses between the neurons within the central nervous system (CNS), and also at the neuromuscular junctions [17]. In this process, a nerve impulse or action potential typically triggers axon uptake of calcium ions via specific Ca2+ channels. The increase in Ca2+ concentration influences the synaptic vesicles that contain neurotransmitters, causing them to fuse with the presynaptic cell membrane and to open up into the extracellular space between an axon and a dendrite: the synaptic cleft. The released neurotransmitters then bind and modulate receptors such as ligand-gated ion channels in the postsynaptic cell, subsequently shifting the local transmembrane potential and thus allowing transmission of an electrical pulse or signal. In this cascade, the ion channels are of particular interest as parasiticide targets, as disruption of their regular function by blockage or prolonged activation may decisively interfere in the signaling process. Characteristic features in arthropod ion channel sequences are envisaged as strategic points for selectivity.

Dieldrin and fipronil have been identified as antagonists for γ-aminobutyricacid-gated chloride channels (GABACls), essentially blocking the influx of chloride ions and consequently counteracting the inhibitory function of the natural neurotransmitter [8, 9, 18]. As a result, depolarization and ensuing hyperexcitation occur, leading to tremors, convulsion, and death. GABACls are essential in both the mammalian and invertebrate CNS, which makes parasiticide selectivity a prerequisite to ensure a relevant therapeutic window [19]. In contrast, glutamate-gated chloride channels (GluCls) are specific to invertebrates and are predominantly located in the skeletal muscle tissue and the CNS. They have recently been associated with the same superfamily of cysteine-loop ligand-gated chloride channels (CysLGCCs) as GABACls, with the added advantage of being species-specific [20, 21]. Notably, ivermectin has been shown to activate glutamate receptors and intensify the chloride influx, giving rise to hyperpolarization followed by ataxia and coma at higher concentrations [7, 20–22]. The sedative effect has also been partially attributed to the activation of GABACls, albeit to a lower degree [23].

Isoxazolines: Novel Ectoparasiticides Acting on GABACls and GluCls

Twenty years since the introduction of the phenylpyrazole fipronil as the most recent addition to the existing portfolio of ectoparasiticides, the demand for new molecular entities addressing the CysLGCCs in a different manner has grown steadily. A clear distinction from known commercially available antagonists is essential, with a novel mode of action or binding site to circumvent cross-resistance concerns. In this regard, structural diversity is key. Previously investigated as highly selective modulators of arthropod ryanodine receptors [24, 25], derivatives of phthalic and anthranilic diamides came into focus when the incorporation of isoxazoline moieties changed the activity profile toward ligand-gated chloride channels [26, 27]. Extensive research on this novel compound class led to the discovery of potent ectoparasiticides by researchers at Du Pont [28, 29] and Nissan [26, 27]. Active ingredients with the international nonproprietary names (INNs) such as afoxolaner (Chapter 13) and fluralaner were later approved by the US Food and Drug Administration (FDA) in short succession for Merial's NexGard (September 2013) [30, 31] and MSD-AH's Bravecto (May 2014) [26, 32–34], respectively (Table 12.2). Structurally related to fipronil to a certain degree with its chloride-substituted phenyl heterocyclic core, the novel compound

Table 12.2 Isoxazoline-derived parasiticides.

R^1	R^2	Ar	Name	References
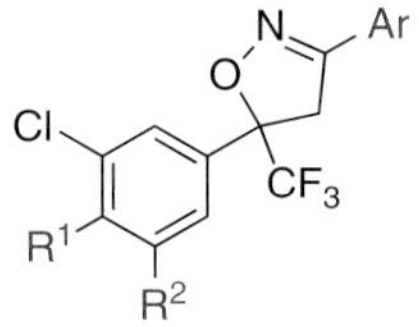				
H	CF_3	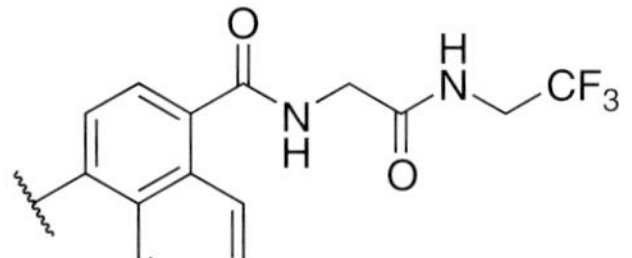	Afoxolaner	[29, 31] (chapter 13)
H	Cl	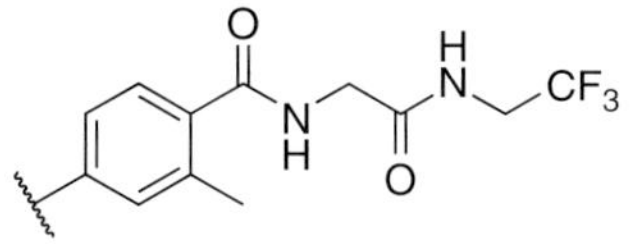	Fluralaner	[26, 33, 34]
Cl	Cl	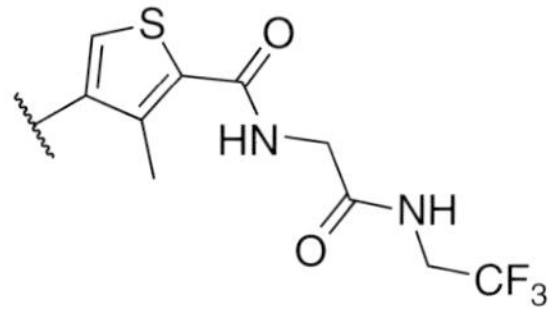	Lotilaner	[35]
F	Cl	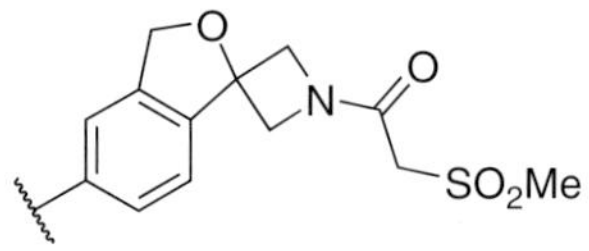	Sarolaner	[36] (Chapter 15)

class contains a nonaromatic isoxazoline and an extended side chain that indicates its link to the preceding diamides. The aryl amide side chain and the substitution pattern on the terminal aromatic unit have consequently been the central areas for further derivatization. With regard to the chiral center within the isoxazoline ring, studies have revealed only the *S* enantiomer of fluralaner to be active [27], with no adverse effects of the respective *R* enantiomer known to date. In this regard, the ectoparasiticides afoxolaner and fluralaner both are distributed as racemic mixtures [27, 31].

Structure and Active Sites of Chloride Channels

As a family, CysLGCCs differ noticeably in complexity in terms of their subunits' composition; however, the homo- or heteromeric ion channels show common structural and functional features and can be generalized to some degree [20, 21]. Embedded in the postsynaptic cell membrane, each unit of the pentameric ion pore that forms the CysLGCCs consists of four distinct helical transmembrane regions (M1–M4) and a characteristic cysteine loop located in the large extracellular N-terminal domain. A schematic representation of the ion channel's α-helical domain is depicted in Figure 12.1, with M2 creating the inner lining of the pore while the five independent helix clusters are linked between M1 and M3 of adjacent subunits. Accordingly, the amino acid sequence in M2 has a pronounced influence on the ion selectivity of the respective receptor. With the emergence of parasiticide resistance, especially against cyclodiene derivatives, research has focused on locating the most conspicuous modifications in the relevant genes that code for the ion channels.

An amino acid exchange of alanine to serine or glycine at position 302 in the M2 helix was observed in a dieldrin-resistant (rdl) GABA subunit gene of *Drosophila melanogaster*, and the single point mutation at this position was later identified as a central cause for resistance development across a number of different species [37–39]. In addition, cross-resistance to fipronil and other GABAergic parasiticides has been observed to varying degrees of intensity, suggesting a high congruency of the binding sites for the different potent GABACl blockers and linking the similarities in mode of action to a common, but not necessarily identical, allosteric-binding site in the channel lumen [23, 40]. By way of better illustration, a homology model of a pentameric GABACl of *C. felis* was generated on the basis of crystallographic data for the respective *Erwinia chrysanthemi* analog (PDB ID: 2VL0; Figure 12.2) [41–43]. The crucial resistance-inducing point mutation located in the transmembrane region is highlighted in dark green with the amino acid residue as a red sphere in Figure 12.2c, the characteristic Cys loop in the extracellular region in light green. In a view along the channel pore axis (Figure 12.2b,c), the channel-pore-lining M2 helices are visible.

Besides voltage clamp electrophysiology studies for investigations on the parasiticides' influence on the chloride ion flux, a multitude of target-site binding assays have been used to evaluate their inhibition and modulation capacity

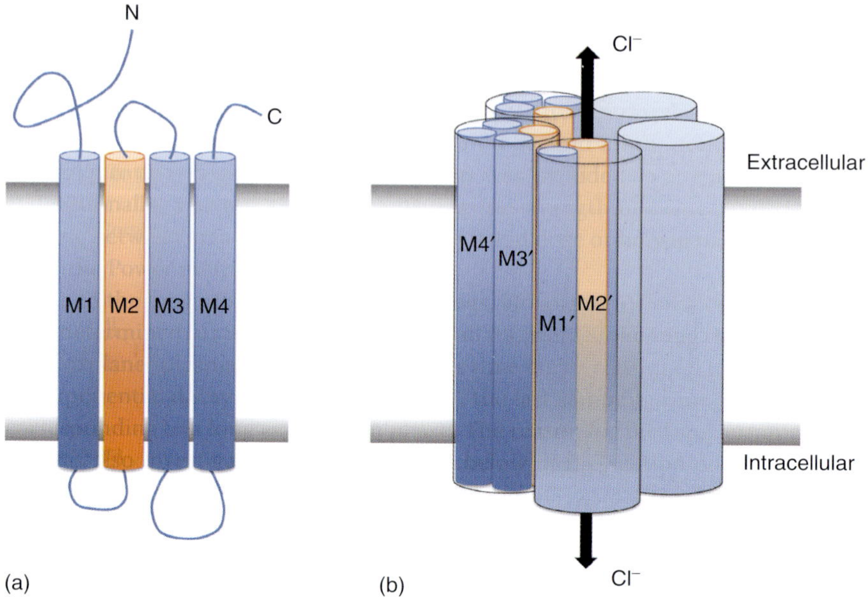

Figure 12.1 Schematic representation of the transmembrane domain of a ligand-gated chloride channel. (a) Four transmembrane helices (M1–M4) that make up one subunit are depicted with their connecting loops. (b) Arrangement of the transmembrane subunits to form a pentameric ion channel with M2 Cl^- helices (orange) lining the pore, M1 and M3 linking the adjacent subunits, and M4 facing the outside of the pore. Helices M3′ and M4′ of the foremost subunit have been omitted for clarity. The composition of helices is illustrated for only three of the five subunits.

[44, 45]. The frequently used radiolabeled receptor channel blocker 4′-ethynyl-4-*n*-propylbicycloorthocarboxylate ([3H]EBOB) shows good correlation between its displacement and the toxicity of GABACl antagonists such as cyclodienes and fipronil, thus further supporting the close link between the parasiticides' binding sites [44, 45]. Noncompetitive and incomplete binding was observed for avermectins, however, emphasizing the distinction in pathology and their conserved sensitivity even in dlr-GABACls [45]. Analogous studies on GluCls suggest an ivermectin-stimulated activation of the ion channel by insertion between the linking M3 and M1′ helices of adjacent subunits in a transmembrane region closer to the N-terminal extracellular domain [46–48]. Crystallization of a GluCl of *Caenorhabditis elegans* in complex with Fab molecules and ivermectin has been reported by Hibbs and Gouaux, defining a binding site and activation mechanism of the avermectin potentiators (PDB ID: 3RHW) [49]. A representation of the ion channel in its proposed open conformation is shown in Figure 12.3, with one ivermectin-binding pocket highlighted in yellow.

Despite the favorable distinction and high parasiticide selectivity for invertebrates, the complex scaffold of the macrocyclic lactones constrains further extensive and profitable derivatization and optimization of this compound class.

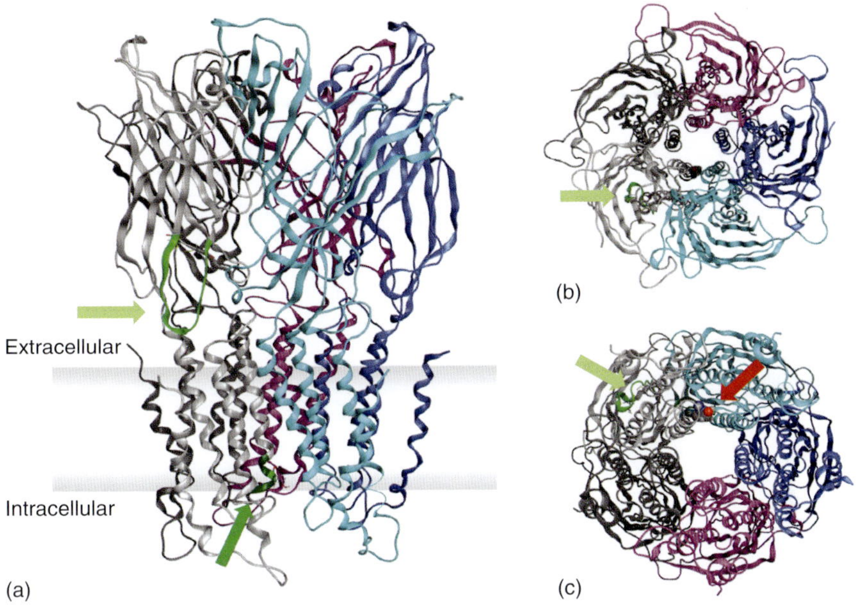

Figure 12.2 Homology model of GABACl in a proposed closed form. Sequences of *Erwinia chrysanthemi* GABACl and dlr-GABACl of *Ctenocephalides felis* were aligned using the BLAST algorithm, and final modeling was performed with the MOE software package [41]. The five individual subunits are shaded in different colors. The Cys loop of one subunit is highlighted in light green (light-green arrows), the resistance-inducing mutation in dark green (dark-green arrow). (a) Protein shown parallel to membrane; horizontal gray bars indicate membrane boundaries. (b) View along the channel pore axis from the extracellular side, with M2 helices visible as the inner pore lining. (c) View along the channel pore axis from the intracellular side, with resistance-inducing residue (red dot) highlighted with a red arrow.

Coupled with the emerging resistance toward GABA blockers, this constraint makes the development of novel modulators of ligand-gated chloride channels pressing. However, a new innovative substrate class has been developed over recent years with an intriguing activity profile.

Isoxazoline Mode of Action and Binding Site

A dose-dependent inhibition of GABACls in *D. melanogaster* by afoxolaner was observed in electrophysiology studies with an IC50 value of 3.7 nM and no detectable cross-resistance effect [29]. Similarly, electrophysiology studies on CysLGCCs of *Musca domestica* confirmed the efficient inhibition of agonist-induced currents in GABACls and GluCls by fluralaner, with IC50 values of 5 and 80 nM, respectively, identifying the GABA receptor as the more sensitive of the two [27]. In addition, it retains its potency even in the dlr mutant of the *M. domestica* chloride channel (A299S). In comparison with fipronil, both isoxazoline

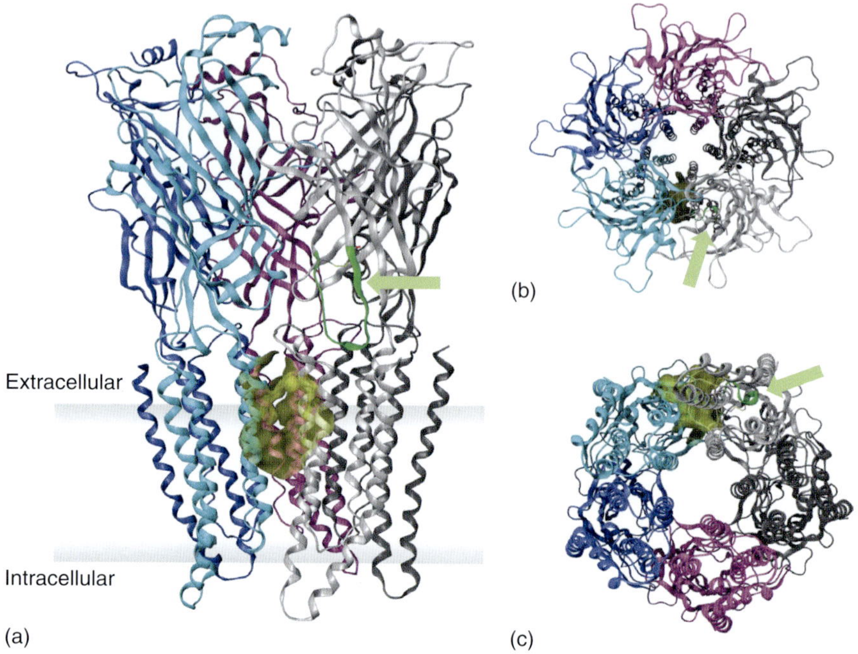

Figure 12.3 GluCl crystal structure of *C. elegans* represented in an activated, open-channel state. These images are based on an original X-ray crystallographic structure (PDB ID: 3RHW) [49] and were produced with the MOE software package [41]. Fab molecules are omitted for clarity. The five individual subunits are shaded in different colors. The surface of one ivermectin-binding pocket is highlighted in yellow. The Cys loop of one subunit is highlighted in light green (light-green arrows). (a) Protein parallel to membrane, horizontal gray bars indicate membrane boundaries. (b) View along the channel pore axis from the extracellular side, with M2 helices visible as the inner pore lining. (c) View along channel pore axis from the intracellular side.

derivatives exhibit markedly higher potency toward GABACls, emphasizing the crucial advancement relative of the aforementioned parasiticides achieved and reviving the susceptibility of CysLGCCs as valid drug targets.

In initial assays on housefly head membrane to elucidate the binding site of isoxazoline derivatives, fluralaner successfully impeded [3H]EBOB binding at subnanomolar concentrations in a manner similar to known antagonists [27]. Ensuing comprehensive evaluation of the specific binding site in housefly head membrane revealed avermectin analogs as effective inhibitors of radiolabeled [3H] fluralaner binding; however, poor displacement by antagonists such as fipronil was detected [50]. The isoxazoline consequently addresses a binding site reasonably distinct from that of CysLGCCl blockers and is closely associated with that of activators, despite being a potent chloride channel blocker. Casida *et al.* summarized this unique sensitivity profile in a comparative binding site study with various radiolabeled ligands including [3H]fluralaner, [3H]EBOB, and avermectin B1a ([3H]AVE) (Table 12.3) [51]. It highlights the exceptional activity of isoxazolines and their distinct target site. Notably, this also implies the absence of common features that might lead to the appearance of cross-resistance and supports the previously reported observations in rdl mutants. No clear assignment of an isoxazoline-binding pocket by X-ray crystallography or other methods has yet been reported. However, based on the evidence provided by the aforementioned binding studies, an interaction with the helical subunits of the transmembrane domain in CysLGCCs between the identified agonist and antagonist sites is conceivable. In comparison, parasiticide-binding regions of GABA blockers and ivermectin relative to each other are depicted in a schematic illustration of a CysLGCC (Figure 12.4). Further in-depth research is required to fully elucidate the inhibition site and mechanism of the novel isoxazoline compound class.

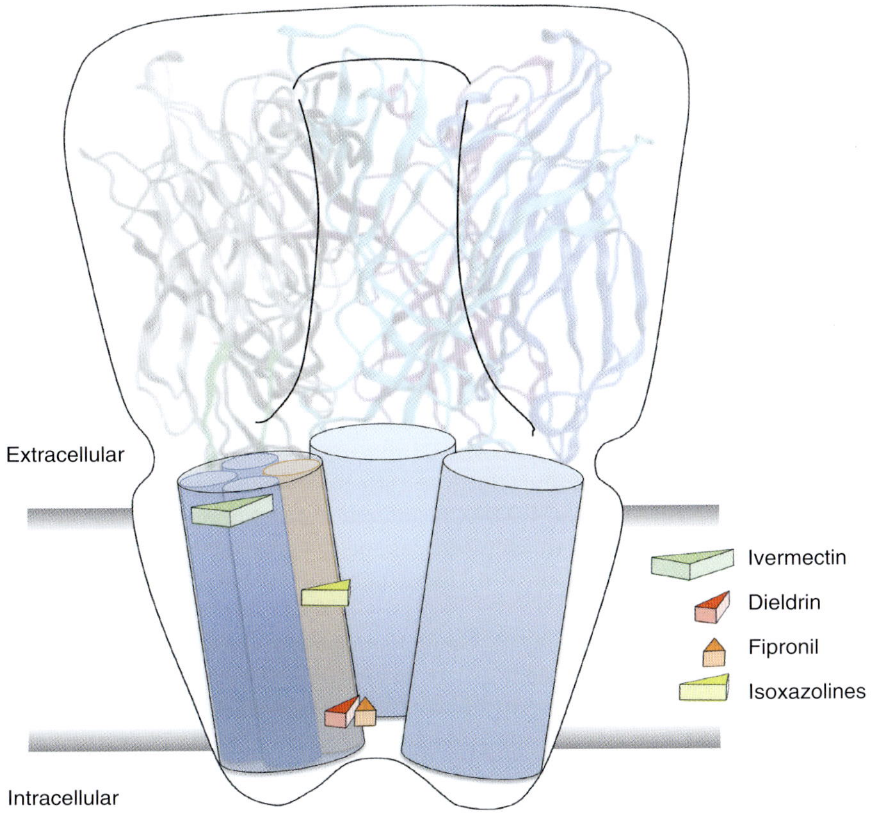

Figure 12.4 Schematic representation of CysLGCCs in a sectional view with only three units of the pentameric transmembrane region depicted for clarity. Positioning of isoxazolines is putative.

Table 12.3 Ectoparasiticide activity (IC50 (nM)) in binding assays on *M. domestica* head membrane.[a)]

Radioligand	Fluralaner	AVE[b)]	Fipronil	References
[^{3}H]fluralaner	0.4	3	PI	[50]
[^{3}H]EBOB[b)]	~40	I	~40	[51]
[^{3}H]AVE[c)]	PI	20	NI	[51]

a) I, inhibition; PI, poor inhibition; and NI, no inhibition.
b) 4′-Ethynyl-4-*n*-propylbicycloorthocarboxylate. Fipronil showed no significant association to the isoxazoline- or avermectin-binding sites, setting it apart from the other two insecticide classes. AVE seems to have a modulating effect on the isoxazoline-binding site, as [3H] fluralaner is displaced; however, fluralaner has only poor activity on [3H]AVE. All three ectoparasiticides displace the more general [3H]EBOB ligand to various degrees, with AVE showing the least activity. The overview is compiled from studies reported by Casida and coworkers [50, 51].
c) Avermectin B1a.

Selectivity and Safety Profile

The ectoparasitic activity of the novel isoxazoline chemotype was evaluated in a series of studies against prominent flea and tick species, such as *C. felis*, *D. variablis*, and *R. microplus*, as well as other pests including an assortment of flies and mosquitoes [27, 33, 52, 53]. Identification, cloning, and functional expression of GABA- and glutamate-receptor genes of *R. microplus* in addition to the expression of rdl and susceptible *C. felis* and *D. melanogaster* GABA receptors have sparked the development of on-target assays, and a comprehensive exploration of the chloride channel inhibiting qualities relative to relevant representative parasiticides ensued [33]. Accordingly, a plethora of studies verify the superiority of the new chemotype, emphasizing the advanced potency over fipronil, dieldrin, and other GABACl antagonists in arthropods. In a direct comparison, the corresponding GluCls are effectively impeded in their activity, but show less sensitivity than GABACls, inferring a clear preference in the targeted receptors. Mammalian chloride channels, as in rat brain membrane, showed no significant response to isoxazoline ectoparasiticides in binding assays [27].

Critically, [3H]EBOB binding in recombinant human GABACl β3 homopentamers, which represent an appropriate model for the analysis of toxicity by displaying generally high sensitivity toward the standard GABAergic antiparasitic agents, remained equally unaffected [51]. In this regard, a high level of specificity for arthropods had been accomplished, justifying the renewed interest in CysLGCCs as antiparasitic drug targets. Fipronil, on the other hand, exhibits less pronounced selectivity against the mammalian GABACls, with IC50 values of 800 and 3.4 nM in rat brain membrane and the human β3 homopentamer, respectively [51].

Isoxazoline Derivatives: Continuous Exploration of the Novel Chemotype

The introduction of isoxazoline derivatives has shown impressive results concerning the selective inhibition of insect and acarid GABACls and GluCls. Moreover, a novel binding site is key to the innovative activity profile, which bypasses the critical cross-resistance observed in other noncompetitive antagonists [51]. With regard to the disclosed isoxazoline selectivity and activity qualities, research is continuously striving to explore the full potential. As such, DuPont researchers reported on isoxazoline derivatives for agricultural applications with small *N* -heterocycles as substitutes for the carboxamide side chain [28]. Structurally reminiscent of the phenylpyrazole fipronil, a range of thus generated compounds exhibit parasiticidal activity against a variety of pests. In addition, one representative was found to efficiently block GABA-induced currents in both wild-type and dlr mutant receptors of *D. melanogaster* [28].

Two other isoxazoline-related compounds have been registered by leading pharmaceutical companies as antiparasitic agents, with the INNs lotilaner [35] by Novartis AH, now Elanco, and sarolaner [36] by Zoetis (Table 12.2, Chapter 15). Both introduce heterocycles as modifications to the phenyl isoxazoline core and adjust the substitution pattern on the terminal phenyl moiety. On September 11, 2015, Zoetis Belgium SA received a positive opinion from the European Medicines Agency and its Committee for Medicinal Products for Veterinary Use, recommending the granting of a marketing authorization for the veterinary medicinal product Simparica chewable tablets – active ingredient sarolaner – intended for the treatment of tick, flea, and mange mite infestations in dogs [54]. In contrast to NexGard and Bravecto, which represent racemates of afoxolaner and fluralaner, respectively, Simparica represents only the biologically active S enantiomer of sarolaner [55, 56]. Searching the patent literature revealed numerous additional isoxazoline-related data and filings coming out of crop science and animal health [57, 58], including isothiazoline compounds (O replaced by S in the name giving ring) [59]. The interest in and research on isoxazolines as effective ectoparasiticides are therefore still unbroken and the high potency in GABACl and GluCls coupled with the encouraging safety profile is an attractive portfolio to justify further investigations.

Conclusions

Acarid and insect ectoparasites such as ticks and fleas pose a serious problem in the agricultural and veterinary sector, as they elicit a severe derogatory effect on crop yield and animal health and welfare. The development of potent antiparasiticides is therefore not only essential but also urgent.

Ligand-gated chloride channels have been prominent targets in the battle against these parasitic species; however, the emerging resistance against various chemotypes, subsequent potency depletion, and slow advancement in this research area have raised concerns over the sustainability of those targets. The outlined recent examples of novel isoxazoline derivatives as potent insecticides mark an inflection point in this development, resurrecting GABA and GluCls as valid drug targets and reviving research interest. Notably, comprehensive exploration and evaluation have led to the development of noncompetitive antagonists with activities as low as subnanomolar against a range of insect and acarid pests while exhibiting low inhibition toward the mammalian GABACl analogs. Unprecedentedly, the isoxazoline blockers were shown to circumvent cross-resistance by addressing a distinct new binding pocket in the chloride channels, setting the stage for an innovative new approach to tackling ectoparasitic infestations. Until now, this progress has culminated in the approval of three orally administered veterinary products: NexGard, Bravecto, and Simparica against ectoparasites on dogs [26, 31, 36]. In addition, combination products such as NexGard Spectra, which combines the ectoparasiticide afoxolaner and the endoparasiticide milbemycin oxime, have been launched [54]. In view of these developments, further derivatives and combinations are certainly in the pipeline to fully explore the isoxazoline scaffold's potential. Notably, although this compound class originates from research in crop science, it was first marketed in the veterinary sector, and no product for the crop science market has been released yet. Similar to other compounds that were initially launched with a veterinary parasitology application, for example, benzimidazoles, praziquantel, ivermectin [60], isoxazolines seem to have the potential to enter the human-host parasitology applications as well, preventing extensive feeding or blood meals by insects, mites, lice, and ticks. If so, isoxazolines might be able to impede arthropod-borne diseases such as Lyme disease, tick-borne encephalitis, or even neglected tropical diseases such as human African trypanosomiasis (sleeping sickness), leishmaniasis, and malaria, for which respective claims have already been made in the patent literature [26, 31, 35, 36, 57, 58]. However, for the time being, the question remains open as to how fast systemic isoxazolines act on the different feeding parasites. It would require the ectoparasite to fall off the host before the disease-causing agent is transmitted, which could be, depending on the parasite and disease, a challenging goal to achieve.

Acknowledgments

T.W. is an Experienced Researcher of the European Commission's Marie Curie Initial Training Network ParaMet (www.paramet.eu). We thank the European Commission for their financial support.

References

1. Popp, J., Pető, K., and Nagy, J. (2013) *Agron. Sustain. Dev.*, **33**, 243–255.
2. Woods, D.J. and Knauer, C.S. (2010) *Int. J. Parasitol.*, **40**, 1177–1181.
3. Ware, G.W. and Whitacre, D.M. (2004) *The Pesticide Book*, 6th edn, Meister Media Worldwide, Willoughby, OH.

4. Shryock, T.R. (2004) *Nat. Rev. Microbiol.*, **2**, 425–430.
5. Casida, J.E. and Durkin, K.A. (2013) *Annu. Rev. Entomol.*, **58**, 99–117.
6. Bermudez, I., Hawkins, C.A., Taylor, A.M., and Beadle, D.J. (1991) *J. Recept. Res.*, **11**, 221–232.
7. Bloomquist, J.R. (2001) in *Biochemical Sites of Insecticide Action and Resistance* (ed. I. Ishaaya), Springer-Verlag, Berlin/Heidelberg, pp. 17–41.
8. Jorgenson, J.L. (2001) *Environ. Health Perspect.*, **109**, 113–139.
9. Cole, L.M., Nicholson, R.A., and Casida, J.E. (1993) *Pestic. Biochem. Physiol.*, **46**, 47–54.
10. Narahashi, T., Zhao, X., Ikeda, T., Salgado, V.L., and Yeh, J.Z. (2010) *Pestic. Biochem. Physiol.*, **97**, 149–152.
11. Campbell, W.C. (2012) *Curr. Pharm. Biotechnol.*, **13**, 853–865.
12. Chabala, J.C., Mrozik, H., Tolman, R.L., Eskola, P., Lusi, A., Peterson, L.H., Woods, M.F., Fisher, M.H., Campbell, W.C., Egerton, J.R., and Ostlind, D.A. (1980) *J. Med. Chem.*, **23**, 1134–1136.
13. Campbell, W.C., Fisher, M.H., Stapley, E.O., Albers-Schönberg, G., and Jacob, T.A. (1983) *Science*, **221**, 823–828.
14. Barragry, T.B. (1987) *Can. Vet. J.*, **28**, 512–517.
15. Lasota, J.A. and Dybas, R.A. (1991) *Annu. Rev. Entomol.*, **36**, 91–117.
16. Schaeffer, J.M. and Haines, H.W. (1989) *Biochem. Pharmacol.*, **38**, 2329–2338.
17. Lummis, S. (1990) *Comp. Biochem. Physiol. C*, **95**, 1–8.
18. Casida, J.E. (1993) *Arch. Insect Biochem. Physiol.*, **22**, 13–23.
19. Ratra, G.S., Kamita, S.G., and Casida, J.E. (2001) *Toxicol. Appl. Pharmacol.*, **172**, 233–240.
20. Sine, S.M. and Engel, A.G. (2006) *Nature*, **440**, 448–455.
21. Lynagh, T. and Lynch, J.W. (2012) *Front. Mol. Neurosci.*, **5**, 60.
22. Rohrer, S. and Arena, J. (1995) in *Molecular Action of Insecticides on Ion Channels, ACS Symposium Series*, vol. **591** (ed. J.M. Clark), American Chemical Society, Washington, DC, pp. 264–283.
23. Bloomquist, J. (1993) *Comp. Biochem. Physiol. C*, **106**, 301–314.
24. Mita, T., Kudo, Y., Mizukoshi, T., Hotta, H., Maeda, K., and Takii, S. (2004) Substituted benzanilide compound and pest control agent. Patent WO 2004018410 A1 (Nissan Chemical Ind., Ltd.).
25. Perspective:Casida, J.E. (2015) *Chem. Res. Toxicol.*, **28**, 560–566.
26. Mita, T., Kikuchi, T., Mizukoshi, T., Yaosaka, M., and Komoda, M. (2005) Isoxazoline-substituted benzamide compound and noxious organism control agent. Patent WO 2005085216 A1 (Nissan Chemical Ind., Ltd.).
27. Ozoe, Y., Asahi, M., Ozoe, F., Nakahira, K., and Mita, T. (2010) *Biochem. Biophys. Res. Commun.*, **391**, 744–749.
28. Lahm, G.P., Cordova, D., Barry, J.D., Pahutski, T.F., Smith, B.K., Long, J.K., Benner, E.A., Holyoke, C.W., Joraski, K., Xu, M., Schroeder, M.E., Wagerle, T., Mahaffey, M.J., Smith, R.M., and Tong, M.H. (2013) *Bioorg. Med. Chem. Lett.*, **23**, 3001–3006.
29. Shoop, W.L., Hartline, E.J., Gould, B.R., Waddell, M.E., McDowell, R.G., Kinney, J.B., Lahm, G.P., Long, J.K., Xu, M., Wagerle, T., Jones, G.S., Dietrich, R.F., Cordova, D., Schroeder, M.E., Rhoades, D.F., Benner, E.A., and Confalone, P.N. (2014) *Vet. Parasitol.*, **201**, 179–189.
30. US Food and Drug Administration (2013) NADA 141-406, www.fda.gov/downloads/AnimalVeterinary/Products/ApprovedAnimalDrugProducts/FOIADrugSummaries/UCM377867.pdf (accessed 7 August 2015).
31. Long, J.K., Selby, T.P., and Xu, M. (2009) Naphthalene isoxazoline compounds for control of invertebrate pests. Patent WO 2009045999 A1 (Du Pont).
32. US Food and Drug Administration (2014) NADA 141-426, www.fda.gov/downloads/AnimalVeterinary/Products/ApprovedAnimalDrugProducts/FOIADrugSummaries/UCM399075.pdf (accessed 7 August 2015).
33. Gassel, M., Wolf, C., Noack, S., Williams, H., and Ilg, T. (2014) *Insect Biochem. Mol. Biol.*, **45**, 111–124.
34. Heckeroth, A.R., Lutz, J., Mertens, C., Williams, H., Zoller, H., and Mita, T. (2009) Isoxazoline compositions and their use as antiparasitics.

Patent WO 2009024541 A2 (Intervet Int. BV).

35. Nanchen, S. (2014) Process for the enantiomeric enrichment of diaryloxazoline derivatives. Patent WO 2014090918 A1 (Novartis AG).
36. Billen, D., Chubb, N.A.L., Curtis, M., Greenwood, S.D.W., Menon, S., Stuk, T., and Vaillancourt, V.A. (2012) Spirocyclic isoxazoline derivatives as antiparasitic agents. Patent WO 2012120399 A1 (Pfizer Inc.).
37. Ffrench-Constant, R.H., Rocheleau, T.A., Steichen, J.C., and Chalmers, A.E. (1993) *Nature*, **363**, 449–451.
38. Ffrench-Constant, R.H., Zhang, H.-J., and Jackson, M. (1995) in *Molecular Action of Insecticides on Ion Channels*, ACS Symposium Series, vol. **591** (ed. J.M. Clark), American Chemical Society, Washington, DC, pp. 192–204.
39. Ffrench-Constant, R.H., Rocheleau, T., Anthony, N., Aronstein, K., and Stilwell, G. (2000) *Annu. Rev. Entomol.*, **45**, 449–466.
40. Bloomquist, J. (1994) *Arch. Insect Biochem. Physiol.*, **26**, 69–79.
41. Chemical Computing Group Inc. (2015) Molecular Operating Environment (MOE), 2013.08, Chemical Computing Group Inc., 1010 Sherbooke Street West, Suite #910, Montréal, QC, H3A 2R7 (Canada).
42. Hilf, R.J. and Dutzler, R. (2008) *Nature*, **452**, 375–379.
43. Nakao, T., Banba, S., Nomura, M., and Hirase, K. (2013) *Insect Biochem. Mol. Biol.*, **43**, 366–375.
44. Deng, Y., Palmer, C.J., and Casida, J.E. (1991) *Pestic. Biochem. Physiol.*, **41**, 60–65.
45. Deng, Y. and Casida, J.E. (1992) *Pestic. Biochem. Physiol.*, **43**, 116–122.
46. Cully, D.F., Paress, P.S., Liu, K.K., Schaeffer, J.M., and Arena, J.P. (1996) *J. Biol. Chem.*, **271**, 20187–20191.
47. Wolstenholme, A.J. and Rogers, A.T. (2005) *Parasitology*, **131**, S85–S95.
48. Wolstenholme, A.J. (2012) *J. Biol. Chem.*, **287**, 40232–40238.
49. Hibbs, R.E. and Gouaux, E. (2011) *Nature*, **474**, 54–60.
50. García-Reynaga, P., Zhao, C., Sarpong, R., and Casida, J.E. (2013) *Chem. Res. Toxicol.*, **26**, 514–516.
51. Zhao, C. and Casida, J.E. (2014) *J. Agric. Food Chem.*, **62**, 1019–1024.
52. Beugnet, F., de Vos, C., Liebenberg, J., Halos, L., and Fourie, J. (2014) *Parasite*, **21**, 42–45.
53. Hunter, J.S. III, Dumont, P., Chester, T.S., Young, D.R., Fourie, J.J., and Larsen, D.L. (2014) *Vet. Parasitol.*, **201**, 207–211.
54. European Medicines Agency (2015) Simparica, Sarolaner, www.ema.europa.eu (accessed 18 October 2015).
55. McTier, T.L., Chubb, N., Curtis, M.P., Hedges, L., Inskeep, G.A., Knauer, C.S., Menon, S., Mills, B., Pullins, A., Zinser, E., Woods, D.J., and Meeus, D.P. (2015) Discovery of sarolaner. 25th International Conference of the World Association for the Advancement of Veterinary Parasitology, Liverpool, Ref. O090/0101.
56. Hedges, L., Collard, W., Merritt, D., Fielder, A., Grover, S., Boucher, J., and Mahabir, S. (2015) The intravenous and oral pharmacokineticsof sarolaner. 25th International Conference of the World Association for the Advancement of Veterinary Parasitology (WAAVP), Liverpool, UK. Ref. P1A216/0519.
57. An, Z., Chen, L., Chen, S., Defauw, J.M., Holmstrom, S.D., Hu, P., Tang, C., White, W.H., Wu, W., and Zhang, Y. (2012) Parasiticidal compounds, methods, and formulations. Patent WO 2012155676 A1 (Eli Lilly & Co.).
58. Akama, T., Balko, T.W., Defauw, J.M., Plattner, J., White, W.H., Winkle, J.R., Zhang, Y.-K., and Zhou, Y. (2013) Isoxazoline derivatives used in the control of ectoparasites. Patent WO 2013078071 A1 (Eli Lilly & Co., Anacor Pharmaceuticals Inc.).
59. Speak, J.D., Perales, J.B., and Rolland, S.C. (2016) Antiparasitic compounds, Patent WO 2016115315 A1 (Avista Pharma Solutions).
60. Geary, T.G., Woods, D.J., Williams, T., and Nwaka, S. (2009) in *Drug Discovery in Infectious Diseases: Antiparasitic and Antibacterial Drug Discovery: From Molecular Targets to Drug Candidates* (ed. P.M. Selzer), Weinheim, Wiley-VCH Verlag GmbH.

13
The Discovery of Afoxolaner: A New Ectoparasiticide for Dogs

Ming Xu, Jeffrey K. Long, George P. Lahm, Wesley L. Shoop, Daniel Cordova, Ty Wagerle, Ben K. Smith, Thomas F. Pahutski, Rafael Shapiro, Michael Mahaffey, Eric J. Hartline, Brandon R. Gould, Molly E. Waddell, Richard G. McDowell, John B. Kinney, Gail S. Jones, Robert F. Dietrich, Mark E. Schroeder, Daniel F. Rhoades, Eric A. Benner, and Pat N. Confalone*

Abstract

The discovery and optimization of a new class of naphthalene isoxazolines led to the identification of afoxolaner as a new ectoparasiticide with excellent effectiveness against fleas and ticks by both topical and oral treatments. The compound is shown to work by inhibition of the GABA-gated chloride channel with no observed cyclodiene cross-resistance.

Introduction

Afoxolaner (**1**) is a new naphthalene isoxazoline with excellent effectiveness against fleas and ticks by both topical and oral treatment (Figure 13.1) [1, 2]. Studies on the mode of action demonstrate that it is a potent inhibitor of the GABA-gated chloride channel but importantly lacks cross resistance with cyclodienes.

Flea control in companion animals has largely been achieved by topical application of fipronil and imidacloprid since the mid 1990s, and spinosad has been demonstrated to have utility as an oral drug for fleas. However, we judged that there remained space in this market for a broadly active, long-lasting, safe oral treatment, especially with a new mode or new site of action, for control of both fleas and ticks. Furthermore, previous flea and tick control agents have often derived from insecticide research for agricultural uses, in which the active ingredient was initially developed for crop protection applications before development of the veterinary utility. When DuPont initiated a program to evaluate compounds from our internal collection for flea and tick control, isoxazoline

*Corresponding author.

Ectoparasites: Drug Discovery Against Moving Targets, First Edition.
Edited by Charles Q. Meng and Ann E. Sluder

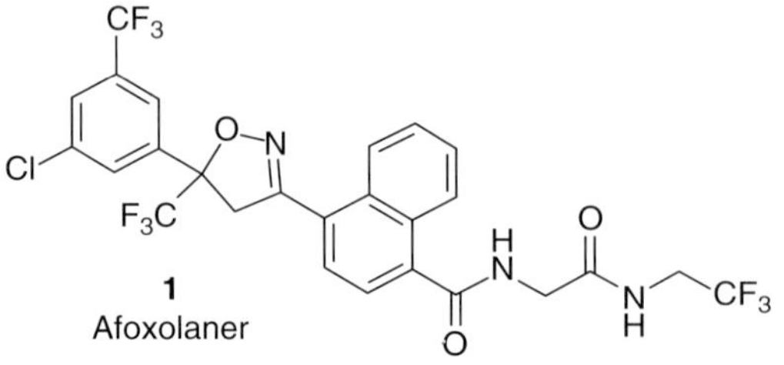

Figure 13.1 Isoxazoline insecticide afoxolaner.

insecticides were an early discovery research program. Optimization of properties for veterinary utility at this stage afforded a unique opportunity to alter the attributes of this chemo-type, and discover a new compound optimized as an orally available antiparasitic drug, effective on both fleas and ticks, with a long duration of action and safety to host animals, rather than accepting compromises that might be present with an existing agricultural agent.

Isoxazoline insecticides, such as Nissan compound **2** (Scheme 13.1), were discovered by Mita *et al.* [3]. These compounds are characterized by broad-spectrum insecticidal activity spanning both agricultural and veterinary utility, particularly useful as ectoparasiticides. While the agricultural utility has not yet been realized, this class has been shown to be useful for the treatment of fleas and ticks on dogs.

The unique chemical features of **2** were attractive. The trifluoromethyl substituent and 3,5-disubstituted aryl group at the 4-position of the isoxazoline, as well as the carboxamide group, appeared to be crucial for activity. In the course of our work in this area, we identified several leads shown in Scheme 13.1, including the 3-cyano-4-triazole of formula **3a**, the pyrimidine of formula **3b,** and the naphthalene of formula **4**, all of which were found to be effective in a cat flea artificial membrane feeding assay (MFA). Further optimization pointed to some of our

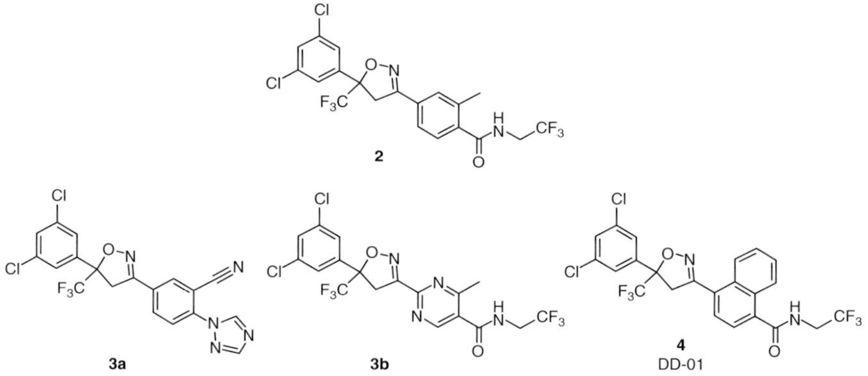

Scheme 13.1 Lead compounds in the discovery of afoxolaner.

best analogs as belonging to the naphthalene class. We herein describe the optimization of the naphthalene chemistry as well as mechanistic studies leading to the discovery of afoxolaner.

Development of Chemical–Biological Structure–Activity Relationships

The synthesis of lead compound DD-01 (**4**) was accomplished as outlined in Scheme 13.2. Mono-lithiation of 1,4-dibromonaphthalene with 1.05 equivalents of *n*-butyllithium followed by addition of dimethylformamide (DMF) afforded aldehyde **6** in 70% yield. Treatment of **6** with hydroxylamine afforded oxime **7**. The chlorooxime of **7** was generated *in situ* by treatment with *N*-chlorosuccinimide, and subsequently reacted with trifluoromethyl styrene **8** in a base-moderated 3 + 2 cycloaddition to afford isoxazoline **9** in an overall yield of 64% for the three steps. Palladium-catalyzed carbonylation and trapping with trifluoroethylamine produced isoxazoline carboxamide **4** (DD-01) in 56% yield. Carboxamides DD-02 through DD-17 of Table 13.1 were prepared in an analogous manner.

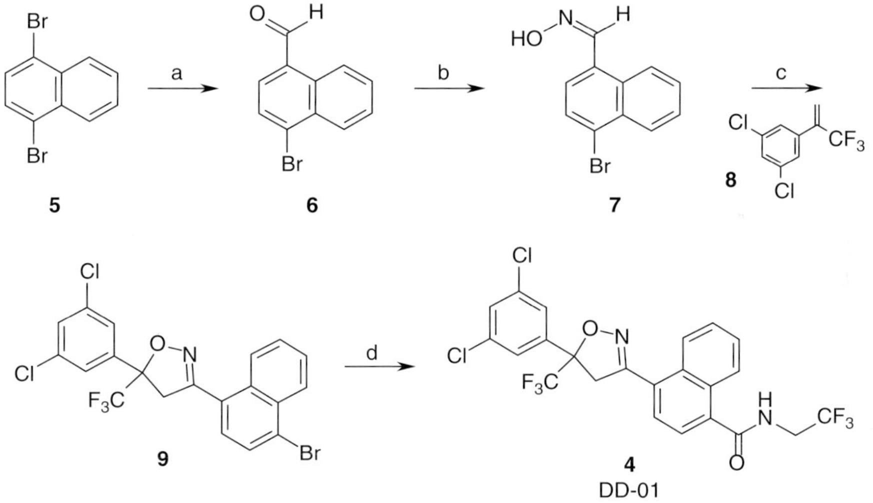

Scheme 13.2 Synthesis of **4**. (a) nBuLi, DMF, THF; (b) NH_2OH, EtOH; (c) NCS, Et_3N, DMF; (d) CO, $PdCl_2dppf$, $NH_2CH_2CF_3$, toluene.

Initial evaluation of naphthalene isoxazolines was done by two primary methods. Inherent potency of compounds was determined in an artificial MFA in which the test substance was diluted in citrated bovine blood and made available for cat fleas to feed upon through a thin membrane. Testing at concentrations of 30, 15, and 3 parts per million (ppm) was used initially, and then compounds showing activity at the lowest rate were titrated to their lowest effective dose in a second round of testing. The MFA data are shown in Table 13.1, with values

Table 13.1 Membrane feeding and oral gavage results for monoamide naphthalene isoxazolines on fleas.

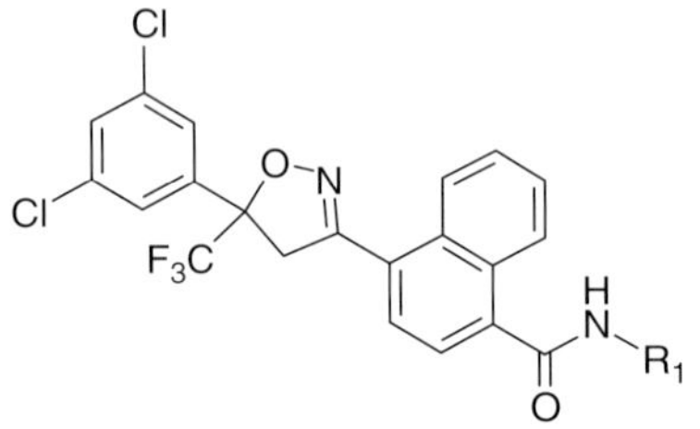

Compound	R_1	MFA[a)]	OG[b)]
DD-01	CH_2CF_3	0.31	86% 2 h
DD-02	Et	0.62	67% 6 h
DD-03	iPr	0.31	17% 2 h
DD-04	s-Bu	0.08	42% 2 h
DD-05	Allyl	0.16	91% 6 h
DD-06	Propargyl	0.16	27% 2 h
DD-07	CH_2Ph	0.31	Not tested
DD-08	CH_2CH_2OMe	0.31	92% 24 h
DD-09	CH_2CN	0.15	15% 2 h
DD-10	$CH_2CH_2SCH_3$	0.08	58% 24 h
DD-11	$CH_2CH_2S(O)CH_3$	0.04	62% 48 h
DD-12	$CH_2CH_2S(O)_2CH_3$	0.04	67% 48 h
DD-13	$CH_2CH_2CH_2SCH_3$	0.08	33% 24 h
DD-14	$CH(Me)CH_2SCH_3$	0.02	42% 48 h
DD-15 (rac)[c)]	$CH(Me)CH_2SCH_3$ (*R*)	No data	92% 24 h
DD-16 (*R*)[c)]	$CH(Me)CH_2SCH_3$ (*R*)	No data	Inactive
DD-17 (*S*)[c)]	$CH(Me)CH_2SCH_3$ (*R*)	No data	88% 72 h

a) Membrane feeding assay (MFA) values indicate the concentration in parts per million that provided >95% control of fleas 72 h after exposure.
b) Oral gavage (OG) assay values indicate the percentage of control at the feeding time point shown.
c) Stereochemistry designated with compound number is that of the isoxazoline 5-position (CF_3 attachment).

indicating the concentration in ppm that provided >95% control of fleas, assessed 72 h after exposure.

Compounds advanced to secondary testing in the MFA were also evaluated in a live mouse assay, in which compounds were formulated and dosed at 10 mg/kg (mpk) by oral gavage (OG) to the test animals, and the mice were then exposed to cat fleas (*Ctenocephalides felis*). Fleas ingested a blood meal from treated mice at a series of time points (2 h, 6 h, 24 h, 48 h, etc.), and the exposed fleas were held and evaluated after 48 h to observe mortality and sublethal effects of each test compound. This was done in consideration that fleas intoxicated but not yet killed by the drug could be more easily groomed off by a treated animal, thus achieving effective real-life control even at sublethal doses. Results of this testing are shown in Table 13.1, as OG for administration by OG, with data expressed as total percent mortality (%M) at a time point. Mortality was generally >95% for earlier feeding time points than those in the table.

Blood samples were also collected at periodic intervals via the mouse tail vein for chemical analysis of serum concentration of test substances to determine a time course profile for all substances tested in the oral assay. Thus, data on inherent potency, duration of action, and pharmacokinetics after oral administration were obtained from the earliest stages of the program. For compounds of high interest, additional testing in mice was performed at doses up to 50 mpk to evaluate the effect on duration of action at higher doses, as well as to provide an initial assessment of mammalian safety. No adverse effects were observed on mice for any of the compounds tested.

One of the first compounds evaluated from this area of chemistry, compound DD-01, showed inherent activity at 0.31 ppm in the MFA, with modest efficacy via OG. However, chemical analysis of blood concentration showed a low, but long, duration of exposure. This key feature attracted further testing of analogous naphthalene isoxazolines, as the class already demonstrated a key attribute for the desired application to animal health care. As might be expected, the data in Table 13.1 show that duration of action in the mouse assay did not track directly with inherent potency, as oral bioavailability, dependent upon multiple ADME (absorption, distribution, metabolism, and excretion) properties, is generally more complex than simple differences of inherent potency.

Various amide substitution patterns were interesting for the combination of their inherent potency and improved oral activity. As seen in Table 13.1, the presence of non-carbon elements such as oxygen (DD-08) and especially sulfur in various oxidation states (DD-10 through DD-14, DD-15, DD-17) correlated with enhanced activity in both the MFA and the *in vivo* mouse assay.

The presence of an additional amide group indicated by the compounds prepared in Scheme 13.3, such as by inclusion of amino acid amides as a heteroatom-containing group, appeared particularly promising, as shown in Table 13.2. This feature made possible the exploration of different amino acid side chains as well as a diverse set of substitution patterns on the terminal amide nitrogen. A number of combinations of alkyl and haloalkyl groups at these positions afforded highly active compounds in the flea assays, with compounds containing the trifluoroethylamino group (DD-23, DD-25) showing especially high potency.

Diamide derivatives of glycine and alanine (Table 13.2) along with a variety of aryl substituents R_4 (Table 13.3), were generally prepared as shown in Scheme 13.3. Cycloaddition of substituted trifluoromethyl styrenes of formula **10** with oxime **7** provided naphthalene isoxazolines **11**. Palladium-catalyzed

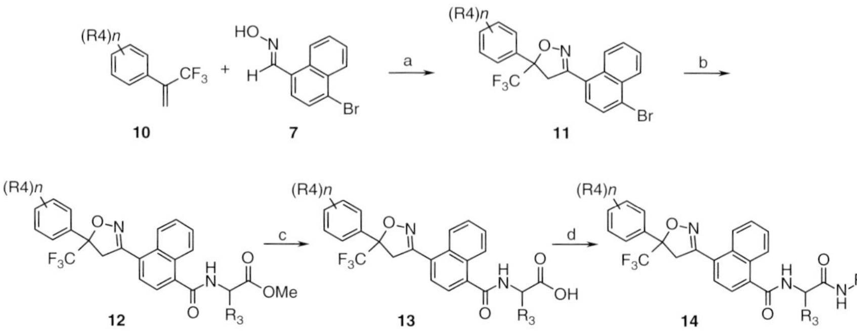

Scheme 13.3 Synthesis of **14**. (a) NCS, Et_3N, DMF; (b) $PdCl_2dppf$, CO, $NH_2CHR_3CO_2Me$, Et_3N, toluene; (c) LiOH, THF/H_2O (1 : 1) then HCl; (d) $(COCl)_2$, DMF, CH_2Cl_2, Et_3N, NH_2R_2.

Table 13.2 Membrane feeding and oral gavage results for diamide naphthalene isoxazolines on fleas.

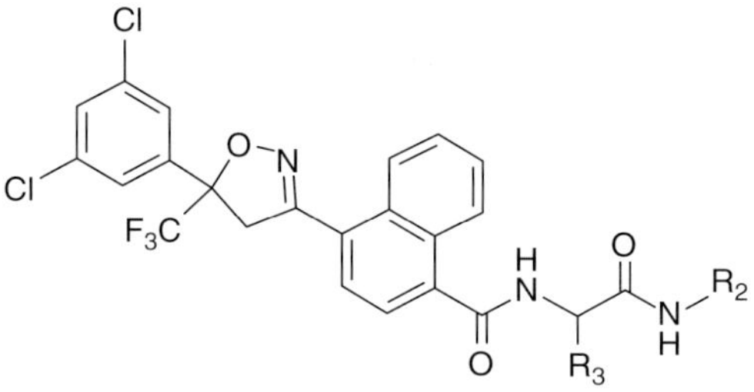

Compound	R_3	R_2	MFA[a)]	OG[b)]
DD-18	H	Et	0.08	100% 6 h
DD-19	H	iPr	0.04	81% 24 h
DD-20	H	CH_2cPr	0.08	54% 6 h
DD-21	H	iBu	0.08	29% 2 h
DD-22	H	tBu	<3	69% 2 h
DD-23	H	CH_2CF_3	0.04	100% 24 h
DD-24	Me (*S*)	CH_2CF_3	0.62	Inactive
DD-25	Me (*R*)	CH_2CF_3	0.04	80% 48 h
DD-26 (*S*)	H	CH_2CF_3	No data	79% 48 h
DD-27 (*R*)	H	CH_2CF_3	No data	8% 48 h

a) Membrane feeding assay (MFA) values indicate the concentration in parts per million that provided >95% control of fleas 72 h after exposure.

b) Oral gavage (OG) assay values indicate the percentage of control at the time point shown.

Table 13.3 Membrane feeding and oral gavage results for diamide naphthalene isoxazolines on fleas.

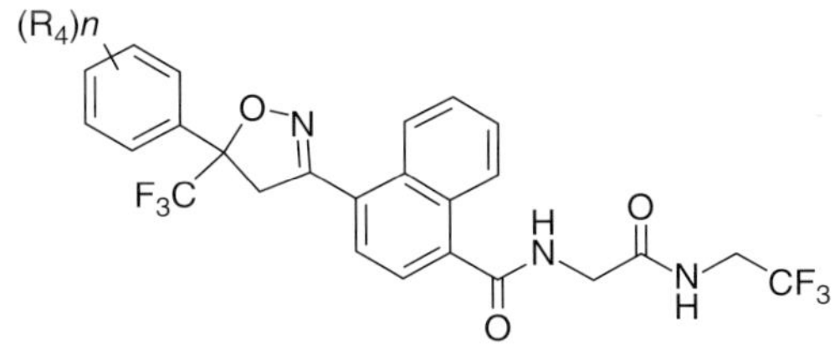

Compound	$(R_4)n$	MFA[a)]	OG[b)]
DD-28	F	3	Inactive
DD-29	Br	0.04	67% 2 h
DD-30	CF_3	3	74% 24 h
DD-31	OCF_3	0.3	86% 6 h
DD-32	3,5-di-F	0.62	Inactive
DD-33	3,5-di-Br	0.04	100% 24 h
DD-34	3,5-di-CF_3	0.04	60% 96 h
DD-35	3-F-5-CF_3	0.3	75% 24 h
DD-36	3-Cl-5-CF_3	0.02	83% 72 h
DD-37	3-Br-5-CF_3	0.04	77% 48 h
DD-38	3-Cl-5-OCF_3	0.16	79% 72 h

a) Membrane feeding assay (MFA) values indicate the concentration in parts per million that provided >95% control of fleas 72 h after exposure.
b) Oral gavage (OG) assay values indicate the percentage of control at the time point shown.

carbonylation and trapping with glycine and alanine methyl esters afforded amides **12**, which were readily converted to diamides **14** by hydrolysis to acid **13** followed by typical amide coupling reactions.

To determine the absolute configuration and activity associated with chirality at the isoxazoline asymmetric center, the diastereomers DD-16 and DD-17 were prepared as shown in Scheme 13.4. Palladium-catalyzed carbonylation of **9,** followed by treatment with methanol, afforded the methyl ester. Subsequent hydrolysis with lithium hydroxide in THF/H_2O (1 : 1) and neutralization afforded carboxylic acid **15** in 85% yield. Acid **15** was treated with oxalyl chloride and then reacted with (*R*)-1-(methylthio)propan-2-amine to afford a mixture of diastereomers **16,** which were separated by chiral column chromatography to afford the (*R,R*) and (*S,R*) diastereomers DD-16 and DD-17 respectively, as determined by X-ray diffraction of DD-16. Activity was determined to largely reside in the (*S,R*) enantiomer DD-17. Of particular note was the determination that the active configuration at the isoxazoline asymmetric center was the *S* enantiomer.

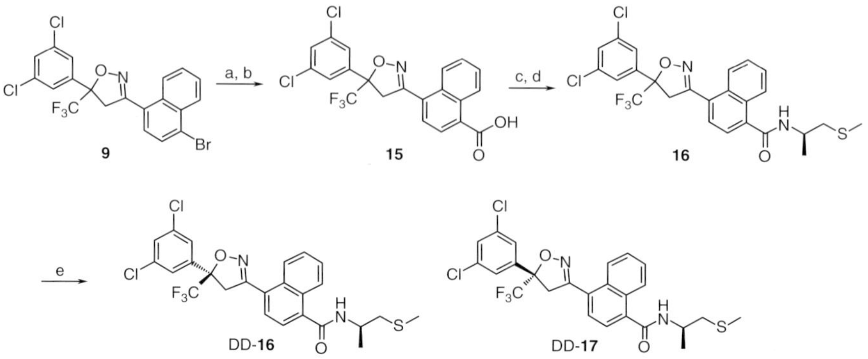

Scheme 13.4 Synthesis of DD-16 and DD-17. (a) CO, $PdCl_2$dppf, MeOH, Et_3N, toluene; (b) LiOH, THF/H_2O then HCl; (c) oxalyl chloride; (d) (*R*)-1-(methylthio)propan-2-amine, Et_3N, CH_2Cl_2; (e) HPLC separation of diastereomers by chiral OJ-RH column with MeOH/CH_3CN as eluent

Early testing of a few compounds in the area demonstrated the potential for tick control with this class of chemistry. A key attribute desired for the veterinary market was to have activity on ticks at an effective dose similar to that used for fleas so that any potential candidate for development could satisfy the market need for control of both through oral administration at an acceptable dosage while retaining a high safety margin for the treated animal. Thus, evaluation of tick activity in a rat model was pursued for a group of compounds showing high efficacy in the mouse/flea assay for a final round of optimization. This work focused on fine-tuning the properties of analogs through a combination of substituent changes on the terminal phenyl group adjoining the trifluoromethyl isoxazoline, alpha-amino acid side chains and amide substituent groups.

This final round of analogs allowed us to identify the best combination of substituents for optimal potency, duration of control, and similar dosage for both fleas and ticks, with safety upon oral administration, balanced against any potential issues of molecular complexity, cost, and other manufacturing considerations. To assess efficacy, rats were treated by OG, but this time at three doses of 0.6, 0.3, and 0.15 mpk, and the treated animals were exposed to ticks for 72 h. Evaluation of tick mortality provided the results summarized in Table 13.4. All compounds demonstrated excellent activity, with compound DD-36 showing the highest levels of control. Compound DD-36 (afoxolaner) was advanced for development based on the combination of this outstanding flea and tick control coupled with the desired combination of additional attributes.

Mode of Action

Previously we demonstrated that similar isoxazolines having a triazole and an ortho-cyano group on the aryl ring had potent insecticidal properties [4]. Pyridine isoxazoline **22** (CPD I) containing the cyano and triazole substituents

Table 13.4 Tick control by oral gavage.

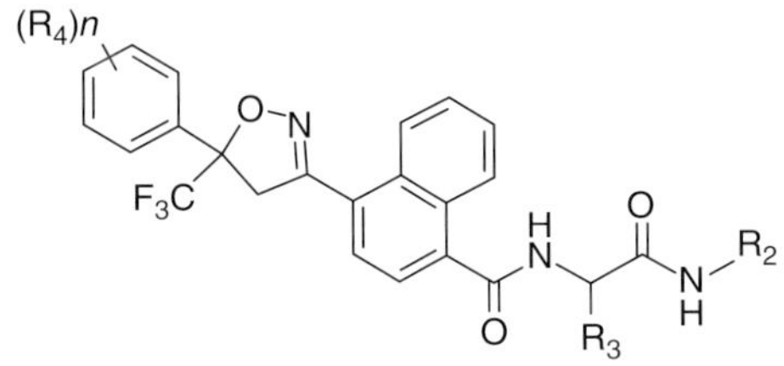

Compound	$(R_4)n$	R_3	R_2	Tick OG[a)]
DD-23	3,5-diCl	H	CH_2CF_3	100% @ 0.3
DD-34	3,5-di-CF_3	H	CH_2CF_3	56% @ 0.15
DD-36	3-Cl-5-CF_3	H	CH_2CF_3	100% @ 0.15
DD-37	3-Br-5-CF_3	H	CH_2CF_3	50% @ 0.15
DD-41	3-Cl-5-CF_3	Me (*R*)	Me	53% @ 0.3
DD-42	3-Cl-5-CF_3	Me (*R*)	Et	89% @ 0.3
DD-43	3-Cl-5-CF_3	Me (*R*)	iPr	100% @ 0.3
DD-44	3-Cl-5-CF_3	Me (*R*)	CH_2CF_3	84% @ 0.6
DD-45	3-Cl-5-CF_3	H	Me	53% @ 0.3
DD-46	3-Cl-5-CF_3	H	cPr	68% @ 0.3

a) Tick oral gavage (OG) assay indicates the percentage of control of ticks at a given dose (mpk) by oral gavage in rats 72 h post-treatment.

was thus useful for mode of action studies owing to its higher water solubility compared with naphthalene analogs. Compound **22** was prepared as outlined in Scheme 13.5. Pyridylboronic acid **17** was coupled with 1,1,1-trifluoro-2-bromopropene using palladium catalysis to afford trifluoropropenylpyridine **18** in 50% yield. A 3 + 2 cycloaddition with oxime **20** afforded isoxazoline **21** in 26% yield. Addition of triazole in the presence of potassium carbonate as base afforded cyano triazole **22** in 93% yield.

Physiological and biochemical studies were then conducted to elucidate the insecticidal mode of action. Poisoning symptoms in American cockroach, *Periplaneta americana*, included periodic wing fluttering, incoordination and leg tremors, indicative of action at a neuronal target site. Among insects, the primary excitatory and inhibitory neurotransmitter receptors are nicotinic acetylcholine receptors (nAChRs) and GABA-gated chloride channels (GABA-Cls), respectively. The isoxazoline insecticide **22** (CPD I) potently inhibited GABA-Cl currents of isolated cockroach neurons with an IC_{50} value of 10.8 nM (Figure 13.2) while lacking significant activity against cockroach nAChRs [2, 5]. A strong correlation between GABA-Cl inhibition and cockroach injection toxicity for a series

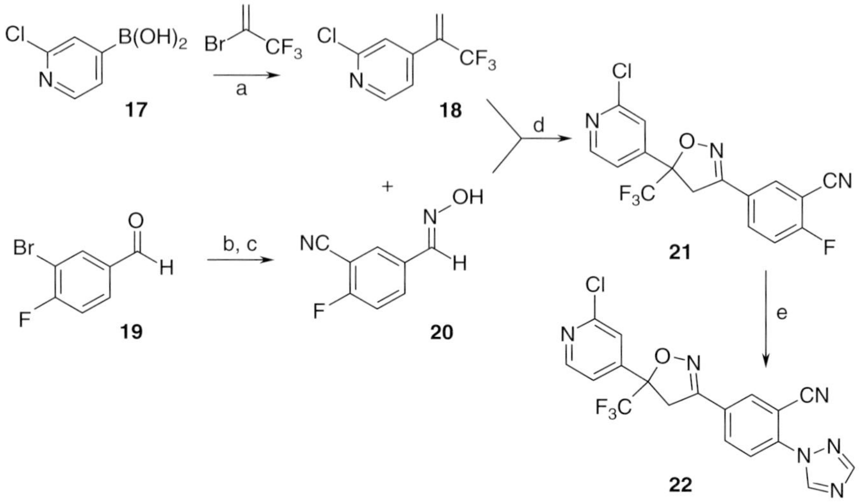

Scheme 13.5 Synthesis of **22**. (a) $Pd(PPh_3)_4$, 4 N KOH, THF/DME, 80 °C; (b) CuCN, NMP 150 °C 15 h; (c) NH_2OH, EtOH, rt, 3 h; (d) NaOCl, THF, 0 °C to rt, 30 min; (e) triazole, K_2CO3, CH_3CN, 80 °C, 18 h.

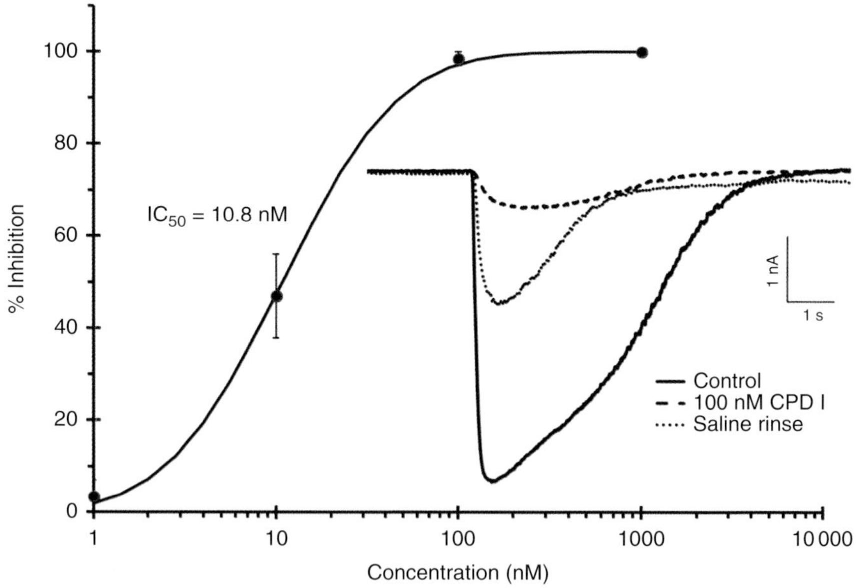

Figure 13.2 CPD I (**22**) inhibits GABA-gated currents in American cockroach, *P. americana,* neurons. Dissociated neurons were clamped at a holding potential of −60 mV and repeatedly stimulated with pulses of 100 μM GABA (inset, solid trace). Perfusion of CPD I (**22**) inhibited the GABA response (inset, dashed trace) in a dose-dependent manner with an IC_{50} = 10.8 nM. Following prolonged saline rinse, a partial recovery of the GABA response (inset, dotted trace) was observed. (Reproduced with permission from Shoop *et al.* [2]. Copyright 2014, Elsevier.)

of isoxazolines confirmed GABA-Cls as the target of this insecticide class. Independently, Ozoe *et al.*, reported the Nissan isoxazoline, A1433, to be a potent inhibitor of ligand-gated chloride channels [6].

Invertebrate GABA-Cls, encoded by the rdl (resistance-to-dieldrin) gene, are the target of cyclodienes and fipronil. It has been established that a single point mutation (A302S) in the fruit fly (*Drosophila*) *rdl* gene confers strong and moderate resistance to cyclodienes and fipronil, respectively [7, 8]. GABA-Cls encoded by *rdl* have also been found in American dog tick, *Dermacentor variabilis*, and in cat flea, *C. felis*, with certain flea strains bearing the A302S mutation [9, 10], although the impact of these mutations on the efficacy of fipronil-based flea and tick control products has not been determined. In a study of flea strains that were homozygous for the *rdl* mutation, a commercial fipronil product was fully effective; and in another study, the *rdl* mutation was present in most field collected and laboratory flea strains, but this had no discernible effect on responses to fipronil [11, 12]. Given that isoxazoline insecticides target GABA-Cls, comparative studies were conducted to determine if the A302S mutation confers resistance to afoxolaner. When tested against *Xenopus* oocytes expressing the wild-type *Drosophila* RDL channel, afoxolaner was found to have an IC_{50} value of 3.7 nM [2]. As shown in Figure 13.3, similar potency was observed with oocytes expressing the mutant (A302S) RDL.

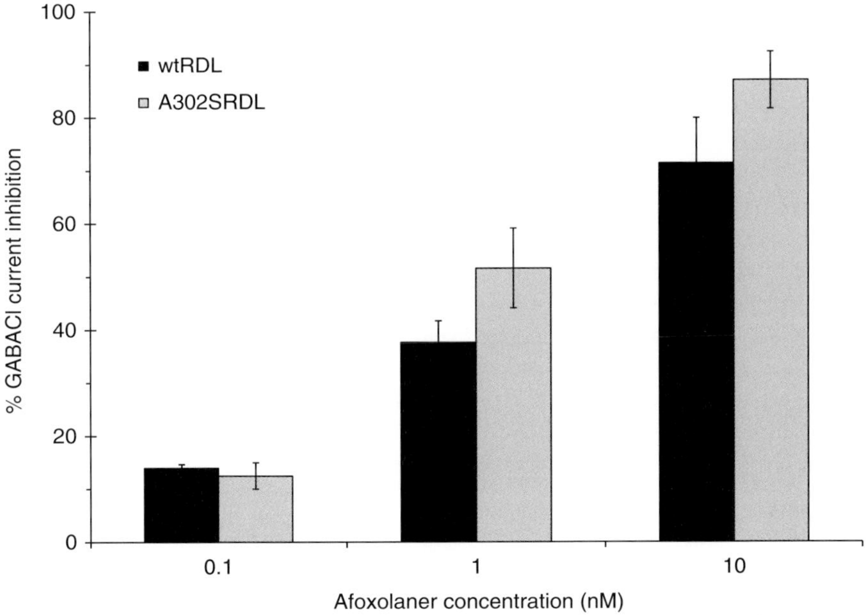

Figure 13.3 Inhibitory effect of afoxolaner, on GABA-gated Cl^- currents recorded from *Xenopus* oocytes expressing either ^{wt}RDL or ^{A302S}RDL(resistant) receptors. Oocytes were recorded using TEVC (two electrode voltage clamp method) with a holding potential of -60 mV. (Reproduced with permission from Shoop *et al.* [2]. Copyright 2014, Elsevier.)

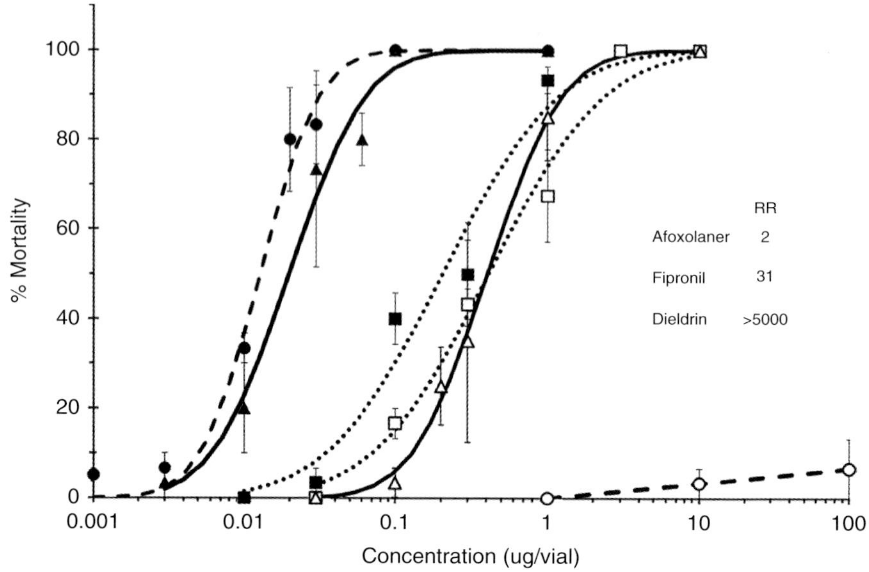

Figure 13.4 Contact toxicity of afoxolaner (square), fipronil (triangle), and dieldrin (circle) against wild-type (Canton-S, closed symbol) and cyclodiene-resistant (Rdl, open symbol) strains of *Drosophila*. Mortality measurements were taken 72 h after flies were transferred to treated glass vials. The resistance ratio (RR) was calculated as Rdl LD_{50}/Canton-S LD_{50} for each compound.

To further test for cyclodiene cross-resistance, afoxolaner toxicity between wild-type *Drosophila* and a mutant strain bearing the A302S mutation (Rdl) was investigated. As predicted, mutant *Drosophila* exhibited strong and moderate resistance to dieldrin and fipronil, respectively. In contrast, afoxolaner exhibited comparable toxicity between the two strains as shown in Figure 13.4. These results confirm those observed in target-based site studies, indicating that afoxolaner binds to the GABA-Cl in a manner distinct from cyclodienes and phenylpyrazoles. This is further supported by preliminary biochemical findings using a radiolabeled isoxazoline [5]. Since cyclodiene resistance among fleas and ticks may be associated with a homologous mutation to the *Drosophila* A302S, one could expect a similar lack of cross-resistance between afoxolaner and cyclodienes among these ectoparasites.

Summary

Afoxolaner is a new naphthalene isoxazoline with excellent effectiveness against fleas and ticks by both topical and oral treatment. This compound works by inhibition of the GABA-gated chloride channel, but there is no observed cross-resistance with cyclodienes. The outstanding efficacy coupled with a favorable pharmacokinetic and safety profile will help make afoxolaner an excellent choice for ectoparasiticide control in dogs.

References

1. Lahm, G.P., Long, J.K., and Xu, M. (2008) Naphthalene isoxazoline invertebrate pest control agents. US Patent 8,410,153, filed Jun. 28, 2008 and issued Apr. 2, 2013.
2. Shoop, W.L., Hartline, E.J., Gould, B.R., Waddell, M.E., McDowell, R.G., Kinney, J.B., Lahm, G.P., Long, J.K., Xu, M., Wagerle, T., Jones, G.S., Dietrich, R.F., Cordova, D., Schroeder, M.E., Rhoades, D.F., Benner, E.A., and Confalone, P.N. (2014) Discovery and mode of action of afoxolaner, a new isoxazoline parasiticide for dogs. *Vet. Parasitol.*, **201**, 179–189.
3. Mita, T., Kikuchi, T., Mizukoshi, T., Yaosaka, M., and Komoda, M. (2004) Isoxazoline-substituted benzamide compound and pesticide. US Patent 8,022,089, filed Jul. 27, 2009 and issued Sept. 20, 2011.
4. Lahm, G.P., Cordova, D., Barry, J.D., Pahutski, T.F., Smith, B.K., Long, J.K., Benner, E.A., Holyoke, C.W., Joraski, K., and Xu, M. (2013) 4-Azolylphenyl isoxazoline insecticides acting at the GABA gated chloride channel. *Bioorg. Med. Chem. Lett.*, **23**, 3001–3006.
5. Cordova, D., Rauh, J.J., Benner, E.A., Schroeder, M.E., Sopa, J.S., Lahm, G.P., Pahutsky, T.F., Long, J.K., Holyoke, C.W., Smith, R.M., Barry, J.D., and Dung M.H. (2010) Isoxazoline insecticides: a novel class of GABA-gated chloride channel blockers. Abstracts of Papers, 12th IUPAC Congress of Pesticide Chemistry, Melbourne, Australia.
6. Ozoe, Y., Asahi, M., Ozoe, F., Nakahira, K., and Mita, T. (2010) The antiparasitic isoxazoline A1443 is a potent blocker of insect ligand-gated chloride channels. *Biochem. Biophys. Res. Commun.*, **391**, 744–749.
7. Ffrench-Constant, R.H., Anthony, N., Aronstein, K., Rocheleau, T., and Stilwell, G. (2000) Cyclodiene insecticide resistance: from molecular to population genetics. *Annu. Rev. Entomol.*, **45**, 449–466.
8. Bloomquist, J.R. (1993) Toxicology, mode of action, and target site-mediated resistance to insecticides acting on chloride channels. *Comp. Biochem. Physiol.*, **106C**, 301–304.
9. Bass, C., Schroeder, I., Turberg, A., Field, L.M., and Williamson, M.S. (2004) Identification of the Rdl mutation in laboratory and field strains of the cat flea, *Ctenocephalides felis* (siphonaptera: Pulicidae). *Pest Manag. Sci.*, **60**, 1157–1162.
10. Zheng, Y., Priest, B., Cully, D.F., and Ludmerer, S.W. (2003) Rdl^{Dv}, a novel GABA-gated chloride channel gene from the American dog tick *Dermacentor variabilis*. *Insect Biochem. Mol. Biol.*, **33**, 595–599.
11. Rust, M.K., Vetter, R., Denholm, A., Blagburn, B., Williamson, M., Kopp, S., Coleman, G., Hostetler, J., Davis, W., Mencke, N., Rees, R., Foit, S., Boehm, C., and Tetzner, K. (2015) Susceptibility of adult cat fleas (Siphonaptera: Pulicidae) to insecticides and status of insecticide resistance mutations at the Rdl and knockdown resistance loci. *Parasitol. Res.*, **114**, S1–S12.
12. Brunet, S., Le Meter, C., Murray, M., Soll, M., and Audonnet, J.-C. (2009) *Rdl* gene polymorphism and sequence analysis and relation to *in vivo* fipronil susceptibility in strains of the cat flea. *J. Econ. Entomol.*, **102**, 366–372.

14
Development of Afoxolaner as a New Ectoparasiticide for Dogs

Laura Letendre, Diane Larsen, and Mark Soll*

Abstract

Afoxolaner was formulated as a nonmeat-based chewable dosing form designed for monthly oral administration and was developed and registered as NexGard®, the first product to effectively control both fleas and ticks on dogs for a month after oral treatment. The studies conducted for the development and registration of NexGard and NexGard Spectra are discussed with a focus on studies to evaluate safety and efficacy in dogs, the target species.

The afoxolaner development program included extensive testing of efficacy and safety, as well as drug disposition studies to define the ADME (absorption, distribution, metabolism, and excretion) profile in laboratory animals and dogs. The relationship between drug exposure (pharmacokinetics) and the safety and efficacy (pharmacodynamics) of the product (PK/PD) was elucidated early in the program and was used to help make key decisions. Studies to define the optimal formulation in terms of bioavailability, palatability, safety, efficacy, and manufacturing scale-up were performed. Efficacy and safety studies in dogs at various doses were conducted to understand the full range of biological activity and benefits of the product for dogs and to select the most appropriate dose level. Toxicity studies in laboratory animals were performed to understand any risks to non-target species and to the environment. Simultaneously, NexGard Spectra®, an endectocide, containing afoxolaner given in a fixed combination with milbemycin oxime in a chewable formulation was developed to treat and control flea and tick infestations and intestinal nematode infections including hookworms, roundworms, and whipworms and for prevention of heartworm disease.

The NexGard development program was thorough and efficient resulting in proven safety and efficacy profiles and a highly acceptable oral dosage form that has been well accepted by veterinarians and pet owners in more than 50 countries around the world, and it is providing protection for millions of dogs against ectoparasites that threaten their health and well-being.

*Corresponding author.

Ectoparasites: Drug Discovery Against Moving Targets, First Edition.
Edited by Charles Q. Meng and Ann E. Sluder

Introduction

Until recently, control of ectoparasites on dogs relied largely on topical application of insecticidal/acaricidal compounds. Although some agents have shown systemic activity, systemic efficacy has been limited to immature stages of fleas, and none of the topical agents provided useful activity against both fleas and ticks. The isoxazolines represent an interesting new class of compounds with useful activity against insects and acarines. Based on preliminary efficacy and safety information, and a novel mechanism of action [1], a naphthalene isoxazoline, afoxolaner, was selected as the lead candidate for further development as a product that could provide extended efficacy against ectoparasites of dogs when administered orally in a palatable dosage form. Afoxolaner was formulated as a non-meat-based chewable dosing form designed for monthly oral administration and was developed and registered as NexGard®, the first product to effectively control both fleas and ticks on dogs for a month after oral treatment. The afoxolaner development program included extensive testing of efficacy and safety, as well as drug disposition studies to define the ADME (absorption, distribution, metabolism, and excretion) profile in laboratory animals and dogs. The relationship between drug exposure (pharmacokinetics (PK)) and the safety and efficacy (pharmacodynamics (PD)) of the product (PK/PD) was elucidated early in the program and was used to help make key decisions. Studies to define the optimal formulation in terms of bioavailability, palatability, safety, efficacy, and manufacturing scale-up were performed. Efficacy and safety studies in dogs at various doses were conducted to understand the full range of biological activity and benefits of the product for dogs and to select the most appropriate dose level. Toxicity studies in laboratory animals were performed to understand any risks to nontarget species and to the environment. Simultaneously, NexGard Spectra®, an endectocide, containing afoxolaner given in a fixed combination with milbemycin oxime in a chewable formulation, was developed to treat and control flea and tick infestations and intestinal nematode infections including hookworms, roundworms, and whipworms and for prevention of heartworm disease [2]. The development program for NexGard Spectra was similar to that of NexGard and additionally included studies to establish non-interaction of the active ingredients and efficacy against intestinal nematodes and heartworms.

The previous chapter described the *in vitro* and efficacy experiments performed to optimize properties of the active pharmaceutical ingredient (API) candidates for veterinary utility, leading to the selection of afoxolaner as the lead candidate with a well-defined mode of action that demonstrated encouraging results for safety and efficacy [1]. The challenge for product development was then to confirm that these results translated into a safe and efficacious product for the target species (dogs and puppies), that is also safe for the pet caregiver, family, veterinarian, and the environment. This chapter outlines the studies conducted for the development and registration of NexGard and

NexGard Spectra with a focus on studies to evaluate safety and efficacy in dogs, the target species.

Study Compliance with Regulatory Requirements

Studies performed in the early stages of development with the aim of establishing proof-of-concept efficacy and safety had fewer animals per group and used prototype formulations. All pivotal PK, safety, and efficacy studies in the target animal were conducted with the final formulation produced under current Good Manufacturing Process (cGMP) conditions and following appropriate regulatory guidelines.

Because the product was to be made available in a fixed dosage form (i.e., a defined size of the product for a specified weight range), it was important to establish the dosing bands and dose of active to be delivered for each of them. Establishing these bands early in development is important, as regulatory authorities generally require testing for efficacy at the minimum dose to be provided within a weight band, and confirmation of safety at multiples of the highest dose to be provided. Table 14.1 gives the four dose bands and afoxolaner amounts in each NexGard beef-flavored chew. Typically, 6–8 animals per group were used for pivotal studies. Pivotal PK, target animal safety, and nonclinical toxicity studies were conducted under Good Laboratory Practice (GLP) guidelines. Pivotal efficacy studies were performed under Good Clinical Practice (GCP) guidelines. Other appropriate International Cooperation on Harmonisation of Technical Requirements for Registration of Veterinary Medicinal Products (VICH), U.S. Food and Drug Administration (FDA), and European Union (EU) guidelines were followed as applicable.

Methods to quantitate drug plasma concentrations were validated early in the development program in rat, cat, and dog plasma. The validated methods were later adapted for use with plasma samples from mice and cattle. The methods used 96-well solid-phase extraction of afoxolaner and milbemycin oxime from 0.25 ml of canine plasma fortified with a proprietary internal standard. Plasma

Table 14.1 Weight bands and active ingredient amounts in the four sizes of NexGard chewable tablets for dogs.

Dog size (kg)	Afoxolaner (mg)
2–4	11.3
>4–10	28.3
>10–25	68.0
>25–50	136.0

samples were analyzed for afoxolaner and, when appropriate, the milbemycin oxime A3 and A4 forms. The extracted analytes were separated by reverse-phase high-performance liquid chromatography (HPLC) and detected using AB Sciex and Waters quadrupole mass spectrometers. The methods have been described in detail [3, 4]. Repeatability (precision), accuracy, assay specificity, stability in plasma and all solutions, and method robustness were verified and passed the FDA guideline criteria for validated bioanalytical methods [5].

All animal procedures used to support studies were reviewed and approved by relevant Institutional Animal Care and Use Committees (IACUC), and animals were handled with due regard for their welfare and in compliance with all local and national regulatory requirements.

Background

Fleas

The cat flea, *Ctenocephalides felis,* is the most common flea species infesting both dogs and cats worldwide [6, 7]. In addition to causing annoyance and discomfort to pets and their owners, fleas are associated with several diseases including flea bite allergy dermatitis in dogs and cats. They are the primary intermediate host of the tapeworm *Dipylidium caninum* [8, 9], can transmit murine typhus (*Rickettsia typhi*) and flea-borne spotted fever (*Rickettsia felis*) [10], and have been implicated in the transmission of some *Bartonella* species such as *Bartonella henselae,* the agent of cat-scratch disease [11].

The optimal flea control program requires the rapid elimination of established flea infestations while providing continued protection against infestation with new fleas emerging from the environment. Therefore, the desirable attributes of an effective flea control product include the ability to eliminate fleas quickly and continuously, and to prevent them from producing viable eggs that infest the environment.

Although the use of newer insecticides has improved flea control, treatment and control of flea infestations remain a major concern for pet owners and veterinarians. New products containing compounds that are fast-acting, long-lasting, and easy-to-administer were needed to complement the existing products on the market.

Ticks

Tick control is an important concern for public health officials, pet owners, and veterinarians [12, 13]. Tick infestations can be a nuisance, and heavy tick infestation can lead to anemia, particularly in young or small dogs. Importantly, certain ticks are vectors of infectious agents such as *Babesia* spp., *Ehrilichia* spp., *Anaplasma phagocytophilum, Borrelia burgdorferi,* and *Rickettsia* spp.

Many species of ticks are capable of infesting dogs, with the spectrum of infestations varying across geographies. Those of primary interest to veterinarians and dog owners include ticks of the genera *Ixodes, Dermacentor, Rhipicephalus, Amblyomma,* and *Haemaphysalis,* and thus these were the genera targeted in the development program of NexGard.

Establishing Proof of Concept in Dogs

Proof-of-concept studies to establish early efficacy and preliminary safety of afoxolaner are reported by Shoop *et al.* [14] and involved treating dogs with the compound in an oral solution. Efficacy against fleas and ticks of doses ranging from 1.5 to 3.5 mg/kg, the impact of prandial state on systemic afoxolaner exposure, and the efficacy and exposure upon repeat monthly dosing were evaluated. Safety was assessed in each study with dogs being periodically checked by a veterinarian. In addition, studies were performed at higher doses of 12.5 and 25 mg/kg and clinical chemistry was monitored. The mean afoxolaner plasma concentration versus time curve following a treatment regimen of 2.5 mg of afoxolaner/kg given in an oral solution monthly for 5 consecutive months provided drug plasma concentrations ranging from 100 to 1000 ng/ml. Steady state was reached by the second dose, with plasma levels not increasing further with multiple doses. The terminal plasma half-life appeared ideal for a monthly product, and the prandial state at the time of dosing did not impact systemic exposure. Effectiveness against fleas was never <99%, and high levels of effectiveness were also shown against ticks (*Dermacentor variabilis*). The results indicated that afoxolaner was a good candidate for further research and development.

Dose Level and Formulation Selection

Dose Level Selection

The objective of dose selection studies is to define a minimal therapeutic dose. This includes not only the dose in weight of drug/body weight but also the required frequency of administration. Exploratory dose characterization studies employing a single dose against fleas (*C. felis*) and ticks (*D. variabilis, Rhipicephalus sanguineus, Amblyomma americanum,* and *Ixodes scapularis*) were performed using preclinical formulations, and these studies revealed that fleas were more sensitive than ticks, with high levels of efficacy obtained against *C. felis* at all dose levels tested. Afoxolaner dose selection studies for ticks evaluated the effectiveness and PK over the dose range of 1.5–3.5 mg/kg in prospective chewable formulations. As an example, the percent efficacy as a function of dose for *R. sanguineus* is shown in Figure 14.1. Although all three dose levels administered in this study (1.5, 2.5, and 3.5 mg/kg) were effective for 1 month, the 1.5 mg/kg dose level provided <90% efficacy on day 30, whereas the 2.5 and 3.5 mg/kg doses provided comparable efficacy that was >90% for the entire month. Therefore, the 2.5 mg/kg dose was chosen as the minimum effective dose to achieve >90% efficacy against *R. sanguineus* for 1 month. Similar results were obtained for *D. variabilis, A. americanum,* and *I. scapularis.* The dose selection studies consistently showed that a single 2.5 mg/kg afoxolaner treatment administered in oral chewable formulation could effectively treat flea and tick infestations for 1 month.

During these early flea and tick efficacy studies, a direct relationship between afoxolaner plasma concentration and percentage of effectiveness relative to control dogs was observed and modeled using a Sigmoidal E_{max} model. The resulting

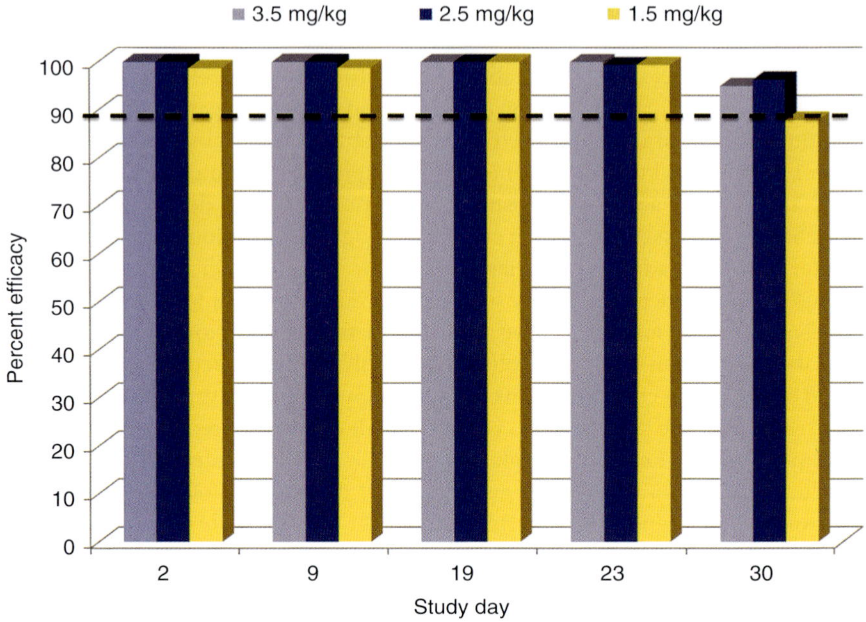

Figure 14.1 Percent efficacy of afoxolaner against *Rhipicephalus sanguineus* following a single oral treatment to dogs at 3 dose levels: 1.5, 2.5, and 3.5 mg/kg.

afoxolaner EC_{90} for fleas was 23 ng/ml and the EC_{90} for ticks was close to 100 ng/ml (Table 14.2). An example of the tick efficacy as a function of plasma concentration is given in Figure 14.2. The single 2.5 mg/kg dose resulted in ≥100 ng/ml mean afoxolaner plasma concentration maintained for 1 month and corresponding to 100% efficacy against fleas and >95% efficacy against all the tick species 28 days after treatment. Understanding the relationship between afoxolaner plasma concentration and efficacy aided in decision making and helped optimize the afoxolaner development program.

Safety is another important consideration when selecting the dose. Afoxolaner administered to dogs at the 2.5 mg/kg dose level had a large margin of safety. Preliminary data also indicated that repeated monthly dosing was not predicted

Table 14.2 Mean ± standard deviation of the effective concentration for 90% efficacy (EC_{90}) determined using a Sigmoid E_{max} model for percent efficacy against *R. sanguineus, D. variabilis*, and *A. americanum* following oral administration of afoxolaner in a chewable formulation to dogs.

	D. variabilis	R. sanguineus	A. americanum
EC_{90} (ng/ml)	110 ± 16	100 ± 14	117 ± 21

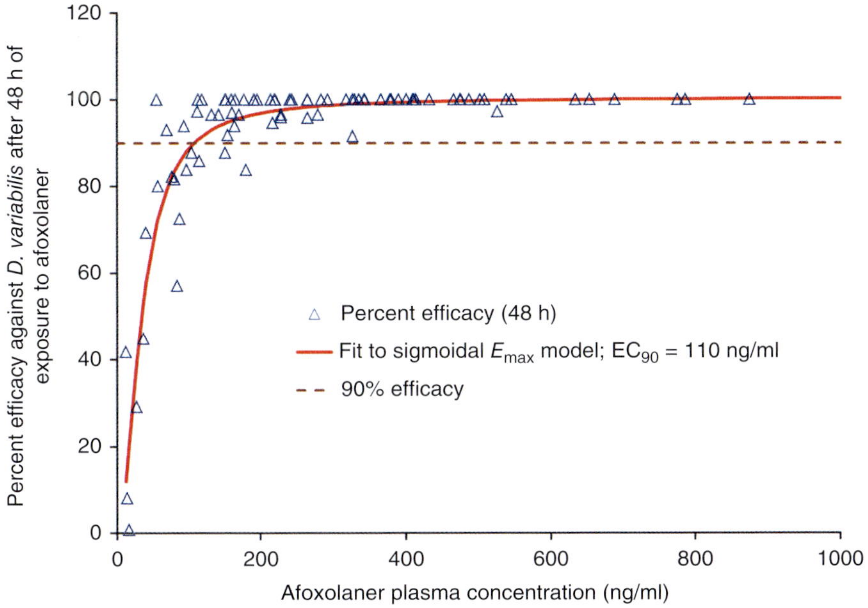

Figure 14.2 Percent efficacy against *Dermacentor variabilis* versus afoxolaner plasma concentration following oral administration of afoxolaner in a chewable formulation to Beagle dogs.

to result in appreciable drug plasma accumulation, and this was later confirmed in multiple-dose PK and safety studies (see subsequent text).

Based on the excellent efficacy, the PK profile and safety data obtained in the dose selection studies, monthly oral treatment with 2.5 mg/kg was selected as the appropriate dose for the afoxolaner oral chewable for dogs. Efficacy and target animal safety studies conducted with the final formulation in the pivotal development program subsequently confirmed the selection of this dose.

Formulation Selection

Formulation of an active ingredient in an appropriate dosage form is a critical step in the development of any new product. Identification of a safe and effective API may not always translate into an effective product if it cannot be formulated as an acceptable product that can be readily applied. Administration compliance is an especially important aspect for effective use of veterinary products, including flea and tick control products where continued suppression of new generations of parasites may be required for an effective control program. Acceptability and palatability of oral formulations are critical attributes when establishing owner compliance, especially for regularly applied products such as parasiticides that require monthly treatments. These dosage forms should be highly palatable to ensure that they are voluntarily and completely consumed by the pet each time they are dosed. The primary goal of the program was therefore to develop a highly palatable and acceptable dosage form that would be well accepted by the great majority of dogs, and that could deliver the target dose rapidly and effectively. Production of the final dosage form needs to be scalable to allow for commercial manufacturing, and the product should be compatible with appropriate packaging materials and stable, with a shelf life of at least 2 years. Early formulation development work involved testing for solubility, polymorphs, solvates, hygroscopicity, particle size, physicochemical properties, hardness (upon drying and over time), and compatibility with potential excipients. Long-term stability of viable formulation candidates was evaluated at specified temperature and humidity conditions to ensure a final product shelf life of >2 years in the climates of all major geographical regions. Dissolution tests were performed and *in vivo* PK and efficacy tests conducted to ensure that the formulations had immediate release *in vitro* and rapid *in vivo* dissolution leading to rapid and complete *in vivo* absorption and fast kill of external parasites. Selected formulation candidates were then tested to determine acceptability to dogs. At an early stage, palatability was determined in a study with placebo formulations given to dogs in the home environment. All formulations were highly palatable, and the final NexGard formulation had the best palatability of all placebos. The final formulation candidates (including the active ingredient afoxolaner) demonstrated excellent palatability, similar to that of HeartGard Plus, the accepted industry gold standard (unpublished data). Excellent palatability and acceptability have been confirmed subsequently in preference tests conducted for comparison with other isoxazoline-containing products [15].

In order to confirm selection of the final formulation, studies were performed to ensure that it could be produced at increasing levels of production up to the final commercial scale. Further work was conducted to ensure that the lead formulation and its excipients were compatible with selected primary packaging. Extensive testing under a range of environmental conditions confirmed long-term stability.

ADME Properties and Pharmacokinetics

ADME and PK studies were performed during the development of afoxolaner with the aim of supporting clinical efficacy and safety in the treated animal. These studies ultimately were used to fulfill the goals outlined in the EMEA (Europe, the Middle East and Africa) Guideline for the Conduct of Pharmacokinetic Studies in Target Animal Species [16], which is now undergoing revision.

Oral Pharmacokinetics in Dogs

The PK properties of afoxolaner are well characterized in dogs. The physicochemical properties, absorption including the effect of prandial state, degree of proportional systemic exposure with dose, bioavailability, multiple-dose kinetics, distribution, protein binding, half-life, principal routes of elimination and metabolism, and lack of interaction between active ingredients were determined during the research and development phases for NexGard and NexGard Spectra. Afoxolaner plasma concentrations were also compared to efficacy and safety data to understand the therapeutic window for each product.

Afoxolaner is a Biopharmaceutics Classification System (BCS) Class II compound, characterized by high permeability and low aqueous solubility measured in defined laboratory cell permeability (Caco 2 and MDR-MDCK) and solubility experiments, and subsequently confirmed in animals. Afoxolaner has a low molecular weight and is lipophilic and hydrophobic. Afoxolaner also remains neutrally charged over a range of physiological pH; afoxolaner is therefore expected to cross cell membranes freely via passive diffusion driven by a concentration gradient and the high afoxolaner Log *P* of 5.5. Absorption of BCS Class II compounds may depend on the rate and extent of dissolution due to the compounds' low aqueous solubility. The high permeability of these compounds results in high bioavailability if dissolution is complete and the drug is in solution. High-permeability compounds may be eliminated primarily by metabolism because they readily cross membranes and access enzymes within the hepatocytes. Lastly, high-permeability compounds are expected to have greater distribution into tissues than low-permeability compounds [17]. Afoxolaner PK properties follow the expectations for a BCS Class II compound [3], although due to high protein binding the majority of the compound does not reach hepatocytes and is excreted unchanged.

The absorption characteristics of afoxolaner were tested when given orally as a solution and in test chew formulations, and confirmed in two GLP studies using the final chewable formulations for NexGard and NexGard Spectra [3, 4]. Afoxolaner absorption was rapid, with maximum afoxolaner plasma concentrations of 1655 ± 332 and 1822 ± 165 ng/ml reached between 2 and 4 h following administration from NexGard and NexGard Spectra, respectively. To determine bioavailability, the oral afoxolaner PK profile was compared to that following intravenous (IV) administration of afoxolaner in a polyethylene glycol (PEG)/ethanol (8:2) solution at a dose rate of 1 mg afoxolaner/kg body weight. Due to the long terminal plasma half-life of afoxolaner (~2 weeks), each dog could not serve as its own control, so the oral and IV treatments were administered to separate dogs, and the average bioavailability was determined. NexGard had an average afoxolaner bioavailability of 73.9% and NexGard Spectra had an average afoxolaner bioavailability of 88.3%.

The effect of prandial state on NexGard absorption was investigated in several studies. For example, in Beagle dogs fed immediately prior to treatment, afoxolaner plasma concentrations reached a mean maximum of 1366 ± 276 ng/ml occurring, for most dogs, 2–24 h following treatment. Dogs that were fasted overnight and for 4 h following treatment had a similar mean maximum afoxolaner plasma concentration of 1453 ± 374 ng/ml occurring 2 h posttreatment. Other PK parameters were also comparable, including half-life and AUCs. The prandial state of the dog prior to treatment does not significantly affect the PK profile of afoxolaner.

The afoxolaner exposure, established by calculating C_{max}, $AUC_{0\text{-}Tlast}$, and $AUC_{0\text{-}Inf}$, increased proportionally with dose, indicating linear PK over the range of 1.0–4.0 mg/kg when afoxolaner chews were dosed orally in a GLP PK/efficacy study with eight dogs per treatment group. Linearity was also demonstrated over the range of 1.0–40 mg/kg when data from a second study delivering 20 and 40 mg/kg to five dogs per treatment group was included in the analysis. Dose proportionality was assessed by calculating the strength of a linear relationship existing between area under the curve (AUC) and dose or between C_{max} and dose using the Power method [18].

Due to the long afoxolaner half-life in dogs, afoxolaner plasma concentrations were determined for 87 days to obtain the full PK profile following treatment. The full afoxolaner plasma concentration versus time curve continued to follow a single exponential decay from day 2 to day 87, the last sampling time for this study, corresponding to a long elimination phase. The reason for the long terminal half-life was also investigated and is discussed below in the section on distribution, clearance, and protein binding.

Multiple-dose afoxolaner kinetics were assessed in dogs following oral 2.5 mg/kg doses of a solution administered three times at 28-day intervals. The maximum afoxolaner plasma concentrations were 699 ± 315, 1150 ± 450, and 908 ± 147 ng/ml observed at approximately 3 h after the first, second, and third doses, respectively. Experimental data confirmed the low accumulation with a ratio of $AUC_{0\text{-}Inf}$ (dose #3) to $AUC_{0\text{-}Inf}$ (dose #1) of approximately 1.0. The accumulation ratio for C_{max} in this study was 1.3 (a 30% increase from dose #1 to dose #3). The half-life was comparable after each of the 3-monthly doses. These parameters indicate that the drug disposition processes are linear upon multiple dosing and that the clearance, distribution, and absorption processes are neither saturated nor induced during regular monthly dosing. The lack of accumulation upon monthly dosing was confirmed in the pivotal target animal safety studies wherein three doses of the final formulation were given monthly followed by three doses given every 2 weeks (see subsequent text).

The mean half-life of afoxolaner administered orally at a dose rate of between 1 and 4 mg/kg across Merial studies is 12.8 ± 5.6 days in adult Beagle and mongrel dogs. For the dog with the longest half-life following monthly oral administrations, NexGard is predicted to have maximum afoxolaner plasma concentrations increasing 2.2-fold at steady state which is reached by the sixth-monthly dose. No further increases in afoxolaner plasma concentration are expected once steady state has been reached. Increases of 2.2-fold do not indicate a safety concern for afoxolaner (see subsequent text).

The PK profile of afoxolaner following oral administration was predictable between dogs and comparable across all studies from which plasma was collected during the chewable development program, encompassing data from >145 dogs.

Distribution and Clearance *of Afoxolaner in Dogs*

Another goal of the afoxolaner development program was to understand the distribution and clearance of afoxolaner in dogs under conditions of use and potential overdose conditions to ensure product safety. Information was compiled from a number of studies to get a full understanding of underlying ADME properties of afoxolaner. Following the 1.0 mg/kg IV dose, the V_{dss} was 2.68 ± 0.55 l/kg and the Cl was 4.95 ± 1.20 ml/h/kg. The single exponential decay of afoxolaner in plasma from day 2 to day 87 suggests that no special tissue depots are present in the dog. This is consistent with the physicochemical properties of afoxolaner which favor passive diffusion into and out of tissues. Active transport, if occurring, was not saturated under the conditions/dose levels tested. PK parameters were consistent across breeds (Beagle, mongrel, and Greyhound dogs) tested following multiple dosing, and following increased dosing of up to 40 mg/kg. No correlation has been found between body weight or age of dog and terminal plasma half-life for Beagle dogs. The V_{dss} indicates moderate tissue distribution within the range of other safe pharmaceutical drugs [19], and afoxolaner has a wide margin of safety with afoxolaner plasma concentrations at steady state following multiple therapeutic doses that are at least 10 times lower than those tested in the NexGard target animal safety study (see subsequent text).

To learn more about hepatic clearance, metabolite identification was performed on plasma samples from dogs and rats following administration of 25, 30, or 100 mg/kg given orally as a solution. A single major metabolite, hydroxylated afoxolaner, was observed in both rat and dog plasma. This metabolite, an oxidation product formed via cytochrome P450 enzymes, is shown in Figure 14.3. The position of the hydroxyl group was determined using mass spectrometry.

The hepatic clearance rate for afoxolaner in dogs is much less than the hepatic blood flow [20] and is primarily responsible for the long half-life of afoxolaner in dogs. To investigate further the factors contributing to low clearance, protein binding was determined via equilibrium dialysis in rat, cat, and dog plasma. Incubations were performed for 2.5 h at 37 °C after which the afoxolaner concentrations were measured in the buffer and plasma compartments. Afoxolaner is highly bound to plasma proteins (>99%) of dogs, cats, and rats, and protein binding is independent of concentration over the range of 200–10 000 ng/ml. The high afoxolaner protein binding (>99%) limits the amount of free drug available to diffuse into hepatocytes or into the kidney. The intrinsic rate of metabolism is also an important aspect of clearance and was measured *in vitro* in rat, cat, and dog hepatic microsomes. In all cases the half-life in microsomes was >120 min, with intrinsic clearance values that were below detection in most cases and all lower

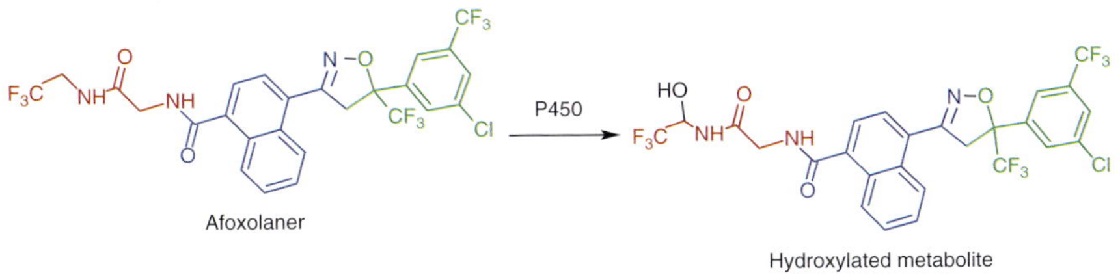

Figure 14.3 Proposed pathway for the formation of the hydroxylated metabolite of afoxolaner.

than 0.007 ml/min/mg of protein. In summary, clearance of afoxolaner is low due to high protein binding and slow intrinsic clearance in all species tested.

Toxicology and Safety of Afoxolaner

Safety Pharmacology Profile of Afoxolaner

Early electrophysiology studies suggested that afoxolaner's mode of action involved a novel, strong, and specific inhibition of insect gamma-aminobutyric acid (GABA)-gated chloride channels [14]. *In situ* studies suggested that central nervous system (CNS) and neuromuscular junction (NMJ), rather than muscle fibers, are the sites of action. Biochemical studies suggested a unique site of GABA binding for some isoxazolines relative to fipronil [21]. Off-target receptor profiling was conducted in over 70 *in vitro* pharmacological assays, which included receptor, ion channel, transporter, enzyme, and second messenger targets. This profiling revealed virtually no cross-reactivity with mammalian GABA receptors and confirmed selectivity for insect GABA receptors and enhanced safety for humans. Subsequent extensive toxicity testing demonstrated that off-target sites were not responsible for the salient toxicity profile in rats, the most sensitive laboratory species (unpublished data).

To appropriately assess the safety profile of afoxolaner, and to meet regulatory requirements, more than 20 acute and repeated toxicity studies were conducted in rodents, rabbits, and dogs. The acute toxicity profile was limited to effects on body weight and food consumption at relatively high dose levels in rats and rabbits. Mortality was not observed in dogs dermally administered upper-limit doses of 2000 mg/kg or in rats orally administered 1000 mg/kg of afoxolaner. In handler safety testing, afoxolaner was neither dermally irritating nor a sensitizer. Similarly, in a genotoxicity test battery, afoxolaner was neither mutagenic (with and without S9, where S9 is the supernatant fraction of the liver homogenate) nor a directly acting genotoxic agent. The repeated-dose toxicity profile in rats was driven by a reduction of feed intake and subsequent effects on body weight, leading to inanition, with effects on organ weight, and clinical and anatomic pathology. Similar clinical effects were observed in rabbits. Developmental and reproductive toxicity was assessed in the rat and rabbit. Afoxolaner was not a selective developmental or reproductive toxicant in these studies.

Safety in Dogs, The Target Species

In vitro cell permeability assays (Caco 2 and MDR-MDCK) demonstrated that afoxolaner is not a substrate for the transporter *p*-glycoprotein. This is important because a subpopulation of dogs have a mutation in the gene encoding *p*-glycoprotein, which results in production of a nonfunctional protein [22]. Phenotypically, this mutation affects absorption and elimination of substrate molecules. From a drug safety perspective, animals with this mutation cannot transport molecules that are substrates for *p*-glycoprotein out of the CNS.

Therefore, these drugs accumulate in the CNS, resulting in signs of toxicity that would not be seen at the same dose in animals without this mutation.

To confirm the results of the *in vitro* study, doses up to 25 mg/kg (10× the target dose) were tested in collies known to have this mutation. The results correlated very well with the results of the receptor binding assay and none of the collies demonstrated any signs of toxicity.

The safety profile of afoxolaner was also tested in a laboratory study in dogs without the *p*-glycoprotein mutation, following regulatory requirements. NexGard was administered six times orally at a dose of 1, 3, or 5× the maximum exposure dose (6.3 mg/kg) in 8-week old Beagle dogs [23]. No afoxolaner-related changes were observed in growth, physical variables, clinical pathology variables, or tissues examined histologically. No clinically or statistically significant health abnormalities related to the administration of afoxolaner were observed.

Safety was also tested in client-owned dogs at the label dose. Multisite trials were conducted in the United States, Europe, Australia, and Japan to evaluate safety of NexGard under field conditions. Afoxolaner was administered in >400 client-owned dogs including different breeds and ages under conditions of normal use. The product was shown to be safe and effective and was not associated with serious adverse reactions.

Pivotal Dose Determination, Confirmation and Field Trials

Following dose selection, and confirmation of the final formulation, a program including multiple studies was conducted at sites around the world to confirm efficacy of the product under controlled conditions employing induced infestations of fleas and ticks, as well as under conditions of natural challenge in the field. The study design for all pivotal ectoparasite dose determination and confirmation studies is very similar. In induced infestation studies, dogs are challenged with an appropriate number of fleas or ticks prior to treatment and weekly thereafter. Fleas or ticks are counted and removed at an appropriate time (usually 24 h for fleas and 48 h for ticks) after treatment or challenge. Based on the systemic mode of action of afoxolaner, ticks could be counted dead or alive, free or attached, and the status of engorgement was included in all experiments, per the WAAVP (World Association for the Advancement of Veterinary Parasitology) guideline by Marchiondo *et al.* revised in 2013 [24]. Mean parasite counts for the treated dogs are compared to the control counts at each time point. Parasite counts were transformed to the natural logarithm (count +1) for calculation of geometric means by treatment group and percent efficacy was calculated using the formula $((C-T)/C)\times 100$, where C = geometric mean for the control group and T = geometric mean for the treated group for each time point.

Dose Confirmation Studies for Fleas

Excellent efficacy was confirmed in all studies using the challenge with *C. canis* and *C. felis* fleas [25, 26]. In these studies, dogs were infested with approximately

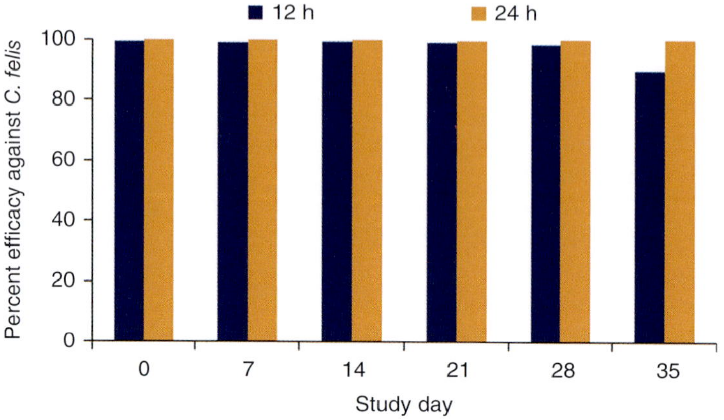

Figure 14.4 Percent efficacy of afoxolaner against *C. felis* following a single oral treatment of NexGard to dogs.

100 fleas on days − 1, 7, 14, 21, 28, and 35. Dogs were treated on day 0 using the chewable tablet sizes listed in Table 14.1, so that the dose was as close as possible to the minimum dose of 2.5 mg/kg. Fleas were counted on day 0 (12 h posttreatment), and on days 1, 7, 14, 21, 28, and 35 (12 and 24 h post-infestation). Efficacy at 12 h was >93% on all count days except two, and efficacy at 24 h post-dose was >99.99% on all count days, confirming a high level of both curative and preventative efficacy, and a rapid onset of action. Figure 14.4 shows the demonstrated efficacy of NexGard against fleas in one of the pivotal studies. Similar results were obtained in flea studies conducted on five continents during the course of the development program.

Two of the flea dose confirmation studies also included assessment of flea egg counts taken at 12 and 24 h on the same days as flea counting. Efficacy was >99.1% on all counting days demonstrating that, due to the rapid speed of flea kill, NexGard prevented fleas from laying eggs, thereby preventing development of further flea infestations. A single treatment therefore also provides control of new flea development for 5 weeks.

Flea studies were also conducted to determine the speed of flea kill following treatment of dogs with a preexisting flea infestation [27]. In a negative control, randomized block design study 80 dogs of both sexes were infested with approximately 100 fleas on day − 1, and on day 0 dogs were treated with NexGard at an average dose of 2.5 mg/kg. Live fleas were counted and removed at 0.5, 2, 4, 8, 12, and 24 h after treatment in study 1 and 0.5 and 2 h after treatment in study 2. Results are given in Figure 14.5. NexGard starts killing fleas within 0.5 h and is >90% effective by 8 h posttreatment.

The results of dose determination studies demonstrated a sustained 12-h count efficacy and ≥99.9% efficacy at 24-h counts against adult *C. felis* for 5 weeks after a single administration. Similarly, studies to determine efficacy against another

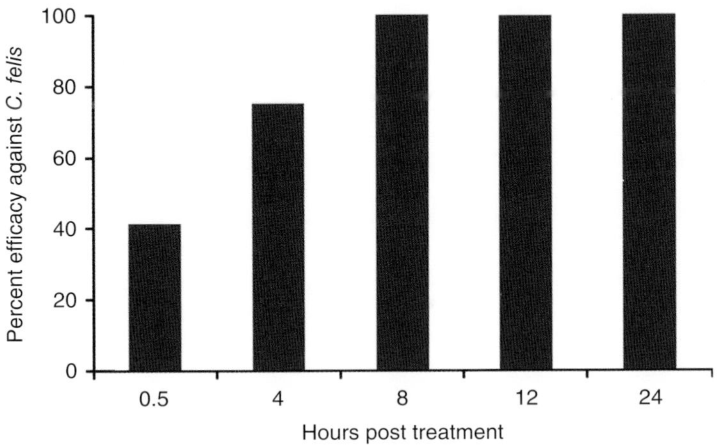

Figure 14.5 Efficacy of NexGard against existing flea (*C. felis*) infestations on dogs.

flea species, *C. canis* showed ≥99% efficacy at the 24-h count for 5 weeks after treatment.

Dose Confirmation Studies for Ticks

Two dose confirmation studies were conducted for each of eight tick species of primary interest on dogs globally. All dose confirmation studies against tick infestations utilized similar study designs. Beagle or mongrel dogs of both sexes were either treated or served as untreated controls in negative-controlled randomized block design studies with 8–10 dogs per group. In all cases, the NexGard chewable product was administered orally to provide an afoxolaner dose as close as possible to the minimum effective dose of 2.5 mg/kg. Because prandial state did not affect afoxolaner exposure in dogs, studies were conducted with dogs that were either fed prior to treatment or fasted overnight prior to treatment and fed 4 h after treatment. Dogs were challenged with an appropriate number of ticks prior to treatment, and again at weekly intervals thereafter and efficacy was determined based on numbers of ticks recovered from treated and control dogs at 48 h (and in some cases 72 h) after treatment or challenge. Mean efficacies for tick species tested following challenge 1 month after treatment are given in Table 14.3.

Field Trials

While induced challenge studies help confirm efficacy and allow for use of relatively small numbers of animals in a trial, the real test of how the product can be expected to perform is obtained by conducting field trials. NexGard was also tested in client-owned animals in several multicentered positive control field trials conducted in the United States, EU, and Japan, with similar good results at the end of a month following the first treatment. Male and female dogs of various ages (e.g., from 3 months to 16 years in the EU) and representing a variety of

Table 14.3 Mean percent efficacy against eight tick species on day 30 or 31 following treatment with NexGard (48- or 72-h counts).

Tick species	Average % efficacy	References
Amblyomma americanum	98	Unpublished data
Ixodes scapularis	95	[28]
Ixodes ricinus	97	[29]
Rhipicephalus sanguineus	97	[30]
Dermacentor reticulatus	96	[29]
Dermacentor variabilis	98	[31]
Ixodes holocyclus	98	Unpublished data
Haemaphysalis longicornis	92	[32]

breeds and crossbreeds were enrolled. The dogs were infested with fleas or ticks at the start of the study, and were randomly allocated to either a NexGard or positive control treatment group. They were treated once at the start of the study (and monthly thereafter in the US study), and evaluated for efficacy at different intervals following treatment. They were maintained at home under their usual conditions, and no premise treatments or other flea or tick control products were applied. Because the dogs were infested with fleas at enrollment, it can be assumed that their home environment was contaminated, and that they would have been subject to continued re-infestation upon returning home. NexGard performed as well as or better than positive control in field trials that included 116 flea cases and 64 tick cases in the EU, 241 flea cases in the United States and 103 flea cases and 102 tick cases in Japan. The results of the field trials are shown in Figure 14.6.

Additional Efficacy Studies

With strong acaricidal activity, it was expected that afoxolaner could provide useful activity against mites as well as ticks on dogs. Mite infestations arising from the mite genera *Demodex, Otodectes, and Sarcoptes* can cause significant skin lesions and compromise the health of dogs. Probe studies indicated that the systemic activity of afoxolaner can provide good efficacy against mange mites, including those that may be active deep in the skin. In a positive-controlled study involving dogs diagnosed with generalized demodicosis, afoxolaner was administered at the recommended dose (at least 2.5 mg/kg) on days 0, 14, 28, and 56. Clinical examinations and deep skin scrapings were performed every month in order to evaluate the effect on mite numbers and the resolution of clinical signs. The percentage reductions of mite counts were 99.2%, 99.9%, and 100% on days 28, 56, and 84, respectively, in the afoxolaner-treated group. The skin condition of the dogs also improved significantly from day 28 to day 84. Mite reductions were significantly higher on days 28, 56, and 84 in the afoxolaner-treated group

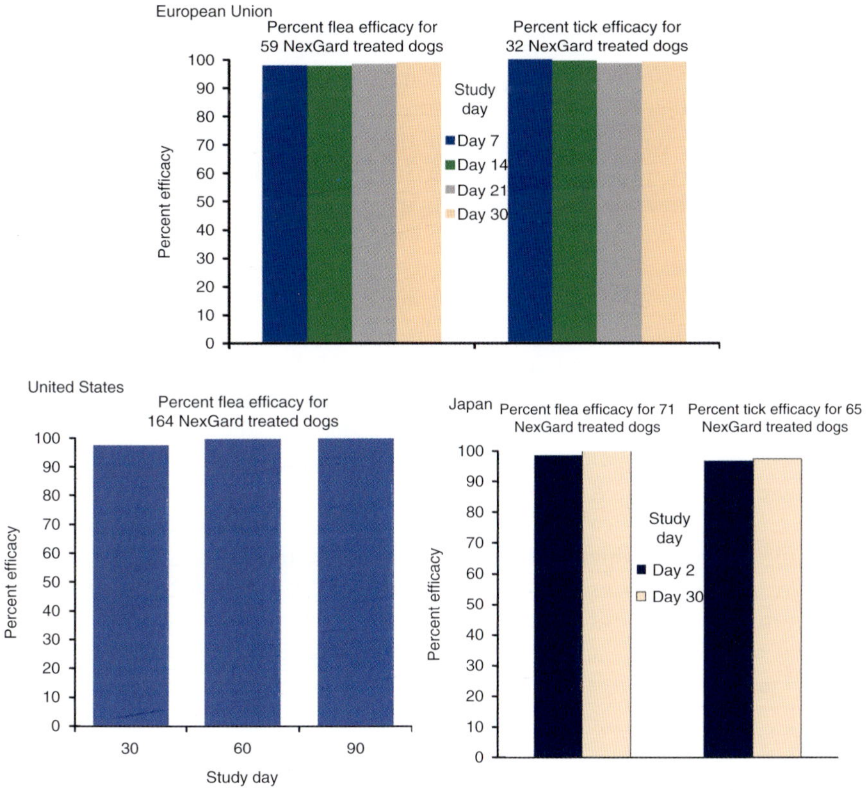

Figure 14.6 Efficacy results from NexGard field trials.

compared to the positive control group, and the results of this study demonstrated that afoxolaner, given orally, was effective in treating dogs with generalized demodicosis within a 2-month period [33].

In a study conducted to evaluate efficacy of afoxolaner against the ear mite *Otodectes cynotis*, a single treatment provided efficacy of >98.5% by 28 days after treatment [34]. Another study was conducted with dogs naturally infested with *Sarcoptes scabiei* [35], and two treatments with afoxolaner at the target dose level on day 0 and again 28 days later eliminated mites (100% efficacy) based on evaluations conducted on days 28 and 56.

As indicated previously, the efficacy of afoxolaner against ticks is not only important because ticks are a nuisance for owners and dogs but they can also transmit dangerous pathogens. A serious tick-borne pathogen affecting humans and dogs is *B. burgdorferi*, the causative organism of Lyme disease. To assess whether NexGard could aid in preventing transmission of pathogens by killing ticks before they transmit disease, a study was conducted using wild caught *I. scapularis* ticks confirmed by polymerase chain reaction (PCR) to be infected with *B. burgdorferi* [36]. Ten dogs were treated orally on day 0 at a dose near the minimum recommended dose of afoxolaner (2.5 mg/kg). Ten control dogs were not treated. On day 28, each dog was infested with approximately 50 adult unfed ticks (67% *B. burgdorferi* infection rate), and any remaining ticks were removed on day 33. To detect *B. burgdorferi* infection, SNAP® 4Dx® tests were performed on serum collected before infestation (all dogs seronegative) and on days 48, 63, 77, and 92. The 10 dogs treated with NexGard® were protected from *B. burgdorferi* infection as they remained seronegative through the end of the study (day 92), while 9 of the 10 untreated control dogs were infected and became seropositive starting on day 48.

Another serious disease pathogen of dogs is *Babesia canis*, transmitted by *D. reticulatus*, which may cause severe hemolysis, leading to death. To test the ability of NexGard to protect dogs from *B. canis*, 16 dogs were infested on days 7, 14, 21, and 28, with 50 adult *D. reticulatus* ticks confirmed to harbor *B. canis* by PCR. Half of the dogs were untreated, while the other half were treated with NexGard on day 0. *B. canis* was transmitted to all untreated control dogs, while all treated dogs remained negative until the end of the study on day 56 [37].

These additional efficacy studies demonstrate further value of the acaricidal effect of afoxolaner, which was shown to have extended efficacy against mange mites including *Demodex*, *Sarcoptes*, and *Otodectes*. In addition to broad activity against multiple species of ticks, treatment with NexGard was also shown to be effective against *Ixodes* and *Dermacentor* ticks soon enough after infestation of dogs to prevent transmission of *B. burgdorferi* and *B. canis*, respectively.

NexGard Spectra®

NexGard Spectra chewable for dogs (approved centrally in the EU in January 2015) is a broad-spectrum parasiticide combination which contains afoxolaner (2.5 mg/kg) and milbemycin oxime (0.5 mg/kg). The product is designed to treat and control a wide spectrum of ectoparasites (by afoxolaner) and endoparasites (by milbemycin oxime) when administered at monthly intervals [2]. It is a unique product that targets fleas, ticks, and gastrointestinal nematodes, and is effective in preventing heartworm disease.

The development of NexGard Spectra was similar to that of NexGard. Dose confirmation studies were performed for efficacy against fleas, ticks [38], intestinal nematodes [39], and for the prevention of heartworm. A target animal safety study was also performed to ensure safety of the combination product [40], and field efficacy and safety studies were conducted at multiple sites in the United States, EU, and Japan.

In addition, studies were conducted to ensure that the compounds administered in this fixed combination did not interact with each other. Exploratory and GLP pivotal studies were performed with groups of dogs receiving NexGard and NexGard Spectra (or similar test formulations) in the same study and the PK parameters were compared [4]. The PK profile of afoxolaner was unchanged by the addition of milbemycin oxime and vice versa [4].

Future Direction

Following the registration of NexGard and NexGard Spectra, other uses for afoxolaner have been investigated. The dose of 2.5 mg/kg is ideal for a monthly oral product for dogs. The relationship between afoxolaner plasma levels and efficacy is well understood; and it was determined that by keeping plasma levels above approximately 100 ng/ml, efficacy will be maintained against fleas and ticks. This knowledge was used to assess the proper dose to maintain afoxolaner plasma concentration above 100 ng/ml for 3 months. A chewable formulation with higher concentrations of afoxolaner may achieve these drug plasma concentrations and would provide a longer acting product for veterinarians and pet owners.

Preliminary studies indicate that afoxolaner is effective in cats for protection against fleas and ticks and in cattle against *Rhipicephalus (Boophilus) microplus*, cattle ticks. Afoxolaner has shown a long duration of action from immediate release formulations in all species tested to date.

Conclusions

In the development of NexGard for dogs, the behavior of the afoxolaner active ingredient was well understood early in the development program. This enabled efficient selection of dose and formulation, ensuring efficacy against the target parasites, and safety for dogs (the patient), the owner, veterinarian, and the environment. Because of extensive work done upfront to understand afoxolaner's safety and efficacy profiles, the pivotal program could be run efficiently to demonstrate the required levels for efficacy for 1 month against all major flea and tick species and safety in use and overdose situations.

Based on its proven safety and efficacy profiles and the highly acceptable oral dosage form, NexGard has been well accepted by veterinarians and pet owners in more than 50 countries around the world, and is providing protection for millions of dogs against ectoparasites that threaten their health and well-being.

Acknowledgments

The authors acknowledge all members of the NexGard and NexGard Spectra development teams and supporting functional groups who contributed to the successful registration of these products.

Disclaimer

This document is provided for scientific purposes only. NexGard® and NexGard Spectra® are registered trademarks of Merial, Inc. Any reference to a brand or trademark herein is for information purposes only and is not intended for any commercial purposes or to dilute the rights of the respective owners of the brand(s) or trademark(s).

References

1. Xu, M., Long, J.K., Lahm, G.P., Shoop, W.L., Cordova, D., Wagerle, T., Smith, B.K., Pahutski, T.F., Shapiro, R., Mahaffey, M., Hartline, E.J., Gould, B.R., Waddell, M.E., McDowell, R.G., Kinney, J.B., Jones, G.S., Dietrich, R.F., Schroeder, M.E., Rhoades, D.F., Benner, E.A., and Confalone, P.N. (2017) The Discovery of Afoxolaner: A New Ectoparasiticide for Dogs, Wiley-VCH.
2. EMA Committee for Medicinal Products for Veterinary Use (2014) CVMP Assessment Report for NexGard Spectra. EMEA/V/C/003842/0000; EMA/695949/2014.
3. Letendre, L., Huang, R., Kvaternick, V., Harriman, J., Drag, M., and Soll, M. (2014) The intravenous and oral pharmacokinetics of afoxolaner used as a monthly chewable antiparasitic for dogs. *Vet. Parasitol.*, **201** (3–4), 190–197.
4. Letendre, L., Harriman, J., Drag, M., Mullins, A., Malinski, T., and Rehbein, S. (2016) The intravenous and oral pharmacokinetics of afoxolaner and milbemycin oxime when used as a combination chewable parasiticide for dogs. *J. Vet. Pharmacol. Ther.* doi: 10.1111/jvp.12332
5. FDA (2001) Guidance for Industry, Bioanalytical Method Validation, http://www.fda.gov/downloads/Drugs/Guidances/ucm070107.pdf (accessed 04 December 2017).
6. Dryden, M.W. and Rust, M.K. (1994) The cat flea: biology, ecology and control. *Vet. Parasitol.*, **52** (1–2), 1–19.
7. Rust, M.K. and Dryden, M.W. (1997) The biology, ecology, and management of the cat flea. *Annu. Rev. Entomol.*, **42**, 451–473.
8. Dunn, A.M. (1978) Cestoda in *Veterinary Helminthology*, 2nd edn, Heinemann, London.
9. Pugh, R.E. (1987) Effects on the development of *Dipylidium caninum* and on the host reaction to this parasite in the adult flea (*Ctenocephalides felis felis*). *Parasitol. Res.*, **73** (2), 171–177.
10. Azad, A.F., Radulovic, S., Higgins, J.A., Noden, B.H., and Troyer, J.M. (1997) Flea-borne rickettsioses: ecologic considerations. *Emerg. Infect. Dis.*, **3** (3), 319–327.
11. Chomel, B.B., Boulouis, H.J., and Breitschwerdt, E.B. (2004) Cat scratch disease and other xoonotic *Bartonella* infections. *J. Am. Vet. Med. Assoc.*, **224** (8), 1270–1279.
12. Dantas-Torres, F., Chomel, B.B., and Otranto, D. (2012) Ticks and tick-borne diseases: a one health perspective. *Trends Parasitol.*, **28** (10), 437–446.
13. Otranto, D. and Wall, R. (2008) New strategies for the control of arthropod vectors of disease in dogs and cats. *Med. Vet. Entomol.*, **22** (4), 291–302.
14. Shoop, W.L., Hartline, E.J., Gould, B.R., Waddell, M.E., McDowell, R.G., Kinney, J.B., Lahm, G.P., Long, J.K., Xu, M., Wagerle, T., Jones, G.S., Dietrich, R.F., Cordova, D., Schroeder, M.E., Rhoades, D.F., Benner, E.A., and Confalone, P.N. (2014) Discovery and mode of action of afoxolaner, a new isoxazoline parasiticide for dogs. *Vet. Parasitol.*, **201** (3–4), 179–189.
15. Carithers, D.S., Halos, L., Crawford, J., Stanford, H., Everett, W.R., and Gross, S.J. (2016) Comparison of preference demonstrated by dogs when offered two commercially available oral ectoparasiticide products containing either afoxolaner (NexGard®) or sarolaner (Simparica™). *Int. J. Appl. Res. Vet. Med.*, **14** (3), 217–222.
16. EMEA (1999) Guideline for the Conduct of Pharmacokinetic Studies in Target Animal Species. EMEA/CVMP/133/99-Final.
17. Wu, C.-Y. and Benet, L.Z. (2005) Predicting drug disposition via application of BSC: transport/absorption? Elimination interplay and development of a biopharmaceutics drug disposition classification system. *Pharm. Res.*, **22**, 11–23.
18. Hummel, J., McKendrick, S., Brindley, C., and French, R. (2009) Exploratory assessment of dose proportionality: review of current approaches and proposal for a practical criterion. *Pharm. Stat.*, **8**, 38–49.

19. Toutain, P.L. and Bousquet-Melou, A. (2004) Volumes of distribution. *J. Vet. Pharmacol. Ther.*, **27**, 441–453.
20. Davies, B. and Morris, T. (1993) Physiological parameters in laboratory animals and humans. *Pharm. Res.*, **10**, 1093–1096.
21. Ozoe, Y., Asahi, M., Ozoe, F., Nakahira, K., and Mita, T. (2010) The antiparasitic isoxazoline A1443 is a potent blocker of insect ligand-gated chloride channels. *Biochem. Biophys. Res. Commun.*, **391**, 744–749.
22. Roulet, A., Puel, O., Gesta, S., Lepage, J.-F., Drag, M., Soll, M.D., Alvinerie, M., and Pineau, T. (2003) MDRA-deficient genotype in Collie dogs hypersensitive to *p*-glycoprotein substrate ivermectin. *Eur. J. Pharmacol.*, **460**, 85–91.
23. Drag, M., Saik, J., Harriman, J., and Larsen, D. (2014) Safety evaluation of orally administered afoxolaner in 8-week-old dogs. *Vet. Parasitol.*, **201** (3–4), 198–203.
24. Marchiondo, A.A., Holdsworth, P.A., Fourie, L.J., Rug, D., Hellmann, K., Snyder, D.E., and Dryden, M.W. (2013) World Association for the Advancement of Veterinary Parasitology (W.A.A.V.P.) second edition: guidelines for evaluating the efficacy of parasiticides for the treatment, prevention and control of flea and tick infestations on dogs and cats. *Vet. Parasitol.*, **194** (1), 84–97.
25. Dumont, P., Gale, B., Chester, T.S., and Larsen, D.L. (2014) Curative and preventive efficacy of orally administered afoxolaner against *Ctenocephalides canis* infestation in dogs. *Vet. Parasitol.*, **201** (3–4), 212–215.
26. Hunter, J.S., Dumont, P., Chester, T.S., Young, D.R., Fourie, J.J., and Larsen, D.L. (2014) Evaluation of the curative and preventive efficacy of a single oral administration of afoxolaner against cat flea *Ctenocephalides felis* infestations on dogs. *Vet. Parasitol.*, **201** (3–4), 207–211.
27. Kunkle, B., Drag, M., Chester, T., and Larsen, D. (2014) Assessment of the onset of action of afoxolaner against existing adult flea (*Ctenocephalides felis*) infestations on dogs. *Vet. Parasitol.*, **201** (3–4), 204–206.
28. Mitchell, E.B., McCall, J.W., Theodore, C.S., and Larsen, D. (2014) Efficacy of afoxolaner against *Ixodes scapularis* ticks in dogs. *Vet. Parasitol.*, **201** (3–4), 223–225.
29. Dumont, P., Blair, J., Fourie, J., Chester, T.S., and Larsen, D.L. (2014b) Evaluation of the efficacy of afoxolaner against two European dog tick species: *Dermacentor reticulatus* and *Ixodes ricinus*. *Vet. Parasitol.*, **201** (3–4), 216–219.
30. Kunkle, B., Daly, S., Dumont, P., Drag, M., and Larsen, D. (2014) Assessment of the efficacy of orally administered afoxolaner against *Rhipicephalus sanguineus* sensu lato. *Vet. Parasitol.*, **201** (3–4), 226–228.
31. Mitchell, E.B., Dorr, P., Everett, W.R., Chester, T.S., and Larsen, D. (2014) Efficacy of afoxolaner against *Dermacentor variabilis* ticks in dogs. *Vet. Parasitol.*, **201** (3–4), 220–222.
32. Kondo, Y., Kinoshita, G., Drag, M., Chester, T.S., and Larsen, D. (2014) Evaluation of the efficacy of afoxolaner against *Haemaphysalis longicornis* on dogs. *Vet. Parasitol.*, **201** (3–4), 229–231.
33. Beugnet, F., Halos, L., Larsen, D., and de Vos, C. (2016) Efficacy of oral afoxolaner for the treatment of canine generalised demodicosis. *Parasite*, **23** (14), 1–8.
34. Carithers, D., Crawford, J., de Vos, C., Lotriet, A., and Fourie, J. (2016) Assessment of afoxolaner efficacy against *Otodectes cynotis* infestations of dogs. *Parasit. Vectors*, **9** (635), 1–5.
35. Beugnet, F., de Vos, C., Liebenberg, J., Halos, L., Larsen, D., and Fourie, J. (2016) Efficacy of afoxolaner in a clinical field study in dogs naturally infested with *Sarcoptes scabiei*. *Parasite*, **23** (26), 1–7.
36. Baker, C.F., McCall, J.W., McCall, S.D., Drag, M., Mitchell, E.B., Chester, S.T., and Larsen, D. (2016) Ability of an oral formulation of afoxolaner to protect dogs from *Borrelia burgdorferi* infection transmitted by wild *Ixodes scapularis* ticks. *Comp. Immunol. Microbiol. Infect. Dis.*, **49**, 65–69.

37. Beugnet, F., Halos, L., Larsen, D., Labuschagne, M., Erasmus, H., and Fourie, J. (2014) The ability of an oral formulation of afoxolaner to block the transmission of *Babesia canis* by *Dermacentor reticulatus* ticks to dogs. *Parasit. Vectors*, **7**, 283.
38. Rehbein, S., Fourie, J.J., de Vos, C., Anderson, A., Larsen, D.L., and Jeannin, P. (2016) Efficacy of oral afoxolaner plus milbemycin oxime chewables against induced infestations with *Dermacentor reticulatus* in dogs. *Parasitol. Res.*, **115** (5), 1845–1851.
39. Rehbein, S., Dorr, P., Bowman, D.D., Crafford, D., Kusi, I., Postoli, R., Yoon, S., Chester, S.T., Dollhofer, D., Visser, M., and Larsen, D.L. (2016) Efficacy of afoxolaner plus milbemycin oxime chewable tablets against naturally acquired intestinal nematodes in dogs. *Vet. Parasitol.*, **217**, 29–35.
40. Drag, M., Saik, J., Harriman, J., Letendre, L., Yoon, S., and Larsen, D. (2017) Safety evaluation of orally administered afoxolaner and milbemycin oxime in eight-week-old dogs. *J. Vet. Pharmacol. Ther.*, **40** (5), 447–453.

15
Discovery, Development, and Commercialization of Sarolaner (Simparica®), A Novel Oral Isoxazoline Ectoparasiticide for Dogs

Debra J. Woods and Tom L. McTier*

Abstract

In this chapter, we discuss the focused efforts by Zoetis scientists to discover and develop a novel isoxazoline, sarolaner (Simparica®[1]), expressly for use in companion animals; highlighting this important advance in the effective treatment and control of ectoparasites on dogs.

Introduction

The canine ectoparasiticides market is very crowded, with multiple topical and oral products available both over the counter and by prescription; with generics only adding to the complexity. Despite this, there was a gap in the market for more effective, broader spectrum, oral ectoparasiticides [1], which was filled by the orally administered novel ectoparasiticidal isoxazoline class [2–4], with the initial entries, afoxolaner and fluralaner, leveraged from crop protection research and discussed in more detail elsewhere in this book. In 2009, Zoetis initiated a program to design an isoxazoline molecule specifically targeting the drug properties required for a veterinary ectoparasiticide (parasite spectrum, potency, duration of activity, and safety), which would demonstrate advantages over older isoxazolines (see section titled "Comparative Flea and Tick Efficacy"), and a lead optimization program was initiated. Achieving high efficacy via oral administration is anticipated to increase pet owner compliance by removing the difficulties of administering topical products, and eliminating the need to temporarily isolate treated animals from children and other pets required with some topical products.

*Corresponding author.

[1] SIMPARICA is a trademark or registered trademark of Zoetis Services LLC in the United States and other countries.

Ectoparasites: Drug Discovery Against Moving Targets, First Edition.
Edited by Charles Q. Meng and Ann E. Sluder

Discovery of Sarolaner

Zoetis medicinal chemists synthesized over 3000 isoxazoline compounds, which were then evaluated by the parasitology team to build an understanding of the structure–activity relationships of the compounds in the Zoetis isoxazoline library. This led ultimately to the identification of sarolaner (Figure 15.1) as the primary candidate for product development [4, 5]. The molecule was optimized specifically for potency against a broad spectrum of ectoparasites, enhanced duration of exposure, and safety in the dog. A 4-substituted fluorine attachment to the 3,5-dichlorophenyl head unit provided outstanding tick potency compared to 4-hydrido-3,5-substituted patterns. The spiroazetidinebenzofuran linker moiety has not previously been described in the parasiticide literature and provides rigidity, potency, and novelty to the molecule. The final optimization generated a methylsulfonylethanone tail, which increased the polar surface area of the molecule, enhancing the pharmacokinetic exposure and ensuring rapid kill of fleas and ticks. To augment safety, the molecule was prepared as the active single *S*-enantiomer; with selection of the pure chiral form decreasing the potential of off-target effects that may result from incorporation of the inactive enantiomer.

The screening strategy that led to the identification of sarolaner is discussed in detail in the paper by McTier *et al.* (2016) [4]. Efficacy screening was conducted initially in an *in vitro* membrane feeding assay against the cat flea, *Ctenocephalides felis felis.* Active hits were progressed to testing in a membrane feeding assay against the soft tick, *Ornithodoros turicata*. Potent *in vitro* hits were then assessed for safety in a mouse symptomatology model and compounds with an acceptable rodent safety profile were progressed to target animal toleration and pharmacokinetic and efficacy studies in the dog. For all studies described in this chapter involving animals, all protocols and procedures were reviewed and approved by institutional animal care and use committees and all procedures were conducted according to state, national, or international regulations. All efficacy calculations are based on geometric means, unless otherwise stated.

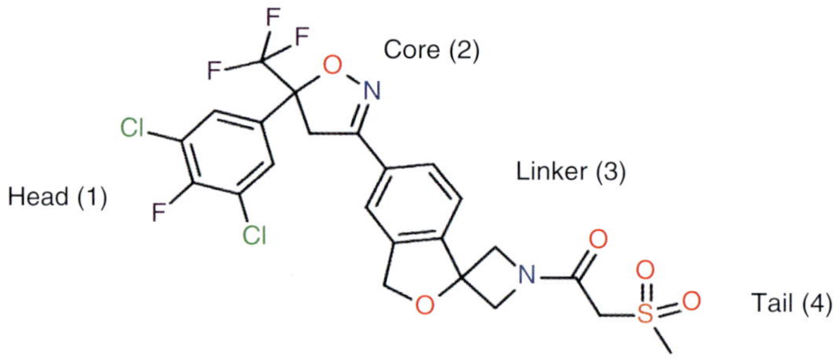

Figure 15.1 Structure of sarolaner (1-(5′-((5S)-5-(3,5-dichloro-4-fluorophenyl)-5-(trifluoromethyl)-4,5-dihydroisoxazol-3-yl)-3′-H-spiro(azetidine-3,1′-(2) benzofuran)-1-yl)-2-(methylsulfonyl) ethanone): (1) phenyl head group, (2) isoxazoline core, (3) spiroazetidinebenzofuran moiety, and (4) methylsulfonylethanone tail.

Table 15.1 Comparative *in vitro* whole parasite efficacy for afoxalaner, fluralaner, and sarolaner fed in blood to *C. felis* and *O. turicata* ($n = 3$).

Compound	Chemistry	Flea feed (*C. felis*) LC 80 (μg/ml)	Soft tick feed (*O. turicata*) LC 100 (μg/ml)
Afoxalaner (MW 625.9)	Racemic	1	0.1
Fluralaner (MW 556.3)	Racemic	1	0.1
Sarolaner (MW 581.4)	*S*-enantiomer	0.1	0.03

In vitro, sarolaner was highly potent against both fleas and ticks in comparison to afoxalaner and fluralaner (Table 15.1), with 10-fold increased potency against fleas and threefold increased potency against soft ticks. *In vivo*, sarolaner was rapidly and well absorbed following oral dosing, with a bioavailability calculated at >85% and a $t_{1/2}$ calculated at 11–12 days; >99.9% was protein bound. Dose proportionality was demonstrated in aqueous solution across doses of 1.25–5 mg/kg. In early mouse symptomatology testing, sarolaner showed no adverse signs associated with isoxazoline toxicosis at 30 mg/kg. Oral administration of ≤10 mg/kg of sarolaner three times at 28-day intervals to adult dogs and ≤20 mg/kg of sarolaner two times at 28-day intervals to 8-week-old dogs was also well-tolerated [4].

With potent *in vitro* activity, high exposure, a long half-life and good toleration in dogs, sarolaner was then progressed to efficacy studies at an exploratory oral dose of 2.5 mg/kg in solution [4]. A >99.9% efficacy was observed against fleas (*C. felis*) and excellent efficacy against three hard-tick species, *Ixodes ricinus* (>99%), *Rhipicephalus sanguineus* (100%), and *Dermacentor reticulatus* (>98%) at all the time points observed through to day 35 (2, 7, 14, 21, 28, and 35 days post-dosing).

Three laboratory studies were subsequently conducted [6], to determine the effective dose of sarolaner compared to a placebo control. In the first study, sarolaner was dosed in a suspension formulation at 1.25, 2.5, or 5.0 mg/kg to dogs infested with *C. felis, R. sanguineus*, and *D. reticulatus*, and then re-infested weekly for up to 8 weeks. Sarolaner delivered 100% flea efficacy for all three doses at 48 h after treatment of the existing infestation and at 24 h after each weekly infestation for 35 days. The lowest dose had efficacy of >98% at time points up to day 56 and also achieved 99.7–100% control of both species of ticks through to day 28, at 48 h posttreatment or after re-infestation. In the second study, dogs were dosed orally with sarolaner suspension formulations at 0.625, 1.25, or 2.5 mg/kg and infested with *Ixodes scapularis* prior to treatment and weekly for 6 weeks, *Amblyomma americanum* (pretreatment and on day 26), *Dermacentor variabilis* (on day 33) and *Amblyomma maculatum* (on day 41). *I. scapularis* was the most susceptible tick species, with the lowest dose delivering >95% efficacy through day 43, at 48 h posttreatment and after re-infestation. On day 35, efficacy against *D. variabilis* was >96% at 1.25 and 2.5 mg/kg, while the 0.625 mg/kg dose provided only 61.4% efficacy. *Amblyomma* spp. were the least susceptible ticks; for *A. americanum*, efficacy at day 28 with the 1.25 mg/kg dose was much lower (88.5%) than that achieved at day 28 for *D. reticulatus* in the first study (100%) and also lower than at day 35 for *D. variabilis* (96.2%). *A. maculatum* efficacy at day 43 was similarly low, with the two lower doses (0.625 and 1.25 mg/kg) resulting in <70% control. This indicated that a minimum dose of between 1.25 and 2.5 mg/kg would be needed to ensure effective monthly treatment and control of fleas and multiple species of ticks following a single treatment. The final dose determination study focused on one of the dose-limiting ticks, *A. maculatum*. Dogs infested with *A. maculatum* on day-1 were dosed with sarolaner in the anticipated commercial tablet (Simparica) at 1.0, 2.0, or 4.0 mg/kg and ticks counted on day 2 and after re-infestation at days 7, 14, 28, and 35. A 100% control of the existing infestations was seen on day 2 at all doses tested. The two highest doses gave >93% control of subsequent challenges for 5 weeks, at 48 h post-infestation. The 4.0 mg/kg dose offered no significant improvement in efficacy over the 2.0 mg/kg dose ($P > 0.05$). The 2.0 mg/kg dose was superior to the 1.0 mg/kg on day 14 ($P = 0.0086$). Efficacy for the 1.0 mg/kg dose fell below 90% at day 28, and did not provide a full month of tick control. Therefore, a single oral treatment of 2.0 mg/kg sarolaner was selected as the dose rate providing effective control of fleas and all major species of ticks infesting dogs for at least 1 month.

As part of the discovery program, the team evaluated the mechanism of action and pharmacology of sarolaner [4], by recombinantly expressing the *C. felis* gamma-aminobutyric acid (GABA)-gated chloride channel RDL (resistance to dieldrin) subunits (CfRDL-A285 and CfRDL-S285) in stable CHO-K1 cell lines. A single amino acid substitution of alanine to serine at position 285 of the CfRDL-S285 subunit confers resistance to dieldrin. Electrophysiology studies were conducted using the IonWorks platform [7], which enabled rapid voltage clamp recordings of up to 384 individual cells, comparable to traditional electrophysiology measurements. Sarolaner potently blocked GABA-induced currents with IC_{50} values of 135 and 136 nM at CfRDL-A285 and CfRDL-S285 receptors, respectively, in agreement with published literature on the isoxazoline mechanism of action [2, 3, 8] and demonstrating that the resistance to dieldrin mutation does not affect sarolaner activity at the GABA receptor. As a comparison, afoxolaner was evaluated in the same assays and demonstrated 3- to 4-fold weaker blockade of GABA-induced currents, with IC_{50} values of 539 and 412 nM at CfRDL-A285 and CfRDL-S285 receptors, respectively [4]. In unpublished data, blockade of glutamate-induced currents has been demonstrated in recombinantly expressed *C. felis* glutamate-gated chloride channels; and also block of both GABA- and glutamate-induced currents in recombinantly expressed *Rhipicephalus microplus* GABA and glutamate-gated chloride channels. Finally (unpublished data), sarolaner exhibits selective blockade of flea GABA-gated chloride channels over recombinantly expressed human GABA-gated $\alpha1\beta2\gamma$ 2 chloride channels (by approximately 244-fold).

Laboratory Studies

Sarolaner was formulated in a palatable chewable tablet, and its activity was determined in controlled laboratory studies at study sites worldwide [9–11].

Flea Efficacy

Five studies were conducted to determine the efficacy of the projected minimum effective dose of 2 mg/kg sarolaner against existing *C. felis* and *Ctenocephalides canis* flea infestations and weekly challenges for 35 days after a single oral dose (Table 15.2, [9]). Four of the studies were run at facilities in the United States (US), Europe, and Australia using laboratory flea colonies regularly enriched with fleas from the field. The *C. canis* fleas were from a laboratory colony in Ireland, established with fleas from the field 4 years prior to the study commencing. *C. felis* studies were run in Arkansas, US, with fleas from a North Carolina laboratory colony, infused with Arkansas field fleas 2 years prior to the

Table 15.2 Summary of dose confirmation laboratory flea efficacy studies [9].

		Geometric mean flea counts (percentage reduction)					
		Day of count					
Flea strain origin	Treatment	Day 1	Day 7	Day 14	Day 21	Day 28	Day 35
EU	Placebo	63.9	70.2	66.5	62.4	79.1	85.5
	Sarolaner	0.0[a] (100)	0.0[a] (100)	0.0[a] (100)	0.0[a] (100)	0.0[a] (100)	0.0[a] (100)
US	Placebo	92.3	96.3	79.5	71.9	62.7	75.6
	Sarolaner	0.0[a] (100)	0.0[a] (100)	0.0[a] (100)	0.0[a] (100)	0.0[a] (100)	0.0[a] (100)
Australia	Placebo	59.2	70.7	83.3	82.3	79.6	74.9
	Sarolaner	0.07 (99.9)	0.0[a] (100)	0.0[a] (100)	0.0[a] (100)	0.0[a] (100)	0.0[a] (100)
KS1 (fipronil-resistant)	Placebo	69	90.5	84.9	94	94.4	87.1
	Sarolaner	0.0[a] (100)	0.0[a] (100)	0.0[a] (100)	0.0[a] (100)	0.15 (99.8)	0.0[a] (100)
Flea Species							
C. canis	Placebo	74.4	95.4	84.3	85.4	75.8	70.9
	Sarolaner	0.0[a] (100)	0.0[a] (100)	0.0[a] (100)	0.0[a] (100)	0.0[a] (100)	0.0[a] (100)

a) Sarolaner significantly lower than placebo ($P < 0.0001$).
Source: Reproduced with permission from Six *et al.* [9]. Copyright 2016, Elsevier.

study; the second study was run in Ireland, with a laboratory colony (from the United Kingdom (UK)), infused with field fleas (from Europe) 9 years prior to the study; the third study was run in New South Wales (NSW), Australia, with fleas from a NSW laboratory colony, infused with field fleas, also from NSW, 2 years prior to the study. A fifth study was run against the KS1 strain of fleas, which has demonstrated reduced susceptibility to many older topical flea control parasiticides.

For all five studies, at the minimum oral dose of 2 mg/kg sarolaner was highly and rapidly effective for both the treatment and control of *C. felis* over the full 35-day duration of the studies, after a single dose. Existing flea infestations were cleared by 24 h post-dosing and efficacy was ≥99.8% at all the time points over the 35 days of the study, at 24 h after re-infestation. Sarolaner, at the same dose rate, was equally effective and persistent against *C. canis* and the insecticide-tolerant *C. felis* strain, KS1. No adverse events related to treatment with sarolaner were observed in any study.

Tick Efficacy

United States (US) Ten laboratory studies were run to determine the efficacy of the minimum effective dose of 2 mg/kg sarolaner against five common tick species that infest dogs in the US, *I. scapularis, D. variabilis, R. sanguineus, A. americanum*, and *A. maculatum* (Table 15.3) [10]. Ticks of each species were acquired from eight different laboratory cultures, originally isolated from the field, with wild caught ticks introduced into each colony every couple of years or generations; one *I. scapularis* study used wild caught adult ticks from South Carolina. Treatment with a single oral dose of sarolaner chewable tablets at 2 mg/kg resulted in ≥99.6% efficacy against existing infestations of all five tick species within 48 h of treatment. For at least 35 days after treatment, the efficacy against weekly posttreatment re-infestations of all tick species was ≥96.9%, at 48 h post-infestation. No adverse events related to treatment with sarolaner were observed in any study.

These studies confirmed that a single oral administration of sarolaner at the minimum effective dose provides rapid treatment (>99.6%) of existing infestations and 5 weeks of control (≥97.1%) against re-infestation by the common US tick species infesting dogs.

Europe Eight laboratory studies were run in Ireland, Germany, and South Africa to determine the efficacy of the projected effective label dose of 2 mg/kg sarolaner against four common tick species that infest dogs in Europe, *D. reticulatus, Ixodes hexagonus, I. ricinus*, and *R. sanguineus* (Table 15.3) [11]. Ticks were sourced from laboratory-maintained colonies with isolates all originating from the field in Europe (UK, The Netherlands, Ireland, Germany, Slovakia); with new ticks being introduced into each colony within the 10 years prior to each study. Treatment with a single oral dose of sarolaner chewable tablets at 2 mg/kg resulted in 100% efficacy, within 48 h, against existing infestations of all tick species except *R. sanguineus*, where the efficacy was 99.7%. For at least 35 days

Table 15.3 Summary of dose confirmation laboratory tick efficacy studies [10, 11].

		Percentage reduction in geometric mean vive tick count					
		Day of count (48-h counts)					
Tick species	**Tick strain origin**	**Day 2**	**Day 7**	**Day 14**	**Day 21**	**Day 28**	**Day 35**
A. americanum	Stillwater, OK, USA	100	99	100	99.6	100	96.9
A. americanum	College Station, TX, USA	99.6	100	100	100	100	100
A. maculatum	College Station, TX, USA	100	100	99.7	100	100	100
A. maculatum	Stillwater, OK, USA	100	100	99.5	100	99.5	99.3
D. variabilis	Greenbriar, AR, USA	100	99.5	99.3	100	100	100
D. variabilis	Henderson, NC, USA	99.7	99.5	98.5	99.7	99.2	99.3
I. scapularis	Stillwater, OK, USA	100	100	100	100	100	100
I. scapularis	Wild caught adults SC, USA	100	100	100	100	100	100
R. sanguineus	Henderson, NC, USA	100	99.7	100	100	99.6	97.1
R. sanguineus	Greenbriar, AR, USA	100	100	100	100	100	100
R. sanguineus	Germany	99.7	100	100	99.6	99.5	99.6
D. reticulatus	Europe (various countries)	100	100	100	100	100	100
D. reticulatus	Ireland/The Netherlands	100	100	99.0	100	99.5	99.5
I. hexagonus	UK	100	ND	ND	ND	ND	100
I. hexagonus	UK	100	ND	ND	100	99.3	ND

(*continued overleaf*)

Table 15.3 (*Continued*)

		Percentage reduction in geometric mean vive tick count					
		Day of count (48-h counts)					
Tick species	**Tick strain origin**	**Day 2**	**Day 7**	**Day 14**	**Day 21**	**Day 28**	**Day 35**
I. hexagonus	The Netherlands	100	100	100	100	100	100
I. ricinus	Germany/Slovakia	100	100	100	100	98.5	100
I. ricinus	Germany	100	100	100	99.4	98.8	97.5

after treatment, the efficacy against weekly re-infestations of all tick species was ≥97.5%. No adverse events related to treatment with sarolaner were observed in any study.

These studies confirmed that a single oral administration of sarolaner at the minimum effective dose provides rapid treatment (>99.7%) of existing infestations and up to 35 days control (>99.6%) of re-infestation by common European tick species infesting dogs.

Flea Speed of Kill and Environmental Flea Control in a Simulated Infested-home Environment

Eradication of fleas from both the animal and its environment is required for effective flea control [12]. Managing fleas in the environment can be achieved with application of an insect growth regulator (IGR), such as lufenuron or methoprene, either alone, or in combination with an adulticide. However, insecticides with a rapid onset of activity can also kill adult fleas before they are able to lay eggs, thus disrupting the flea life cycle and thereby reducing the infestation levels in the environment [13]. Laboratory studies were run to evaluate the speed of kill of sarolaner against existing and posttreatment flea infestations, the effects on flea reproduction and the efficacy in a flea-infested, simulated home environment [14].

Two studies were conducted to determine the speed of kill sarolaner (at the minimum effective dose of 2 mg/kg) against both existing and re-infestations of *C. felis* on dogs; at 1, 2, 3, 4, 8, and 12 h post-dosing or weekly flea re-infestations up to 35 days post-dosing. Both these studies confirmed that sarolaner began killing fleas by 3–4 h posttreatment or after re-infestations for up to 1 month, and achieved ≥98% control of fleas by 8 h posttreatment or after re-infestation for 28 days following a single oral 2 mg/kg treatment. Complete kill was attained within 12 h of infestation over 35 days; therefore, monthly sarolaner treatments would be expected to affect the ability of fleas to reproduce, because female fleas need at least 24 h on the dog prior to the start of egg laying [15].

A third study evaluated the effects of a single oral 2 mg/kg sarolaner treatment on flea reproduction. No flea eggs were recovered a day after treatment or following weekly re-infestations until 35 days after treatment.

In the final study, dogs were held in a flea-infested simulated-home environment and the effects of monthly 2 mg/kg sarolaner treatment on flea infestations were evaluated over a 3-month period. Flea infestation reductions in sarolaner-treated dogs relative to placebo-treated dogs on days 14, 30, and 44, respectively, were 95.6%, 98.6%, and 99.6% and 100% on days 60, 74, and 90. The small numbers of fleas detected up to day 44 are likely a result of fleas emerging from eggs laid pretreatment, as the flea life cycle can take up to 8 weeks to complete [12]; the complete clearance of fleas during the final 4 weeks of the study confirms the rapid speed of kill of adults before they could lay eggs during the treatment period.

Overall, a rapid and consistent speed of kill, starting between 3 and 4 h post-treatment, was shown to completely stop flea egg laying for 1 month following a single oral 2 mg/kg treatment with sarolaner. The simulated infested-home-environment study confirmed that monthly treatment of the host with sarolaner provided excellent control of existing environmental flea infestations.

Tick Speed of Kill

As highlighted in Chapter 1, in the "Ectoparasiticides" section, transmission of diseases by vectors, especially ticks, to dogs is a major concern; consequently, decreasing the ability of a vector to attach and/or feed with an effective ectoparasiticide will reduce the risk of disease transmission. Key to success in reducing the potential risk of contracting tick-borne diseases is selection of products with rapid speed of kill; defined as the time required to kill ticks that are already attached when the product is initially administered, or to kill ticks after re-infestation after the product has already been applied. Acaricides with rapid onset of efficacy are also more effective in reducing the irritation and reducing the debilitating effects due to blood loss or tick toxicosis. An initial attachment and feeding of at least 24–48 h is required before transmission of most tick-borne pathogens can occur, a period during which reactivation of tick-borne pathogens takes place [16]. The transmission of the tick-borne pathogens may be prevented if the infected ticks are killed within that period of time [17].

Three laboratory studies were conducted to evaluate the speed of kill of the minimum effective oral dose of 2 mg/kg sarolaner against induced infestations of three tick species (*I. scapularis, I. ricinus* and *A. maculatum*) for 5 weeks post-treatment [18].

Live tick counts were reduced significantly at 8 h compared to placebo counts for all tick species, indicating that sarolaner started killing existing infestations of ticks very quickly, resulting in rapid, effective control with efficacy of 90.1% against *I. ricinus*, 98.8% against *I. scapularis*, and 99.2% against *A. maculatum* within 12 h posttreatment, and 100% efficacy within 24 h posttreatment against all three tick species. This rapid speed of kill was sustained with ≥95.7%, ≥98.7%, and ≥89.6% efficacy against *I. scapularis, I. ricinus,* and *A. maculatum*, respectively, at 24 h after re-infestation through at least day 28. There were no adverse events observed in any of the studies that were considered related to sarolaner treatment.

This rapid speed of kill and consistent and persistent efficacy for a full month against ticks following oral administration demonstrates that sarolaner will effectively reduce the chances of ticks surviving and/or feeding for the critical period required for the transmission of most tick-borne diseases. Used in a control program, Simparica will reduce the impact of tick infestation with the additional potential to reduce the risk of dogs contracting tick-borne diseases.

Flea and Tick Field Studies

In the field, dogs are continuously exposed to re-infestations by fleas and ticks from the environment. Ectoparasiticides should therefore provide not only immediate rapid efficacy but also persistent efficacy after a single administration until the end of the treatment period to protect the dogs from re-infestation. Reducing the ability of a vector to attach and/or feed, with an effective ectoparasite control program, will reduce the risk of disease transmission. The laboratory studies described in the section "Laboratory Studies", earlier in this chapter, demonstrate that sarolaner has these characteristics, but how does this translate to efficacy in the home environment?

The efficacy and safety of sarolaner was evaluated for the control of fleas on dogs in a randomized, controlled clinical study run in 19 general veterinary practices across the United States [19]. Around 479 dogs from 293 households were enrolled and tablets administered orally once a month for 3 months. Dogs were randomly assigned to treatments with either sarolaner (Simparica) chewable at a minimum dose of 2.0 mg/kg (range 2–4 mg/kg), or an approved comparator product, Comfortis® (oral spinosad, Elanco), at the label dose. Most sarolaner tablets (91.5%) were taken by free choice from the hand or in food; only 8.5% of doses had to be pilled. This compared with 81.1% free choice and 18.9% pilling with spinosad tablets.

After the first treatment, live flea counts were reduced by >99% after 14 days (the first efficacy time point) and continued to reduce through the study. Treatment success rates, based on the number of dogs with ≥90% reduction in fleas, were statistically superior for sarolaner compared with spinosad ($P \leq 0.025$) on days 14 and 30 (97.7 vs 90.0% and 98.3 vs 69.0%, respectively). Following the second and third treatments, efficacy was similar for the two products. This confirmed that the excellent efficacy of sarolaner against flea infestations demonstrated in the laboratory studies [9, 14] was corroborated in the general dog population under typical use conditions, with treatment administered by the owner. The rapid reduction of flea infestations in these dogs reflects the rapid onset of activity of sarolaner demonstrated in speed of kill studies [14] indicating that fleas are likely to be killed before they are able to lay eggs [14, 20]. This effected a reduction in the flea populations in the environment, rapidly killing newly emerged fleas on the dogs before they laid eggs and contributed to re-infestation of the environment. The >99% reduction of fleas within 14 days of the first treatment is consistent with the results obtained in a simulated-home, flea-infested environment study [14].

Prior to the first treatment, almost 90% of the sarolaner-treated dogs had pruritus, the most common sign associated with flea allergy; this was reduced to only 9% of dogs by the end of the study. Similarly 49.9% of papules pre-study were reduced to 1.9%; 96.2% erythema to 13.2%; 67.9% scaling to 15.1%, and alopecia from self-trauma was reduced from 69.8% pre-study to 5.7% post-study. This shows that treatment with sarolaner tablets, causing a rapid reduction in flea infestations, led to a concurrent rapid resolution of the clinical signs of flea allergic dermatitis (FAD). Sarolaner was well tolerated by dogs in the study.

The efficacy and safety of sarolaner (Simparica), was also evaluated for the control of fleas and ticks on dogs in a randomized, multicentered clinical study conducted at veterinary clinics in Belgium, Hungary, Italy, France, and the United Kingdom [21]. In the flea study, 285 primary and 137 supplementary dogs, harboring natural infestations of ≥5 live fleas, were enrolled and treated. In the tick study, 181 primary and 48 supplementary dogs, harboring natural infestations of ≥3 live attached ticks, were enrolled and treated. Dogs were randomly assigned to treatments with either sarolaner (Simparica) chewable at a minimum dose of 2.0 mg/kg (range 2–4 mg/kg), or an approved comparator product; spinosad (Comfortis Chewable Tablets, Elanco) was used in the flea study; and fipronil (Frontline® Spot on, Merial) was used as positive control in the tick study. Tablets were dosed orally once a month for 3 months. In the two studies, sarolaner chewable tablets were well accepted, with 93% taken by free choice from the hand; this compared with 84.2% for Comfortis tablets. Sarolaner chewable tablets may be administered with or without food, so the high palatability and ease of administration of sarolaner chewable tablets should increase owner compliance, a key contributory factor for the success of companion animal ectoparasite treatment [22]. Efficacy for the European field studies is reported as percentage reduction in posttreatment arithmetic mean flea or tick counts compared to pretreatment counts, arithmetic means, following European regulatory guidelines.

In the flea study, sarolaner treatment caused a rapid decrease in the incidence of infestations with >95% of dogs being flea-free at all posttreatment time points, starting at day 14. The incidence of *C. felis* infestations was higher for spinosad-treated dogs with >10% of dogs having up to 4 or 39 fleas on days 14 and 30, respectively (<90% of dogs were flea–free), and 2.3% and 4.3% of dogs with up to 17 and 19 fleas on days 60 and 90, respectively. There were no treatment-related adverse events in sarolaner-treated dogs.

For *C. canis*, sarolaner treatment was also effective, with only a single dog with a single flea found from day 30 onward, similar to observations in dogs treated with spinosad. Thirty primary dogs in the sarolaner-treated group and 12 in the spinosad-treated group were identified as having FAD at enrolment. The clinical signs of FAD improved in all dogs following administration of treatment in both groups. Incidence of any one of the clinical signs of FAD ranged from 50% to 100% at enrolment; by completion of the study, the incidence of FAD signs had dropped to 0–8.3% in both groups.

In the tick field study, 52.8% of dogs were infested at enrollment with *I. ricinus*, 41.1% with *R. sanguineus*, 23.9% with *D. reticulatus*, and 6.7% with *I. hexagonus*. Some dogs harbored infestations of more than one tick species. In general, the initial incidence of each species was similar for each treatment group. In the sarolaner-treated group, efficacy on posttreatment days 14, 30, 60, and 90 was 97.4%, 97.6%, 99.8%, and 100%, and in the fipronil group, 94.1%, 88.5%, 89.9%, and 98.1%. Sarolaner was non-inferior to fipronil at all time points, and superior to fipronil on days 30 and 60. The high efficacy of sarolaner against *D. reticulatus* during the study period is noteworthy, as this species is the main vector of *Babesia canis*, a potentially fatal disease in dogs. *I. hexagonus* and *I. ricinus*, the most prevalent tick species infesting dogs in Europe, are the main vectors of *Borrelia burgdorferi sensu lato* which causes Lyme disease in dogs and people. At 30 days after the second treatment, neither of these species was found on any sarolaner-treated dogs and no live ticks of any species were found after the third treatment.

By contrast, fipronil had relatively poor efficacy against both *Ixodes* species, at these time points, with a maximum of 145 *I. hexagonus* recovered from almost 6% of fipronil-treated dogs and a maximum of 59 *I. ricinus* recovered from nearly 10% of fipronil-treated dogs at day 60, and a maximum of 10 live ticks found on over 6% of dogs 30 days after the third treatment. By reducing the ability of the ticks to remain attached and/or feed, sarolaner will reduce the risk of disease transmission. There were no treatment-related adverse events in sarolaner-treated dogs.

From these three studies it can be concluded that oral sarolaner, administered at monthly intervals at a minimum dosage of 2 mg/kg, was safe and highly effective against natural infestations of fleas and ticks on dogs in geographically separate locations. The flavored, chewable tablets were highly palatable and clinical signs associated with FAD improved in dogs treated with sarolaner.

Mite Efficacy

Efficacy Against *Demodex* Infestations in Dogs

Demodicosis is a debilitating and often life-threatening inflammatory disease in dogs with an impaired or underdeveloped immune system, caused by sensitivity to large numbers of *Demodex* spp. mites, and is difficult to control with existing therapies [23]. A laboratory study was therefore run in dogs to evaluate the efficacy and safety of sarolaner, at the minimum dose of 2 mg/kg, against natural infestations of *Demodex* spp.

Naturally infested dogs were allocated to blocks of two by descending mite count and randomly assigned to treatment with either sarolaner or the positive control, topical imidacloprid plus moxidectin (Advocate®/Advantage® Multi Spot-on solution for Dogs, 100 mg imidacloprid/25 mg moxidectin/ml, Bayer), with eight dogs assigned to each group. Dogs in the sarolaner group were dosed orally with a single tablet that was trimmed to provide the target dose of 2 mg/kg of sarolaner on days 0, 30, and 60. Positive control dogs were dosed topically once weekly from day 0 to day 81 with the appropriate band dose providing ≥10 mg imidacloprid/≥2.5 mg moxidectin/kg, per label directions for dogs with severe generalized demodicosis. There were no adverse events related to treatment with sarolaner.

A single oral dose of 2.0 mg/kg resulted in >97% reduction in *Demodex* mite counts at 14 days and >99% at 29 days with a concurrent improvement in the clinical signs of demodicosis. Following a second monthly treatment, no live mites were recovered from any sarolaner-treated dog. This level of efficacy compared favorably to that of the commercial comparator (topical imidacloprid plus moxidectin), as a dog in the comparator group continued to host live mites even after administration of 11 weekly treatments.

Sarolaner administered at three consecutive monthly doses of 2 mg/kg in dogs with generalized demodicosis was highly effective in eliminating mites and resolving clinical signs of the disease. In regions where it is approved, Simparica offers an attractive alternative to existing therapies for this debilitating parasite infestation.

Efficacy Against *Otodectes cynotis* Infestations in Dogs

Otodectic mange is caused by *Otodectes cynotis*, an obligate parasite which inhabits the vertical and horizontal ear canals of dogs and cats. Liquid aural treatments (including antibiotics, antifungals, steroids, and parasiticides) are utilized for treatment, but require regular reapplication; monthly topical spot-on treatments are also available and effective [24]. An attractive therapeutic option would be a systemic product with persistent activity.

A laboratory study was conducted in dogs to evaluate the efficacy and safety of sarolaner, at the minimum dose of 2 mg/kg against induced aural infestations of *O. cynotis*. Dogs were assigned to single- or two-dose treatments, with either placebo or sarolaner oral tablet, trimmed to provide the target dose of 2 mg/kg, with eight dogs per treatment or placebo group. A single oral 2 mg/kg dose of sarolaner resulted in a 98.2% reduction in live mite counts; with two doses of sarolaner, administered 1 month apart, resulting in a 99.5% reduction in ear mites compared to placebo-treated controls. There were no adverse events related to treatment with sarolaner.

Against an induced infestation of *O. cynotis* in dogs, oral sarolaner reduced mite counts by >98% after a single dose and by >99% after two monthly doses. In regions where it is approved, Simparica offers an attractive alternative to existing therapies for this parasite infestation.

Efficacy Against *Sarcoptes scabiei* Infestations in Dogs

Sarcoptes scabiei var. *canis* is a highly contagious and zoonotic parasite and is one of most common mites infesting dogs worldwide, causing severe pruritus, with potential for concomitant secondary bacterial and yeast infections. Licensed treatments are primarily topical. These include selamectin and moxidectin/imidacloprid containing spot-on products; plus, in some countries amitraz dip, with adjunctive therapy including shampooing/bathing with products to rehydrate the skin and treat seborrhea. This may also have the regrettable effect of reducing the efficacy and/or persistence of the topical treatments. Milbemycin oxime, delivered orally, is also approved in some regions, but has to be dosed every other day. Having an oral monthly treatment option would therefore be a convenient and effective option for many patients.

A laboratory study was run to evaluate the efficacy of 2 monthly doses of sarolaner, at the minimum dose of 2 mg/kg, for the treatment of sarcoptic mange in dogs. In addition, a multicentered field study was conducted in veterinary patients to confirm the efficacy and safety of this treatment and dosing regimen against *S. scabiei* [25]. In the laboratory study, infested dogs were assigned to a sarolaner treatment group or to a placebo group, based on pretreatment mite counts. In the field study, dogs were dosed with the Simparica chewable tablets to provide the recommended minimum dose of 2 mg/kg (range 2–4 mg/kg). The imidacloprid/moxidectin spot-on treatment (Advocate) was applied topically according to its label directions to deliver 10–25 mg/kg imidacloprid and 2.5–6.25 mg/kg moxidectin.

Sarolaner achieved 100% parasitological cure in both studies following 2 monthly administrations, with clinical signs of sarcoptic mange markedly improved in the treated dogs. The efficacy of sarolaner was non-inferior to the topically applied positive control in the field study; however, parasitological cure was not achieved in one imidacloprid/ moxidectin-treated dog after 2 monthly treatments, whereas mites were eliminated from all sarolaner-treated dogs. Sarolaner tablets were also highly palatable with 90.5% acceptance by free choice within 1 min of offering. There were no adverse events related to treatment with sarolaner.

Sarolaner, administered orally twice at monthly intervals at the minimum label dosage of 2 mg/kg, was safe and achieved complete parasitological cure in dogs with natural infestations of *S. scabiei*. The clinical signs of sarcoptic mange also improved without topical or systemic concomitant treatment, confirming that Simparica offers a convenient, efficacious therapy for sarcoptic mange, in regions where it is approved.

Prevention of Tick-borne Disease Transmission

Reducing the ability of a vector to attach and/or feed with an effective ectoparasiticide program will reduce the risk of disease transmission. Currently, recommendations for pet owners are to reduce tick populations in the dogs' environment, frequently examine and remove any ticks found on the dog, and to use an approved product year-round to protect dogs against unexpected exposure to ticks [26].

A laboratory study was therefore run to evaluate the efficacy of sarolaner (Simparica) to prevent transmission of *B. burgdorferi* and *Anaplasma phagocytophilum* from infected wild-caught *I. scapularis* to dogs [27].

Dogs were allocated to one of three oral treatment groups (eight dogs per group): placebo administered on days 0 and 7, or sarolaner at 2 mg/kg administered on day 0 (28 days prior to tick infestation) or on day 7 (21 days prior to tick infestation). On day 28, all dogs were infested with 25 female and 25 male wild-caught adult *I. scapularis* that had 57% prevalence for *B. burgdorferi* and 6.7% for *A. phagocytophilum*. Tick counts were conducted on days 29 and 30. On day 33, all ticks were counted and removed. There were no adverse reactions to treatment with sarolaner.

The live tick count reductions at day 29, day 30, and day 33 were 86.3%, 100%, and 100% for the group treated with sarolaner 21 days prior to infestation, and 90.9%, 97.1%, and 100% for the group treated with sarolaner 28 days prior to infestation. Transmission of *B. burgdorferi* to all eight placebo-treated dogs was confirmed by positive antibody, polymerase chain reaction (PCR), and/or culture. Similarly, transmission of *A. phagocytophilum* was confirmed by the presence of antibodies in four of eight placebo-treated dogs. By contrast, treatment with a single dose of sarolaner prevented transmission of *B. burgdorferi* from infected ticks to dogs infested 21 or 28 days after treatment. Prevention of transmission of *A. phagocytophilum* was demonstrated in all sarolaner-treated dogs.

As highlighted earlier, attachment and feeding of at least 24–48 h is required before transmission of most tick-borne pathogens can occur [16]; the transmission of the tick-borne pathogens may be prevented if infected ticks are killed, or feeding halted, within that period of time [17]. After a single dose of sarolaner, the rapid speed of kill, within 24 h after treatment and with persistent high efficacy for a full month, indicates that treatment with sarolaner is likely to effectively reduce the chances of ticks surviving and/or feeding for this vital time period. The use of sarolaner in a tick-control program should therefore reduce the risk of dogs becoming infected with tick-borne diseases, as well as minimize the negative health effects accompanying tick infestation.

This study demonstrated that a single oral dose of sarolaner (Simparica) was successful at preventing transmission of *B. burgdorferi* and *A. phagocytophilum* from infected wild-caught *I. scapularis,* to dogs when challenged 28 days after treatment.

Comparative Flea and Tick Efficacy

Sarolaner (Simparica) is the most recently approved product in the isoxazoline class of parasiticides for dogs, with proven rapid onset of activity and sustained persistence against flea and tick infestations. There are many topical and oral products already available, with differing profiles with respect to spectrum of activity, speed of kill, and duration and consistency of efficacy. The efficacy of sarolaner was compared with other commonly used flea and tick control products, both topical and oral, over the approved dosing intervals. These comparisons allow an evidence-based evaluation of the relative onset of effect, speed of action, and duration and consistency of protection these products provide against flea and ticks throughout the dosing interval.

Comparative Flea Studies Against Topical and Oral Products

Rapid kill of fleas is desirable to relieve both the immediate irritation caused by fleas as well as to reduce associated allergenic responses (FAD) and the risk of transmission of flea-borne pathogens [28]. Rapid speed of kill of adult fleas on the host is also important in the control of infestations, as female fleas do not begin producing eggs until 24–48 h after they start feeding [16]. Killing fleas before they lay eggs will, over time, effectively control an environmental infestation.

Three studies were conducted to evaluate the speed of kill and persistence of sarolaner against induced infestations of fleas, when compared to afoxolaner (NexGard®, Merial) [29], fluralaner (Bravecto®, Merck) [30] and spinosad/milbemycin oxime (Trifexis®, Elanco) [31]. Summary data can be found in Table 15.4.

In the afoxolaner and spinosad/milbemycin oxime studies, dogs were treated with either a placebo tablet; a Simparica chewable tablet to deliver sarolaner at

Table 15.4 Flea efficacy: comparative speed of kill [29–31].

	Sarolaner geometric mean live flea counts Significantly ($P < 0.05$) lower than comparator[a)] – yes/no					
Count day	**Day 0**	**Day 7**	**Day 14**	**Day 21**	**Day 28**	**Day 35**
NexGard®						
8 h[b)]	Yes	Yes	Yes	Yes	Yes	Yes
12 h	No	No	No	No	Yes	Yes
24 h	No	No	No	No	No	No
Trifexis®						
8 h	No	No	No	Yes	Yes	Yes
12 h	No	No	No	No	No	Yes
24 h	No	No	No	No	No	Yes
Count day	**Day 0**	**Day 14**	**Day 29**	**Day 44**	**Day 274**	**Day 90**
Bravecto®						
8 h[c)]	No	No	No	No	Yes	Yes
12 h	No	No	No	No	No	No
24 h	No	No	No	No	No	No

a) No comparator was significantly lower than sarolaner.
b) Count times after treatment on day 0 and after each subsequent weekly infestation. Dogs treated once on day 0.
c) Count times after treatment on day 0 and after each subsequent infestation. Sarolaner administered on days 0, 30, and 60; Bravecto® administered on day 0 only. Counts conducted prior to treatment on days 30 and 60.

the minimum label dose of 2 mg/kg (range 2–4 mg/kg); or NexGard (afoxolaner at 2.5–6.8 mg/kg); or Trifexis (spinosad at 30–60 mg/kg plus milbemycin oxime at 0.2–0.4 mg/kg). Live fleas were counted at 8, 12, and 24 h posttreatment and at successive re-infestations on days 7, 14, 21, 28, and 35. In the sarolaner and afoxolaner comparative study [29], a single oral dose of sarolaner delivered ≥98.8% efficacy within 8 h of treatment or successive weekly re-infestations of fleas to day 35 and by 12 h posttreatment fleas were virtually eliminated from all dogs; efficacy was 100% at all other time points (Table 15.4). Significantly higher numbers of live fleas were recovered from afoxolaner-treated dogs at 8 h on all days and at 12 h on days 28 and 35 ($P < 0.05$). In the sarolaner and spinosad/milbemycin oxime study [31], a single oral dose of sarolaner delivered ≥ 94.0% efficacy within 8 h of treatment or successive weekly re-infestations of fleas to day 35 (Table 15.4). By 12 h, fleas were eliminated from all dogs and they remained free of fleas at 24 h. Significantly higher numbers of live fleas were recovered from spinosad/milbemycin oxime-treated dogs at 8 h from day 21 to day 35 ($P \leq 0.0085$), and at 12 and 24 h on day 35 ($P \leq 0.0002$). There were no adverse reactions to treatment.

In the third study [30], dogs were treated with either a placebo tablet; the appropriate strength Simparica chewable tablet to provide sarolaner at the recommended label dose of 2 mg/kg (range: 2–4 mg/kg) on days 0, 30, and 60; or Bravecto on day 0 only (fluralaner at 25–50 mg/kg). On days 30 and 60, dogs in the fluralaner group were administered placebo. Three monthly doses of sarolaner provided ≥97.6% efficacy (by arithmetic means) within 8 h of treatment or at weekly re-infestations of fleas for 3 months. By 12 h posttreatment or re-infestation, 100% fleas were eliminated from all dogs. Significantly higher numbers of live fleas were recovered from fluralaner-treated dogs at 8 h posttreatment on days 74 and 90 ($P \leq 0.0043$) with efficacies of only 80.7% and 72.6%, respectively (Table 15.4). There were no adverse reactions to treatment.

Sarolaner (Simparica) was demonstrated to deliver ≥94.0% efficacy within 8 h of treatment or post-weekly flea infestations, across all three studies. By 12 h posttreatment or re-infestation, fleas were eliminated or virtually eliminated at all the time points. Toward the end of the treatment period, sarolaner maintained the rapid and consistent speed of kill, whereas the speed of kill of the comparator treatments declined.

Comparative Tick Studies

Comparative Speed of Kill of Sarolaner and Afoxalaner Against Induced Infestations of *I. scapularis* *I. scapularis*, or black-legged (deer) tick, is the primary vector for *B. burgdorferi* and *A. phagocytophilum*, which cause Lyme borreliosis and granulocytic anaplasmosis in humans and dogs [28]. They also secrete a salivary neurotoxin that may cause tick paralysis in dogs and humans. While tick efficacy for regulatory claims are based on evaluation at 48 h posttreatment or re-infestation [32], the speed of kill is key in disrupting or preventing feeding and thereby reducing the risk of pathogen transmission, which occurs after the infected tick has attached and been feeding for 24–48 h, for many pathogens [16, 17].

In this study, dogs were infected at day 2 and treated with either a placebo tablet, a sarolaner chewable tablet (range 2–4 mg/kg), or afoxolaner tablet (at 2.5–6.8 mg/kg) [33]. Live ticks were then counted at 8, 12, and 24 h posttreatment and successive re-infestations on days 7, 14, 21, 28, and 35. Sarolaner reduced tick counts significantly versus placebo from day 0 to day 21 at 8 and 12 h, and on day 35 at 12 h ($P \leq 0.0174$), while afoxolaner was significantly lower at 8 h only on days 0 and 14 ($P \leq 0.0309$), and at 12 h only on day 0 ($P < 0.0001$). Significantly more live ticks were recovered from afoxolaner-treated dogs than from sarolaner-treated dogs at 24 h post-infestation from day 14 to day 35 ($P \leq 0.0278$). At 24 h, efficacy of afoxolaner declined to <80% from day 21 to the end of the study, while efficacy for sarolaner was >95% for the full 35 days, at 24 h. There were no adverse reactions to treatments.

Sarolaner had a faster speed of kill against *I. scapularis* than afoxolaner, an effect that was markedly more evident toward the end of the monthly treatment period. This therapy should provide highly effective and reliable control of ticks over the entire treatment interval, and reduce the risk of tick-borne diseases, including Lyme disease and anaplasmosis.

Comparative Speed of Kill of Sarolaner and Imidicloprid/Permethrin Against Induced Infestations of *D. reticulatus* *D. reticulatus* infests dogs throughout Europe and is a vector for *B. canis*, a lethal protozoan disease in dogs. Rapid kill of the tick before *B. canis* can be transmitted is desirable.

In this study, dogs were dosed with either a placebo tablet, sarolaner at 2–4 mg/kg or a topical application of Advantix® Spot-on solution for dogs (imidacloprid + permethrin at 10–25 mg/kg imidacloprid and 50–125 mg/kg permethrin) [34]. On days 0, 7, 14, 21, 28, and 35, live ticks were counted 8, 12 (±0.5) and 24 (±0.5) h posttreatment or following each weekly re-infestation. The efficacy of sarolaner was ≥75.6% (89.6%), by arithmetic (geometric) mean tick counts, within 8 h of treatment, with tick counts significantly lower than placebo and imidacloprid/permethrin-treated dogs ($P < 0.0001$). Imidacloprid/permethrin caused no significant reduction ($P \geq 0.3990$) at 8 or 12 h posttreatment. Sarolaner killed all ticks within 24 h of treatment, while the efficacy of imidacloprid/permethrin was only 48.1%. Following weekly re-infestations, sarolaner reduced the tick counts significantly versus placebo within 8 h on days 7, 14, and 35 ($P \leq 0.0239$), and at 12 and 24 h ($P \leq 0.0079$) until day 35. Sarolaner efficacy was ≥95.8% within 24 h for 35 days. Significantly more live ticks ($P \leq 0.0451$) were recovered from imidacloprid/permethrin-treated dogs than from sarolaner-treated dogs at 24 h after infestation on all days. There were no sarolaner-related adverse reactions during the study.

Oral Simparica had a faster and more consistent speed of kill against *D. reticulatus* compared to topically applied Advantix. With rapid and consistent efficacy within 24 h for a full 5 weeks after a single oral dose, Simparica provides effective, reliable control of *D. reticulatus*, reducing the risk of transmission of *B. canis*.

Comparative Speed of Kill of Sarolaner, Afoxolaner, and Fluralaner Against Induced Infestations of *A. americanum* The lone star tick, *A. americanum*, infests dogs and cats in North America transmitting *Ehrlichia chaffeensis* and *Ehrlichia ewingii*, which cause monocytic and granulocytic ehrlichiosis in dogs and humans, and *Cytauxzoon felis* which causes cytauxzoonosis in cats. *A. americanum* was shown to be one of the dose-limiting parasites for Simparica [6].

Two studies were run to evaluate the speed of kill of sarolaner, compared to afoxolaner and fluralaner, against induced infestations of *A. americanum* [35, 36]. In the first study, dogs were treated with sarolaner (2–4 mg/kg), afoxolaner (2.5–6.8 mg/kg), or a placebo and live ticks counted at 8, 12, and 24 h posttreatment and following re-infestations on days 7, 14, 21, 28, and 35. A single oral dose of sarolaner delivered 100% efficacy within 24 h of treatment, and consistently provided >90% efficacy against successive weekly re-infestations with ticks to day 28. Significantly more live ticks were recovered from afoxolaner-treated dogs than from sarolaner-treated dogs at 24h after infestation from day 7 through day 35 ($P \leq 0.0247$). At 24h, efficacy of afoxolaner dropped to <90% from day 14 to the end of the study. No sarolaner-related adverse reactions were observed during the study.

In the second study, dogs were treated with placebo or sarolaner at the label rate (2–4mg/kg) on days 0, 30, and 60 or with fluralaner (25–56mg/kg) once according to label instructions on day 0. Live ticks were counted at 8, 12, and 24h posttreatment and following re-infestations on days 14, 28, 42, 58, 76, and 90. No sarolaner-related adverse reactions were observed during the study. Monthly oral dosing of sarolaner delivered >95% efficacy within 24h of treatment, with consistent >70% efficacy against subsequent re-infestations with ticks within 24h over the whole treatment period. Significantly more live ticks were recovered from fluralaner-treated dogs than from sarolaner-treated dogs at 24h after re-infestation from day 42 onwards. At 24h, efficacy of fluralaner had declined to ≤20% from day 42 through to the end of the study on day 90.

These studies demonstrated that sarolaner had a faster speed of kill than both afoxalaner and fluralaner, highlighting that a single oral dose of Simparica provides effective, reliable control of *A. americanum*, and reducing the risk of transmission of tick-borne diseases.

Comparative Speed of Kill of Sarolaner, Afoxolaner, and Fluralaner Against Induced Infestations of *R. sanguineus* The brown dog tick, *R. sanguineus sensu lato*, infests dogs globally and is the major vector of *E. canis*, which causes canine monocytic ehrlichiosis; and *Babesia vogeli*, the causative agent of canine babesiosis, and a number of other important pathogens.

Two studies were run to evaluate the speed of kill of sarolaner, compared to afoxolaner and fluralaner, against induced infestations of *R. sanguineus* [37, 38]. The study design was identical to the *A. americanum* study described, apart from infestation with *R. sanguineus*. There were no treatment-related adverse reactions observed during the studies.

In the first study, sarolaner provided, by geometric means, >94% efficacy within 8h of treatment, and >99% after 12 and 24h. With subsequent weekly re-infestations of ticks, sarolaner achieved ≥91.7% efficacy to day 35 at 24h. Sarolaner also significantly reduced tick counts versus placebo on days 0 and 28 at 8h ($P \leq 0.0390$), on days 0 to 14 and 28 at 12h ($P \leq 0.0142$), and on all days at 24h ($P < 0.0001$). By comparison, tick counts for afoxolaner were significantly lower than placebo at 8h on days 0 and 28 ($P \leq 0.0117$), at 12h on day 0 ($P < 0.0001$), and on all days at 24h ($P \leq 0.0078$). Significantly more live ticks were recovered from afoxolaner-treated dogs than from sarolaner-treated dogs 8 and 12h posttreatment ($P \leq 0.0286$), at 12h after re-infestation on days 7 and 28 ($P \leq 0.04630$), and at 24h post-re-infestations from day 7 to day 35 ($P \leq 0.0119$). At 24h, afoxolaner efficacy was <90% from day 7 onwards, and declined to <45% by day 35, while efficacy for sarolaner was >90% for 35 days.

In the second study, both sarolaner and fluralaner significantly reduced live ticks within 8h posttreatment against an existing infestation with *R. sanguineus*, and killed all ticks within 24 h. After re-infestation, however, sarolaner provided ≥98.5% reduction within 24 h on all days except days 74 and 95 ($P < 0.0001$), compared to fluralaner which provided ≥95.5% reduction only until day 44. Geometric mean live tick counts for sarolaner were significantly lower ($P \leq 0.0415$) at 24 h posttreatment and after re-infestation than those for fluralaner on all days, except on days 0, 14, and 28 ($P \geq 0.0678$).

These studies demonstrated that sarolaner had a significantly faster speed of kill than both afoxalaner and fluralaner against the brown dog tick, *R. sanguineus.*

Summary of Comparative Studies These comprehensive comparative studies reflect the success of the Zoetis strategy to design and develop the novel isoxazoline, sarolaner (Simparica), with properties expressly optimized for use in companion animals.

Commercialization of Sarolaner

Sarolaner was discovered and developed by Zoetis, with approvals being achieved in the European Union and New Zealand in late 2015 and Australia, Brazil, Canada, Ecuador, Nicaragua, and the United States by the end of 2016. Further submissions are in review.

Acknowledgments

We acknowledge the commitment and efforts of Dr. Steven Maeder and the Zoetis colleagues who enabled the successful development, approval and launch of Simparica. Thanks also go to Dr. Maeder for his helpful comments in reviewing the final manuscript.

References

1. Beugnet, F. and Franc, M. (2012) Insecticide and acaricide molecules and/or combinations to prevent pet infestation by ectoparasites. *Trends Parasitol.*, **28**, 267–279.
2. Shoop, W.L., Hartline, E.J., Gould, B.R., Waddell, M.E., McDowell, R.G., Kinney, J.B., Lahm, G.P., Long, J.K., Xu, M., Wagerle, T., Jones, G.S., Dietrich, R.F., Cordova, D., Schroeder, M.E., Rhoades, D.F., Benner, E.A., and Confalone, P.N. (2014) Discovery and mode of action of afoxolaner: a new isoxazoline parasiticide for dogs. *Vet. Parasitol.*, **201**, 179–189.
3. Gassel, M., Wolf, C., Noack, S., William, H., and Ilg, T. (2014) The novel isoxazoline ectoparasiticide fluralaner: selective inhibition of arthropod gamma-aminobutyric acid- and -glutamate-gated chloride channels and insecticidal/acaricidal activity. *Insect Biochem. Mol. Biol.*, **45**, 111–124.
4. McTier, T.L., Chubb, N., Curtis, M.P., Hedges, L., Inskeep, G.A., Knauer, C.S., Menon, S., Mill, B., Pullins, A., Zinser, E., Woods, D.J., and Meeus, P. (2016) Discovery of sarolaner: a novel, orally administered, broad-spectrum,

isoxazoline ectoparasiticide for dogs. *Vet. Parasitol.*, **222**, 3–11.

5. Curtis, M.P., Vaillancourt, V., Goodwin, R.M., Chubb, N.A.L., Howson, W., McTier, T.L., Pullins, A., Zinser, E.W., Meeus, P.F.M., Woods, D.J., Hedges, L., Stuk, T., Price, J.E., Koch, J.D., and Menon, S.R. (2016) Design and synthesis of sarolaner, a novel, once-a-month, oral isoxazoline for the control of fleas and ticks on dogs. *Bioorg. Med. Chem. Lett.*, **26**, 1831–1835.
6. McTier, T.L., Six, R.H., Fourie, J.J., Pullins, A., Hedges, L., Mahabir, S.P., and Myers, M. (2016) Determination of the effective dose of a novel oral formulation of sarolaner (Simparica™) for the treatment and month-long control of fleas and ticks on dogs. *Vet. Parasitol.*, **222**, 12–17.
7. Schroeder, K., Neagle, B., Trezise, D.J., and Worley, J. (2003) IonWorks HT: a new high-throughput electrophysiological measurement platform. *J. Biomol. Screen.*, **8**, 50–64.
8. Ozoe, Y., Asahi, M., Ozoe, F., Nakahira, K., and Mita, T. (2010) The antiparasitic isoxazoline A1443 is a potent blocker of insect ligand-gated chloride channels. *Biochem. Biophys. Res. Commun.*, **391**, 744–749.
9. Six, R.H., Geurden, T., Packianathan, R., Colgan, S., Everett, W.R., Grace, S., Hodge, A., Mahabir, S.P., Myers, M.R., Slootmans, N., and Davis, K. (2016) Evaluation of the effectiveness of a novel oral formulation of sarolaner for the treatment and control of fleas on dogs. *Vet. Parasitol.*, **222**, 18–22.
10. Six, R.H., Everett, W.R., Young, D.R., Carter, L., Mahabir, S.P., Honsberger, N.A., Myers, M.R., Holzmer, S., Chapin, S., and Rugg, J.J. (2016) Efficacy of a novel oral formulation of sarolaner (Simparica™) against five common tick species infesting dogs in the United States. *Vet. Parasitol.*, **222**, 28–32.
11. Geurden, T., Becskei, C., Grace, S., Strube, C., Doherty, P., Liebenberg, J., Mahabir, S.P., Slootmans, N., Lloyd, A., and Six, R.H. (2016) Efficacy of a novel oral formulation of sarolaner (Simparica™) against four common tick species infesting dogs in Europe. *Vet. Parasitol.*, **222**, 33–36.
12. Rust, M.K. and Dryden, M.W. (1997) The biology, ecology, and management of the cat flea. *Annu. Rev. Entomol.*, **42**, 451–473.
13. Jacobs, D.E., Hutchinson, M.J., and Ryan, W.G. (2001) Control of flea populations in a simulated home environment model using lufenuron: imidacloprid or fipronil. *Med. Vet. Entomol.*, **25**, 73–77.
14. Six, R.H., Becskei, C., Carter, L., Gale, B., Young, D.R., Mahabir, S.P., Chapin, S., and Myers, M.R. (2016) Evaluation of the speed of kill, effects on reproduction, and effectiveness under a simulated infested-home environment of sarolaner (Simparica™) against fleas on dogs. *Vet. Parasitol.*, **222**, 43–48.
15. Dryden, M.W. and Broce, A.B. (2002) Integrated flea control for the 21st century. *Compend. Contin. Educ. Vet.*, **24** (Suppl), 36–39.
16. Little, S.E. (2007) Changing paradigms in understanding transmission of canine tick-borne diseases: the role of interrupted feeding and intrastadial transmission. 2nd Canine Vector-Borne Disease (CVBD) Symposium, Mazara del Vallo, Sicily, Italy, 30–34.
17. Kidd, L. and Breitschwerdt, E.B. (2003) Transmission times and prevention of tick-borne diseases in dogs. *Compend. Contin. Educ. Vet.*, **10**, 742–751.
18. Six, R.H., Geurden, T., Carter, L., Everett, W.R., McLoughlin, A., Mahabirm, S.P., Myers, M.R., and Slootmans, N. (2016) Evaluation of the speed of kill of sarolaner (Simparica™) against induced infestations of three species of ticks (*Amblyomma maculatum, Ixodes scapularis, Ixodes ricinus*) on dogs. *Vet. Parasitol.*, **222**, 37–42.
19. Cherni, J.A., Mahabir, S.P., and Six, R.H. (2016) Efficacy and safety of sarolaner (Simparica™) for the treatment and prevention of natural flea infestations on client-owned dogs in the US. *Vet. Parasitol.*, **222**, 43–48.
20. Dryden, M.W., Payne, P.A., Lowe, A., Mailen, S., Smith, V., and Rugg, D.

(2007) Efficacy of a topically applied formulation of metaflumizone on cats against the adult cat flea, flea egg production and hatch, and adult flea emergence. *Vet. Parasitol.*, **150**, 263–267.

21. Becskei, C., De Bock, F., Illambas, J., Mahabir, S.P., and Six, R.H. (2016) Efficacy and safety of a novel oral isoxazoline, sarolaner (Simparica™), in the treatment of naturally occurring flea and tick infestations in dogs presented as veterinary patients in Europe. *Vet. Parasitol.*, **222**, 49–55.
22. Halos, L., Beugnet, F., Cardoso, L., Farkas, R., Franc, M., Guillot, J., Pfister, K., and Wall, R. (2014) Flea control failure? Myths and realities. *Trends Parasitol.*, **30**, 228–233.
23. Mueller, R.S., Meyer, D., Bensignor, E., and Sauter-Louis, C. (2009) Treatment of canine generalized demodicosis with a 'spot-on' formulation containing 10% moxidectin and 2.5% imidacloprid (Advocate, Bayer Healthcare). *Vet. Dermatol.*, **20**, 441–446.
24. Six, R.H., Clemence, R.G., Thomas, C.A., Behan, S., Boy, M.G., Watson, P., Benchaoui, H.A., Clements, P.J.M., Rowan, T.G., and Jernigan, A.D. (2000) Efficacy and safety of selamectin against *Sarcoptes scabei* on dogs and *Otodectes cynotis* on dogs and cats presented as veterinary patients. *Vet. Parasitol.*, **91**, 291–309.
25. Becskei, C., De Bock, F., Illambas, J., Cherni, J.A., Fourie, J.J., Lane, M., Mahabir, S.P., and Six, R.H. (2016) Efficacy and safety of a novel oral isoxazoline, sarolaner (Simparica™), for the treatment of sarcoptic mange in dogs. *Vet. Parasitol.*, **222**, 56–61.
26. http://www.capcvet.org/capc-recommendations/ (accessed 23 June 2016).
27. Honsberger, N.A., Six, R.H., Heinz, T.J., Weber, A., Mahabir, S.P., and Berg, T.C. (2016) Efficacy of sarolaner in the prevention of *Borrelia burgdorferi* and *Anaplasma phagocytophilum* transmission from infected *Ixodes scapularis* to dogs. *Vet. Parasitol.*, **222**, 67–72.
28. Blagburn, B.L. and Dryden, M.W. (2009) Biology, treatment and control of flea and tick infestations. *Vet. Clin. North Am. Small Anim. Pract.*, **39**, 1173–1200.
29. Six, R.H., Liebenberg, J., Honsberger, N., and Mahabir, S.P. (2016) Comparative speed of kill of sarolaner (Simparica™) and afoxolaner (NexGard®) against induced infestations of *Ctenocephalides felis* on dogs. *Parasit. Vectors*, **9**, 90.
30. Six, R.H., Liebenberg, J., Honsberger, N., and Mahabir, S.P. (2016) Comparative speed of kill of sarolaner (Simparica™) and fluralaner (Bravecto®) against induced infestations of *Ctenocephalides felis* on dogs. *Parasit. Vectors*, **9**, 92.
31. Six, R.H., Everett, W.R., Myers, M.R., and Mahabir, S.P. (2016) Comparative speed of kill of sarolaner (Simparica™) and spinosad plus milbemycin oxime (Trifexis®) against induced infestations of *Ctenocephalides felis* on dogs. *Parasit. Vectors*, **9**, 93.
32. Marchiondo, A.A., Holdsworth, P.A., Fourie, L.J., Rugg, D., Hellmann, K., Snyder, D.E. *et al.* (2013) World Association for the Advancement of Veterinary Parasitology (W.A.A.V.P.) 2nd. ed.: guidelines for evaluating the efficacy of parasiticides for the treatment, prevention and control of flea and tick infestations on dogs and cats. *Vet. Parasitol.*, **194**, 84–97.
33. Six, R.H., Young, D.R., Myers, M.R., and Mahabir, S.P. (2016) Comparative speed of kill of sarolaner (Simparica™) and afoxolaner (NexGard®) against induced infestations of *Ixodes scapularis* on dogs. *Parasit. Vectors*, **9**, 79.
34. Becskei, C., Geurden, T., Erasmus, H., Cuppens, O., Mahabir, S.P., and Six, R.H. (2016) Comparative speed of kill after treatment with Simparica™ (sarolaner) and Advantix® (imidacloprid + permethrin) against induced infestations of *Dermacentor reticulatus* on dogs. *Parasit. Vectors*, **9**, 104.
35. Six, R.H., Everett, W.R., Chapin, S., and Mahabir, S.P. (2016) Comparative speed of kill of sarolaner (Simparica™) and afoxolaner (NexGard®) against induced infestations of *Amblyomma americanum* on dogs. *Parasit. Vectors*, **9**, 98.

36. Six, R.H., Young, D.R., Myers, M.R., and Mahabir, S.P. (2016) Comparative speed of kill of sarolaner (Simparica™ Chewables) and fluralaner (Bravecto®) against induced infestations of *Amblyomma americanum* on dogs. *Parasit. Vectors*, **9**, 399.
37. Six, R.H., Young, D.R., Holzmer, S.J., and Mahabir, S.P. (2016) Comparative speed of kill of sarolaner (Simparica™) and afoxolaner (NexGard®) against induced infestations of *Rhipicephalus sanguineus* s.l. on dogs. *Parasit. Vectors*, **9**, 91.
38. Becskei, C., Geurden, T., Liebenberg, J., Cuppens, O., Mahabir, S.P., and Six, R.H. (2016) Comparative speed of kill of oral treatments with Simparica™ (sarolaner) and Bravecto® (fluralaner) against induced infestations of *Rhipicephalus sanguineus* on dogs. *Parasit. Vectors*, **9**, 103.

16
Isoxazolines: Preeminent Ectoparasiticides of the Early Twenty-first Century

Alan Long[*]

Abstract

The isoxazoline chemical series represents the largest and arguably the most exciting new class of insecticidal and acaricidal molecules introduced for animal health uses in the twenty-first century. This chapter details the origins of the chemical class as described by the patent literature through the end of 2016, including commentary on their use against particular pest species and descriptions of the syntheses and purifications utilized to create the various analogs. An account is made regarding the number of analogs synthesized and studied, along with a representative measure of the constitutive properties. Finally, the molecular structure of a typical isoxazoline is deconstructed into pieces to demonstrate what changes have been made and how those changes have initiated new and related chemical series.

Introduction

Origins

Although so far only commercialized for flea and tick control on dogs and cats, the isoxazoline chemical class was first identified in the agrochemical sector; primarily by efforts that grew out of the diamide insecticidal programs at Nihon Nohyaku, Nissan, and DuPont [1]. Scheme 16.1 illustrates how a single starting point (**1**) provided both of the commercial phthalic and anthranilic diamide insecticides flubendiamide (**2**) and chloranthraniliprole (**3**), as well as the orally active flea and tick control isoxazoline molecules fluralaner (**4**) and afoxolaner (**5**).

With respect to the creation of the isoxazoline parasiticidal chemical series, perhaps the most significant finding came from Nissan, who showed that the critical heptafluoro isopropyl substituent found in the insecticidal diamides (such as flubendiamide, **2**) could be replaced with heterocycles or other linking

[*]Corresponding author.

Ectoparasites: Drug Discovery Against Moving Targets, First Edition.
Edited by Charles Q. Meng and Ann E. Sluder

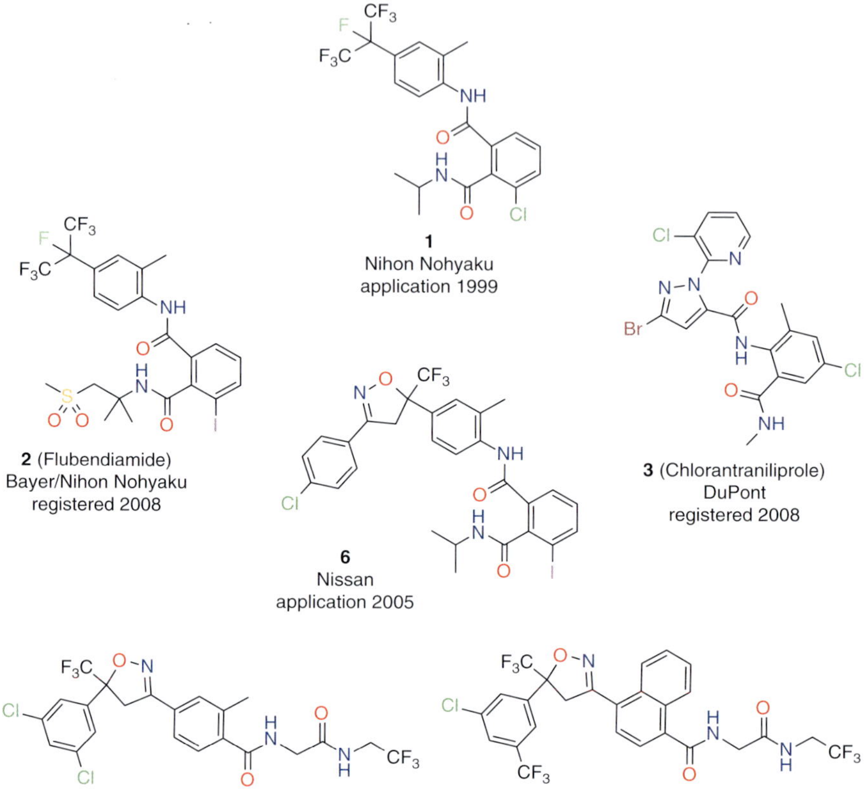

1
Nihon Nohyaku
application 1999

2 (Flubendiamide)
Bayer/Nihon Nohyaku
registered 2008

3 (Chlorantraniliprole)
DuPont
registered 2008

6
Nissan
application 2005

4 (Fluralaner)
Nissan
application 2005

5 (Afoxolaner)
DuPont
application 2009

Scheme 16.1 Diamide starting point resulting in divergent compounds.

functional groups [2]. A key example is shown in compound **6**, and a subsequent reorientation of the isoxazoline moiety provided a new chemical series, albeit one no longer exerting its effects by agonizing the ryanodine receptor as the diamides do [1].

Mode of Action

Similar to fipronil, the isoxazoline compounds appear to operate via the inhibition of GABA (gamma-aminobutyric acid)- and glutamate-gated chloride channels [3]. The spectrum of biological potency is broad and includes arthropods, either insects or acarians, of many orders including Leptidoptera, Diptera, Hemiptera, Blattaria, Coleoptera, Trombidifomes, Thysanoptera, Parasitiformes, Ixodida, and Siphonaptera. As is the case for fipronil, it is hypothesized that the high level of safety observed for the isoxazolines can be at least partially attributed to their lack of effect on GABA-gated chloride channels in mammals [4].

Companion Animal Applications

To date, four isoxazolines have been commercialized. Afoxolaner (**5**) was created by DuPont [5] and subsequently developed by Merial into the first isoxazoline sold for flea and tick control on dogs. Afoxolaner (**5**) was brought to market in 2013, as a prescription product under the tradename of NexGard® in a 2.5 mg/kg minimum dose in a soft chewable form designed for monthly administration. (See Chapters 13 and 14 for an accounting of the discovery and development of Afoxolaner.) Fluralaner (**4**), discovered by Nissan [6] and developed by Intervet/Merck, came to the market soon after in 2014 in a 25 mg/kg tablet form (~10× the minimum monthly dosage) designed for administration every 3 months under the name Bravecto®. In 2016, fluralaner was also registered as a topical formulation for use in dogs and cats at the doses of 25–56 and 40–94 mg/kg, respectively (Bravecto® spot-on). Zoetis, the third company to market an isoxazoline for flea and tick control on dogs, introduced in 2015 sarolaner (**7**) (created by Pfizer [7]) in a 2 mg/kg tablet form for monthly treatments under the name Simparica®. In addition to these three isoxazolines, Elanco received a registration of Credelio® in Australia and Europe in 2017, using Lotilaner (**8**) (discovered by Novartis [8]) as the active ingredient in a 15 mg/kg tablet form presumably meant for monthly administration. It is important to observe that while the first two isoxazolines were brought to market as racemic mixtures, the subsequent entries were developed and marketed as the single enantiomers (Figure 16.1).

Agrochemical Applications

Although the parasiticidal isoxazoline chemical class originated from research in companies having efforts focused on the agrichemical industry, to date no examples have been commercialized for crop protection. Perhaps this is due to the longer timelines associated with getting a new API (active pharmaceutical ingredient) developed and registered for use in food crops relative to use in companion animals, with 8–12 years being the new norm [9]. By using the patent application publication date as a (somewhat artificial and possibly misleading) means to delineate the times marking the end of the discovery phase and the beginning of development and registration work, one can estimate that the flea and tick products were brought to market on the order of 5 years following discovery, and it has now been >11 years since the seminal Nissan patent application [6] was published.

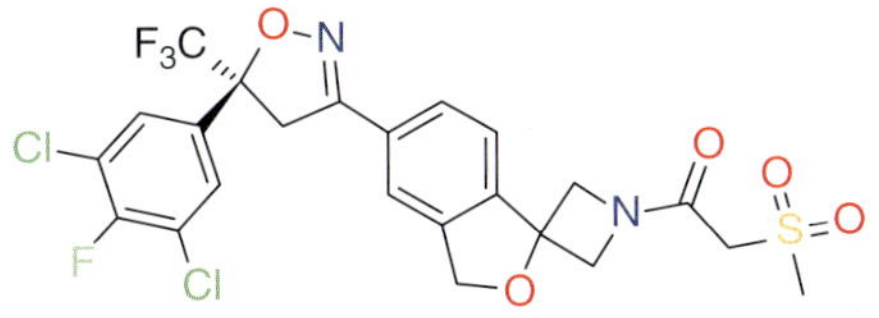

7 (Sarolaner)

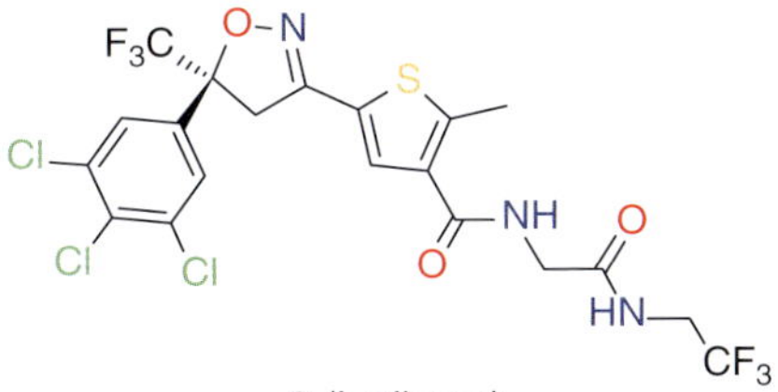

8 (Lotilaner)

Figure 16.1 Other commercial isoxazolines for use against fleas and ticks.

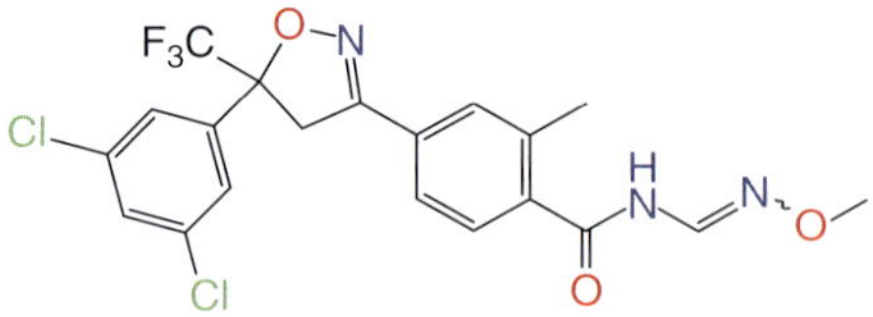

9 (Fluxametamide)

Figure 16.2 Potential first commercial isoxazoline for crop protection.

With that in mind, fluxametamide (**9**) is the ISO name approved in April 2015 for, potentially, the first isoxazoline for use in food crop protection. This compound first appeared in an international patent application by Nissan (Figure 16.2) in 2007 [10]. Fluxametamide (**9**) has been reported to be active against the cabbage moth, common cutworm, beet armyworm, oriental tea tortix, corn earworm, western flower thrip, melon thrip, Lewis spined bug (shield/stink bug), brown rice planthopper, silver leaf whitefly, green peach aphid, Japanese mealy bug, cucurbit leaf beetle, serpentine leaf miner, two-spotted spider mite, and the pink citrus rust mite. As one might expect, fluxametamide (**9**) has also been reported to have activity against the cat flea and the American dog tick.

Production Animal Applications

Based on the patent literature, there appears to be only one application describing the use of isoxazolines in cattle. Researchers at Merial demonstrated efficacy against the cattle tick (*Rhipicephalus microplus*) following treatment with a long-acting injectable formulation of afoxolaner (**5**) [11]. Similarly, a single patent application exists demonstrating efficacy in a poultry study. Intervet describes the administration of fluralaner (**4**) in drinking water to control the northern fowl (*Ornithonyssus sylviarum*) and red (*Dermanyssus gallinae*) mites in chickens [12].

In addition to activity against a range of terrestrial organisms, it is reasonable to expect potency against an expanded set of aquatic parasites for the isoxazoline chemical class in the future. Efficacy against one economically important sea louse, *Lepeophtheirus salmonis*, was first demonstrated for existing members of the chemical class by Novartis [13]. It not only showed potency against the copepodid stage using *in vitro* screens but also demonstrated efficacy against pre-adult and adult lice on salmon using injection, bath, and in-feed experiments. In 2016, a close, and racemic, analog of lotilaner (**8**) was shown to be active against the copepodid life stage at the sub-ppb level *in vitro*, while racemic lotilaner (**8**) itself was inactive at the tested dose [14].

During the same time period, scientists at Pfizer also demonstrated potency against the copepod life stage of *L. salmonis* using their azetidinyl-substituted phenyl isoxazolines [15]. Later Zoetis (formerly Pfizer Animal Health) was able to show superior potency *in vitro* against the pre-adult and adult stages for its spirocyclic molecule sarolaner (7) when compared to emamectin [16].

Synthesis

The large-scale production of a parasiticide, pesticide, agrochemical, APIs, or any drug for that matter, has several important requirements. First, the overall process must be robust, meaning that all of the chemical reactions, isolations, purifications, and material transfers used must perform in a reliable and consistent manner. Second, the reagents, solvents, and intermediates used must be inexpensive and readily available from multiple suppliers. Third, the chemistry reactions and associated equipment must be safe, and this includes the methods employed to dispose of waste generated during the process. Only when these factors are all put into place can the process chemists deliver the parasiticide compound (to the respective formulation production group) on a commercial development scale.

The vast majority of isoxazoline molecules have been synthesized by one of two general routes (Scheme 16.2). In the aldol condensation/cyclization method, an addition-dehydration reaction between an aromatic ketone and (typically) a trifluoromethyl aromatic ketone affords an α,β-unsaturated ketone with the potential for having differentially substituted aromatic rings. This linear intermediate is then cyclized in a process that presumably involves the *in situ* formation of an amino-ether followed by a base-mediated cyclization to afford the desired 3,5-*bis*-aryl-5-trfiluoromethyl isoxazoline; a mechanistic study by researchers at the Nagoya Institute of Technology suggested the alternative oxime is not a viable reaction intermediate [17]. The second general protocol for the preparation of parasiticidal isoxazolines begins with the reaction of an aromatic aldehyde with hydroxylamine. The resulting oxime is then oxidized to the nitrile oxide, most typically done in an *in situ* process involving *N*-chlorosuccinimide wherein the imidoyl chloride is not isolated. In the presence of the requisite alkenyl (styrenyl) partner, the nitrile oxide undergoes a facile [3 + 2] cycloaddition to yield the corresponding isoxazoline.

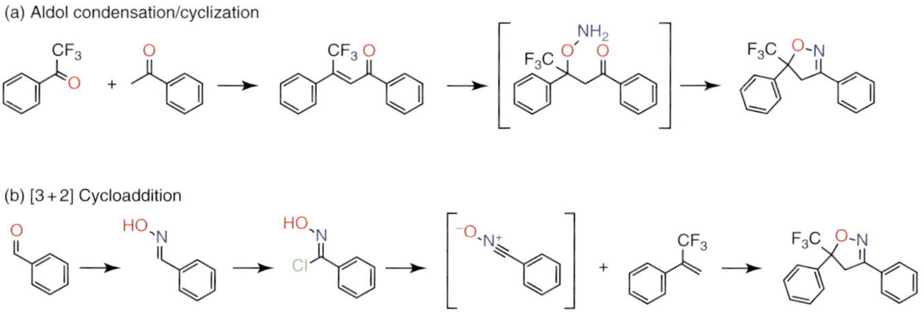

Scheme 16.2 Two general pathways for the synthesis of parasiticidal isoxazolines.

In principle, both of the aforementioned synthetic processes can be performed in a stereoselective manner. In practice, it appears that only the aldol condensation/cyclization method has realized enantiomerically enriched isoxazoline products of value to the animal health and agrochemical businesses. Publications from Nagoya Institute of Technology [17], Nissan [18], Syngenta [19] and Zoetis [20] describe the use of cinchona alkaloid-derived catalysts to achieve asymmetric syntheses of parasiticidal isoxazolines. The elegance of the chiral catalysis method lies in the overall efficiency: essentially only the desired enantiomer is produced which, in turn, minimizes wasted material and simplifies purification of the desired product. Sumitomo has also described an asymmetric synthesis of parasiticidal isoxazolines, but uses optically active thioureas as the chiral catalysts as opposed to the quinine-based variants [21]. Companies including BASF [22], Syngenta [23], and Anacor and Eli Lilly [24] have utilized chiral separation via chromatography to provide optically active isoxazolines. In principle, chiral chromatography could be used to purify a parasiticide for the animal health or agrochemical industries. However, in practice this can be cost-prohibitive and is used primarily to provide resolved material for research purposes. In Scheme 16.3 is shown a chiral resolution process developed by Novartis [25], which uses a fractional crystallization to isolate the desired enantiomer of the requisite carboxylic acid intermediate used to synthesize lotilaner (**8**). A significant aspect of the Novartis process is the recovery and recycling of the undesired enantiomer.

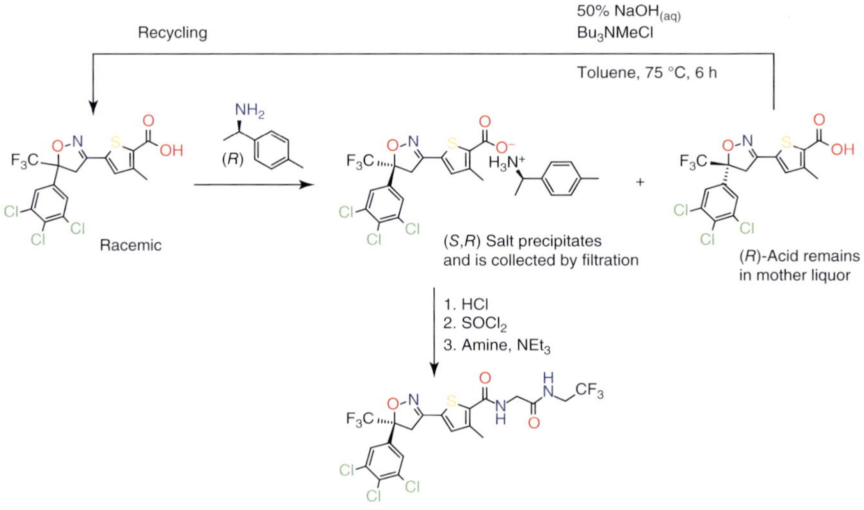

8 (Lotilaner)

Scheme 16.3 Fractional crystallization and recycling processes in the synthesis of Lotilaner.

The Hoveyda chemistry group from Boston College has very recently disclosed a *formal* synthesis of fluralaner (**4**) in enantiomerically pure form, which utilizes an alternative route (Scheme 16.4). In their treatise, the researchers illustrate the careful design of a chiral catalyst to affect the addition of an allyl group into the requisite trifluoromethyl ketone **10** to afford tertiary alcohol **11** in high yield and good enantiomeric purity [26]. Routine chemical manipulations provided ketone **12**, which was subsequently converted to the key isoxazoline intermediate **13** using a three-step ring closure based on a procedure developed by the Shibata group at the Nagoya Institute of Technology [27]. Standard chemistry reactions can then be used to convert the methyl ester into a desired final amide such as **14** (active enantiomer of fluralaner). Overall, the Hoveyda group's process is high yielding, but so far has only been described on a laboratory (small) scale.

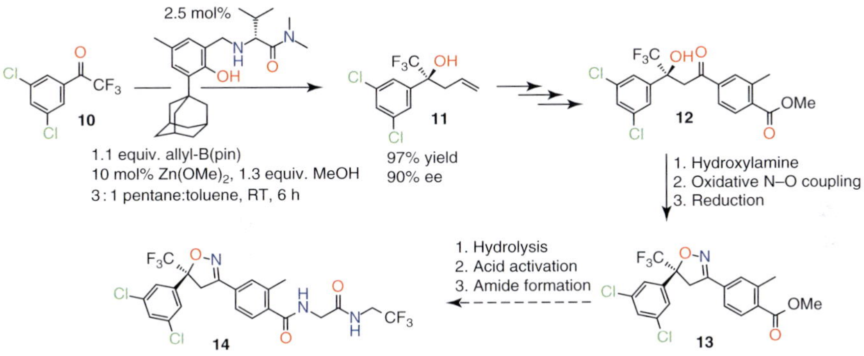

Scheme 16.4 An alternate route to access enantiomerically enriched antiparasitic isoxazolines.

Physicochemical Properties

The physicochemical properties of a compound or series are important as they can help researchers to make predictions and define studies (e.g. stability, dosing levels, and regimen) and interpret *in vivo* results such as potency, bioavailability, and persistency. Relative to small molecules used as human health pharmaceuticals, such as the majority of antibacterials, for instance [28], the parasiticidal isoxazolines highlighted in this chapter are larger and more lipophilic compounds, with halogenation designed to provide both improved potency and stability towards metabolic enzymes present in both the host and the target parasites. As a result of the said halogenation, both a broadened spectrum of biological activity and a predictable pharmacokinetic profile can be effected, in turn leading to efficacious products with the desired duration of activity or environmental persistency.

Figures 16.3–16.5 illustrate select constitutive physicochemical properties for parasiticidal isoxazolines and highly related analogs taken from the patent literature. The examples were retrieved from international patent applications describing new chemical series (vs methods-of-use or process chemistry filings), and efforts were made to select a single, representative analog from each application that was listed as biologically active in the assays mentioned therein. This amounted to 134 examples meant to represent what could total approximately 50 000–100 000 molecules made in the efforts to find parasiticidal isoxazolines and/or close derivatives. To provide the reader with additional context, also

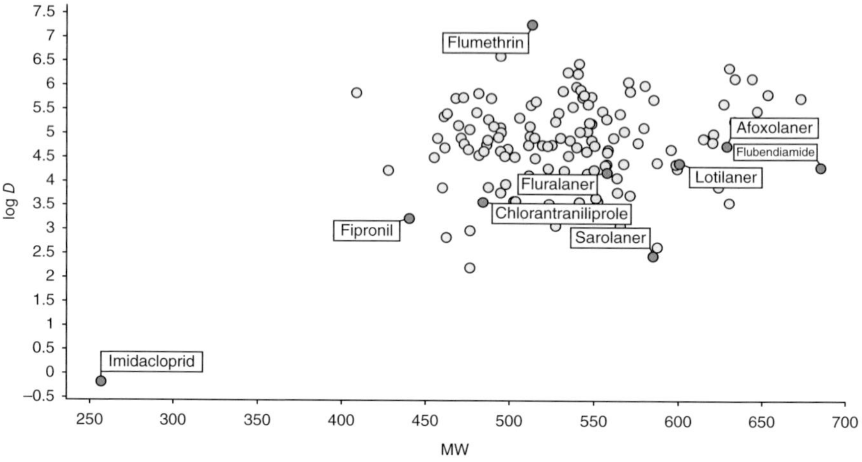

Figure 16.3 Lipophilicity and molecular weight of representative compounds.

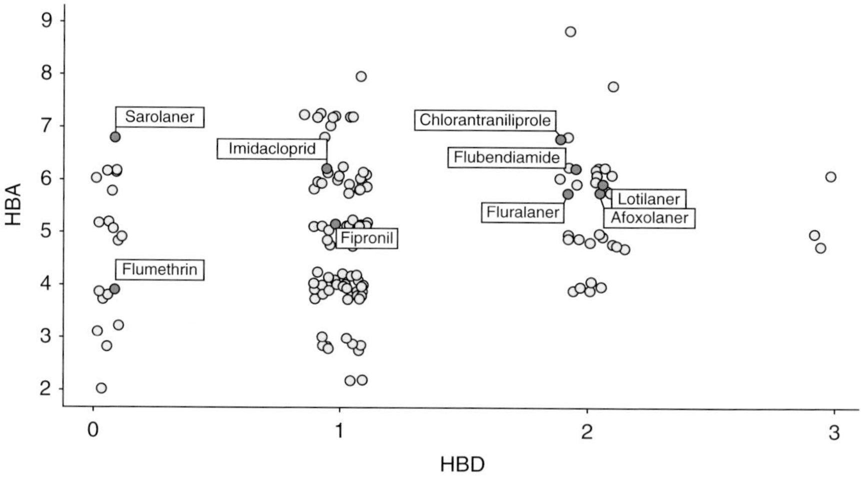

Figure 16.4 Hydrogen bond acceptors and donors for representative compounds.

included in the plots are the commercial insecticides flubendiamide (**1**), chlorantraniliprole (**3**), imidacloprid, flumethrin, and fipronil.

Approximately two-thirds of the examples presented have molecular weights (MWs) in excess of 500 Da and the very large majority of the remaining analogs have masses of >450 g/mol, meaning that the series as a whole contains members that are larger than most small-molecule human health drugs designed for oral absorption [29]. In a similar vein, calculations[1] on this set of examples show the

[1] StarDrop® software (by Optibrium, version 6.1) was used to calculate log *D*, log *S*, molecular weight, total polar surface area, and the number of hydrogen bond acceptors and donors.

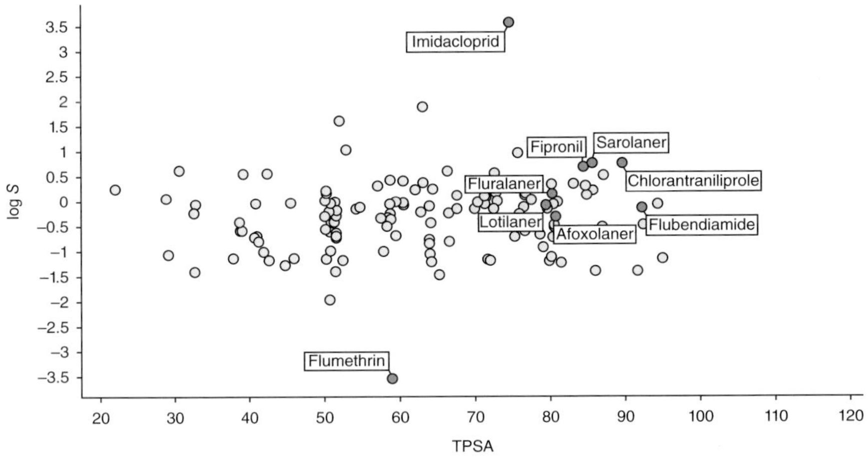

Figure 16.5 Calculated solubility and polar surface area for representative compounds.

analogs to possess poor aqueous solubility and to have an average log D of ~4.8, and a lipophilicity or "greasiness" that is also at the high end of what is desired for a small molecule intended for oral absorption. While the number of hydrogen bond acceptors (many instances of N, O, or S) ranges from 2 to 9, the number of hydrogen bond donors (mostly amide N–H bonds for this set) is relatively fixed and only spans 0–2 for the vast majority of the members considered. However, when viewed in the context of commercial parasiticides for both crop protection and domesticated animal protection, the isoxazolines and related structural analogs appear constitutively similar to other classes with the obvious exception of the smaller and highly polar neonicotinoids (imidacloprid as a typical example).

Patent Literature

A very large number of isoxazoline-containing compounds exist in the literature, with over 32 000 being reported in more than 300 patents and patent applications, the very large majority of which are focused on insectidal and/or acaricidal activity. This number includes only those analogs that, in fact, contain an isoxazoline decorated with a trifluoromethyl group and are flanked by two cyclic groups as is shown in Figure 16.6; between 17 000 and 18 000 were actually made and/or tested, with the remainder disclosed as prophetic examples[2]. Due to the high level of interest in this series from both the animal health and agrochemical sectors, at least 18 different companies have submitted patent applications, and there are at

[2] SciFinder® software (by Chemical Abstracts Services) sub-structure search performed on the general structure shown in Figure 16.6.

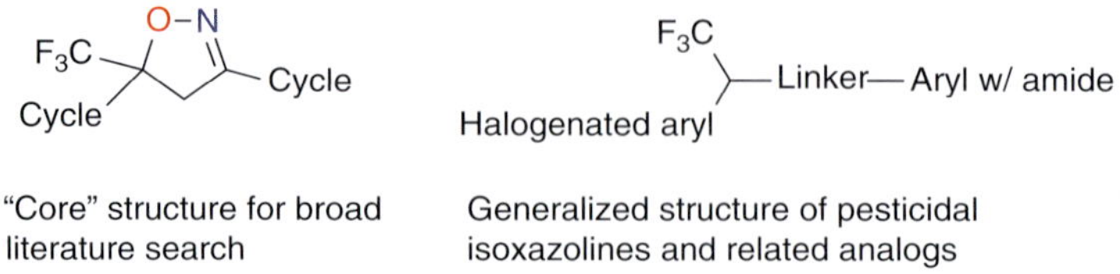

Figure 16.6 General structures for literature analyses.

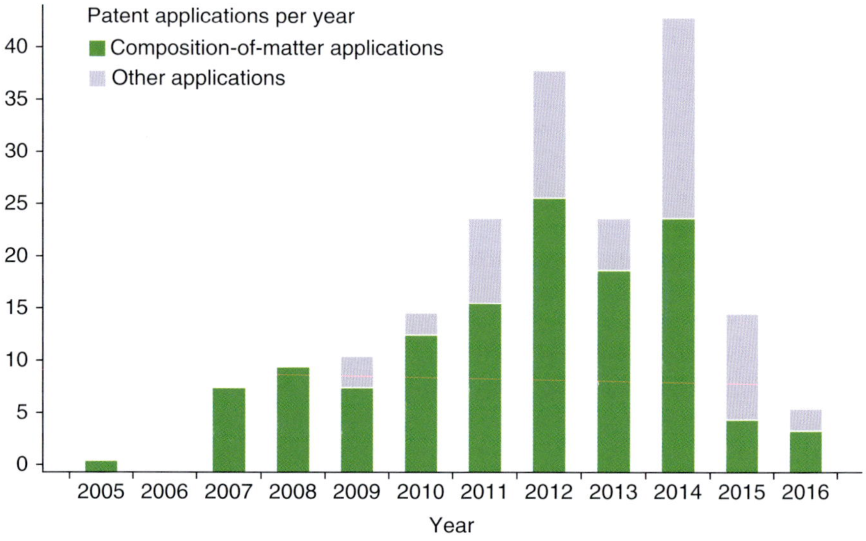

Figure 16.7 Patent applications 2005–present.

least five cases wherein multiple companies disclosed the same, new, compound within 6 months of each other. The reader will note, however, that approximately half of the patent literature mentioned describes methods of use, such as combinations or formulations of isoxazolines, is instead process chemistry related, or even simply includes isoxazoline analogs mentioned in a long listing of named parasiticides. In Figure 16.7 is shown a graphical representation of the number of parasiticidal-isoxazoline-related patents per year, differentiating between those describing new molecules/series and those describing usage of existing APIs. With that in mind, the actual number of patent applications describing new chemical entities for the isoxazoline series and their highly structurally related parasiticidal analogs, is approximately 145[3] as of late 2016.

[3] The actual number of patent applications filed regarding new chemical series for isoxazoline or isoxazoline-related compounds may be closer to 150, as a small number may have been missed due to filings outside of the PCT system or language/keyword/structure issues.

A review of the biological data reported in the patent literature reveals that the isoxazoline group itself is not critical (speculation) for activity; examples are illustrated in the following section on structural variations. It is this author's hypothesis that the isoxazoline functional group serves as a preferential linking group between the critical halogenated aryl group with the requisite geminal trifluoromethyl group and the other aryl group substituted with a preferred amide group. While more difficult to quantify and somewhat subjective, the number of structurally related derivatives (as compared to isoxazolines) with activity against arthropods appears to approximate the number of described isoxazolines. Several examples are given in the following section covering the structural variations.

Structural Variations

A Geminal Relationship

A point of consistency across the very large number of analogs synthesized and tested is that the geminal trifluoromethyl and phenyl groups on the 5-position of the isoxazoline are present in the large majority of analogs published (see Figure 16.8). In fact, ~99% of the compounds retrieved using the "core" structure in Figure 16.6 contain a (substituted) phenyl group at the 5-position, with the remaining 1% largely comprised of (substituted) pyridyl groups. The other half of the germinal relationship, the trifluoromethyl group, is likewise present in ~99% of the published analogs. Approximately half of the companies working in this area have explored other substitutions such as -CF_2H, -CFH_2, -CF_2CF_3, -$CF_2CF_2CF_3$, -CF_2Cl, -CCl_3, or -CH_3, but based on the number of analogs made, none appear to have seriously pursued these analogs. Presumably this is for a lack of potency, but is possibly due to other considerations such as cost and perhaps unfavorable biological exposures. Beyond how it is presented in the "core"

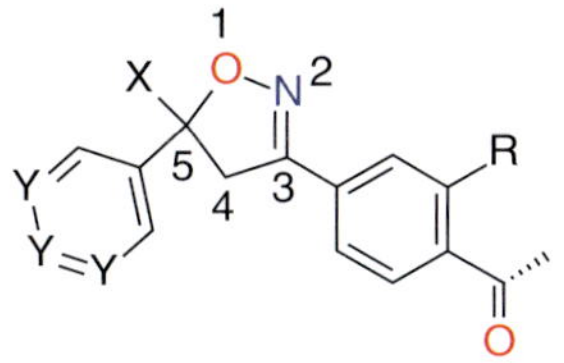

R = H, CH_3, Br, Cl, CN, or CF_3

X = CF_3, or,
X = CF_2H, CFH_2, CF_2CF_3, $CF_2CF_2CF_3$, CF_2Cl, CCl_3, or CH_3

Y = CH, C-CF_3, C-halogen, or,
Y = N

Figure 16.8 Common substitution patterns on the isoxazoline core.

structure in Figure 16.6, the geminal trifluoromethyl-phenyl substitution pattern is maintained across nearly all of the analogs that follow in this section.

An Aromatic Backbone

Switching to the other side of the isoxazoline, the 3-position, again, a substituted phenyl ring appears to be the most commonly chosen group to tether the terminal amide (or other group) to the isoxazoline. While the largest number of analogs were made with R = hydrogen as shown in Figure 16.8, other substitutions *meta* to the isoxazoline have been found to be favorable. For instance, the methyl group present in fluralaner (**4**) also appears in another nearly 3000 exemplified compounds including the analogous commercial compound lotilaner (**8**). It may be more appropriate to say this substitution pattern is *ortho* to the carboxyl group (typically an amide); molecular modeling shows the methyl group to impart an approximate 40–50° twist from planarity[4] to the adjacent amide group. Regardless of whether this structural twist is involved in the binding of the compounds to the molecular target, the *ortho* methyl substitution is a highly preferred moiety presumably for reasons of efficacy/potency.

In addition to the methyl group situated *ortho* to the carboxyl (or other) functional group, a large number of analogs were made and tested which contain either the chloro (~2000 compounds) or the cyano (>600 compounds) group at the same position. The trifluoromethyl group was also introduced at this position, but to a much lesser extent with only about 100 analogs appearing in the patent literature.

As was shown in Figure 16.1, a phenyl ring at the 3-position of the isoxazoline ring is not the only option for achieving potent molecules. For example, afoxolaner (**5**) has a naphthalene ring in lieu of the phenyl and lotilaner (**8**) takes advantage of a thiophene as a phenyl bioisostere [30]. The "extra" ring (specifically the "*ortho*"-methine) in the naphthalene ring system imparts a similar conformational effect to the *ortho*-methyl substitution found in fluralaner (**4**) and related analogs mentioned. Novartis has also described derivatives with *in vivo* activity wherein the *ortho*-methyl thiophenyl "core" is replaced by the analogous furanyl unit [30]; a similar level of conformational strain is presumed to operate in these non-phenyl analogs also.

Besides the naphthyl, thiophenyl, and furan replacements for the phenyl group, at least 10 companies [6, 10, 15, 31–40] have engaged in the synthesis and testing of analogs using heterocycles affixed to the 3-position of the isoxazoline. Included in this group of core linkers are pyridines, pyrimidines, and various orientations of quinolines and isoquinolines as shown in Figure 16.9. DuPont systematically addressed all of the positional isomers for the potential quinolines/isoquinolines and arrived at a compound with potency against multiple crop pests that was demonstrated to be superior to the commercial product indoxacarb (Activyl®) [41].

[4] Maestro® software (by Schrödinger Inc., version 10.5) was used to measure the dihedral angle after performing a series structural minimizations.

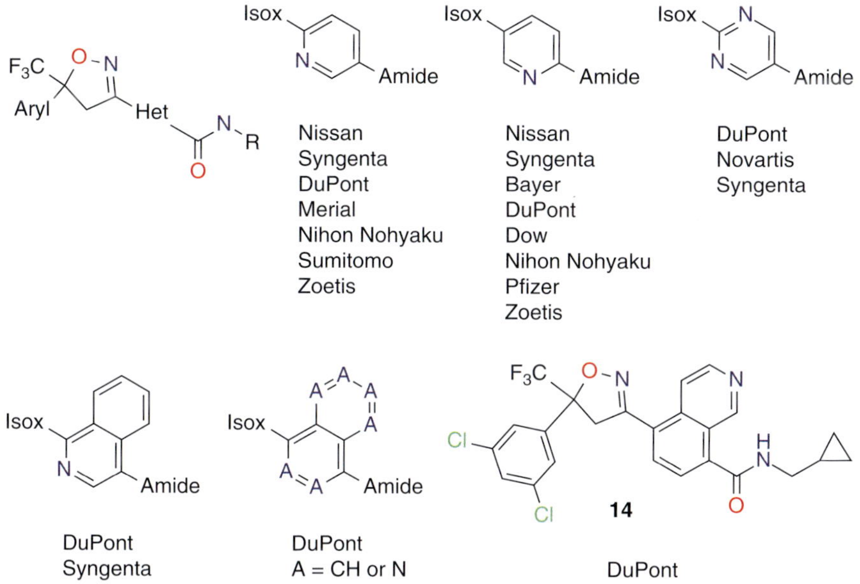

Figure 16.9 Phenyl group replacements.

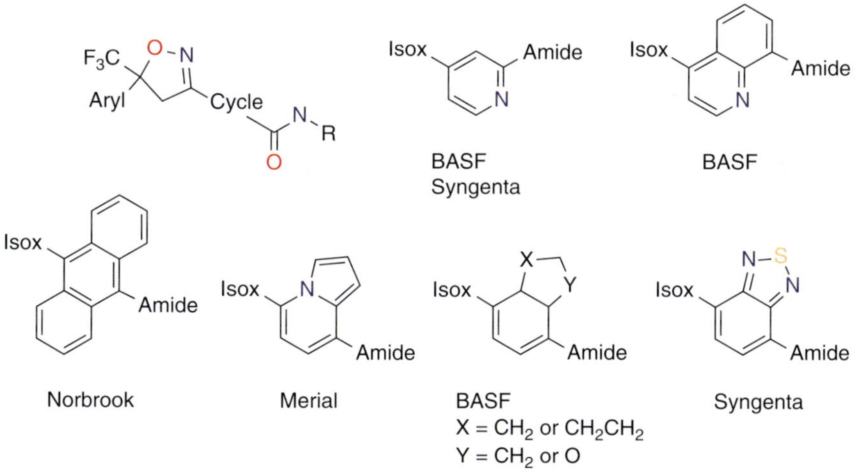

Figure 16.10 Phenyl group replacements.

As one might guess, the number of potential replacements for the central, linking phenyl group is large. Beyond the linear (*para*-substituted) pyridines and quinolines present in a number of other applications, researchers at BASF evaluated the nonlinear heterocyclic linkers such as the pyridyl and quinolinyl cores shown in Figure 16.10 [42]. Norbrook disclosed and claimed activity against ectoparasites for compounds containing anthracenyl cores decorated with the amide and aryl moieties found in fluralaner (**4**) and afoxolaner (**5**) [43]. Merial published examples with animal health activity for analogs containing the indolizine core in lieu of a phenyl [35]. Syngenta revealed a set of substituted benzothiadiazoles in 2010 [32g, 44b]. Finally, BASF has patented several related series of analogs incorporating indanyl, dihydrobenzofuranyl, tetrahydronaphthalenyl, and dihydrobenzopyran linkers [22, 45].

Biologically Active Amides

Under the current focus, a variety of structural searches revealed that over half of the analogs presented in published patent applications describe compounds wherein an amide substitution is located opposite the isoxazoline (across the phenyl ring). The percentage of amide-containing analogs appears much lower, however, when the isoxazoline moiety is replaced with other groups such as dihydropyrrole or pyrrolidine. This could be attributed to a number of factors including a loss of potency or spectrum, or may be reflective of desires of the animal health and agrochemical companies' need to find, define, and defend new intellectual property space. Whatever is the case, the fact remains that all of the commercial pesticides (animal health and potentially crop pest) contain an amide group directly across the aromatic ring from the (*para* to) isoxazoline.

The relative synthetic ease of introducing an amide group into a molecule translates to a very large number of possible analogs, too many to cover in this review. However, shown below in Figure 16.11 are several examples which feature prominently, either in the commercial molecules and/or in a number of published patent applications. In general, it appears that the amides chosen for flea and tick applications for pets have a lower polarity than those intended for crop protection. This is likely due to a number of factors which primarily include the safety profile in mammals, water miscibility, and reduced environmental persistency for

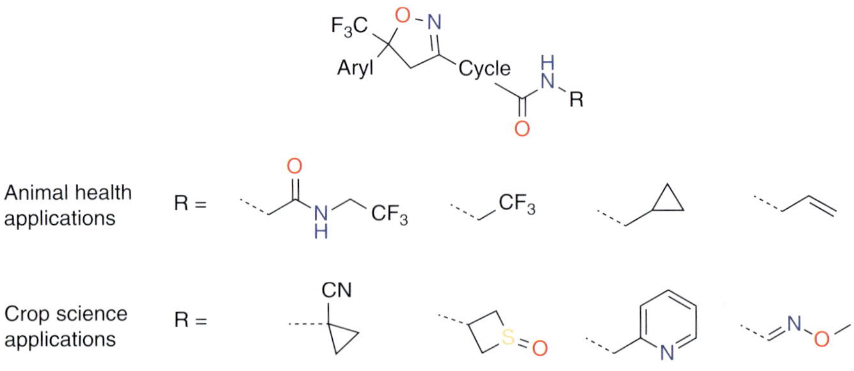

Figure 16.11 Common amide substitutions.

crop pest applications, as well as structure–activity relationship considerations for potency against the differing pest species.

The amides as depicted in Figure 16.11 are not the only biologically active moieties found in this broad set of molecules. Following the success of the initial isoxazolines, "reversed" amides, or one wherein the central aryl group is bound to the amide nitrogen versus to the carboxyl carbon, were demonstrated to also have potency against a variety of crop pests. Chemists from Nissan were the first in 2007 to publish the reversed amides [38b]; and within months, Japanese teams from Nihon Nohyaku [39] and Bayer [37b] also disclosed highly similar molecules. The somewhat more sophisticated examples of the reversed amide such as the ones illustrated by the 2009 series by the Nippon Soda company [46] (Figure 16.12), may have served as inspiration for the later design of the potent flea and tick molecule sarolaner (**7**). Also shown below in Figure 16.12, are other examples of amido-related functional groups found in insecticidally active isoxazolines, such as the hydrazide [47], the semicarbazide [48], and the amidoxime [49] substitution patterns. Similar to the oxime found in fluxametamide (**9**) which presumably helps avoid long-term environmental persistence, BASF has put forth examples of the semicarbazone [50] and amino-carbamate [51] moieties as linkers between the phenyl isoxazoline "cores" and the active groups (such as trifluoromethyl) at the termini.

Other Active Substitutions

Beyond amides, reversed amides, hydrazides, and the like, there have been approximately 2000 examples put forth which incorporate a heterocyclic group in lieu of the amido-substitution. The earliest examples were the "azoles" (pyrroles, pyrazoles, imidazoles, triazoles, and tetrazoles) coming in 2007 from Nissan for crop protection [38c] and DuPont for crop pest/flea control [40a] and then later in 2008 from Bayer [37b]. In 2010, Novartis utilized a wide variety of heterocycles, each plausibly meant to serve as a linking group between the central aryl and the terminus of the known, active amide substitution pattern. The example given in Figure 16.13 was progressed to a gerbil model of tick (*Rhipicephalus sanguineus*) efficacy and shown to have a 96% efficacy using a topical dose of 100 mg/kg bodyweight [52]. A somewhat larger departure from the azoles, Anacor developed a set of benzoxaboroles and together with Elanco, demonstrated 1 month's efficacy in dogs for the (*S*)-enantiomer (**15**) against the American (*Dermacentor variablis*) and brown (*R. sanguineus*) dog ticks and the cat flea (*Ctenocephalides felis*) following oral administration at 50 mg/kg [53]. The lead AN8030 (**15**) was progressed toward a development candidate (**16**) which showed a similarly high level of efficacy in dogs for 1 month post oral administration at only one-half the dose (25 mg/kg bodyweight) [54].

Other functional groups have been investigated as replacements for the amide moiety *para* to the isoxazoline substituent. One such example is the thio group, whether in the thiol, sulfide, sulfoxide, sulfone, or sulfoximine oxidative states. Excluding any examples that still belong to the ryanodine receptor agonist precursors, approximately 1100 isoxazoline compounds are described in 21 patent

Color Fig. 16.12

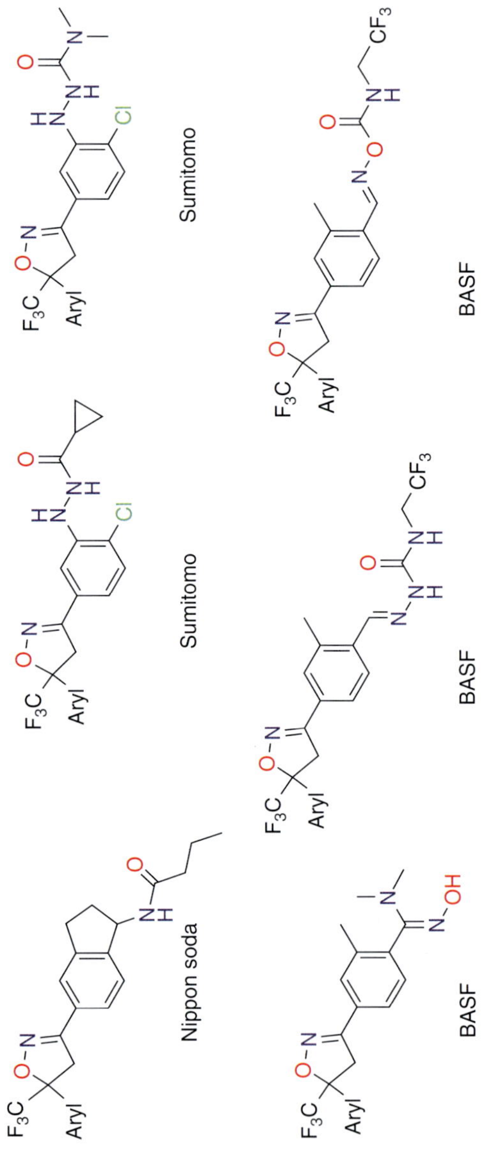

Figure 16.12 Amide-related functional groups in isoxazoline analogs.

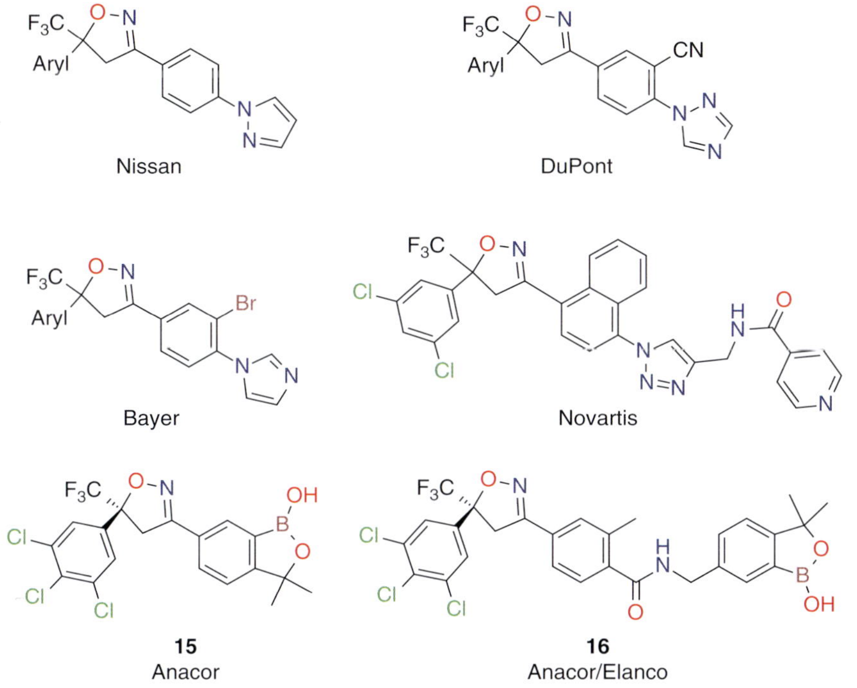

Figure 16.13 Alternative substitutions *para* to the isoxazoline.

applications from 2007 to 2015[5]. Shown in Figure 16.14 are examples from Nissan [55a], DuPont [40a, 55b], Nihon Nohyaku [39], Dow [55c], and Syngenta [32e]. An example from Bayer [55d] describes fused heterocycles with alternative oxidative states for the sulfur.

Ketones are another alternative functional group explored *para* to the isoxazoline substitution on the phenyl ring (Figure 16.15). While present in a number of human health pharmaceuticals and a few anthelmintics, outside of rotenone, the ketone functional group appears to be largely absent from use in *ecto* parasite control agents. As less than 100 total isoxazoline-related compounds appear in only 6 patent applications (see Footnote 5), the supposition is that relative to the amide moiety, this substitution pattern is poorly active against crop and/or animal health pests or there are toxicity or stability issues stemming from the carbonyl functional group. Although several examples of ketonic-isoxazolines can be found in applications submitted by Bayer [55d], BASF [50, 51], Nippon Soda [56], and Syngenta [57], most of these appear to be single compounds and/or synthetic intermediates of other analogs. It would appear that only BASF [58] has systematically explored the use of ketones in the isoxazoline series.

5 Sub-structure searches performed in Scifinder and refined for document type = patent.

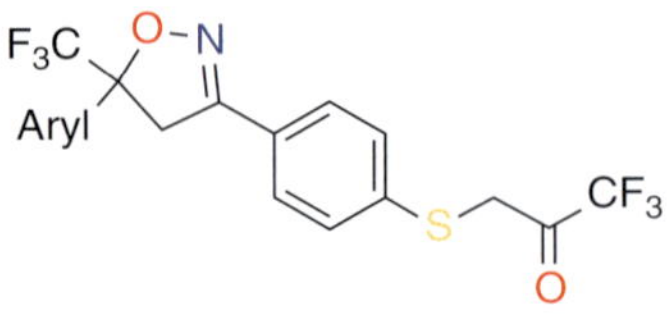

Nissan

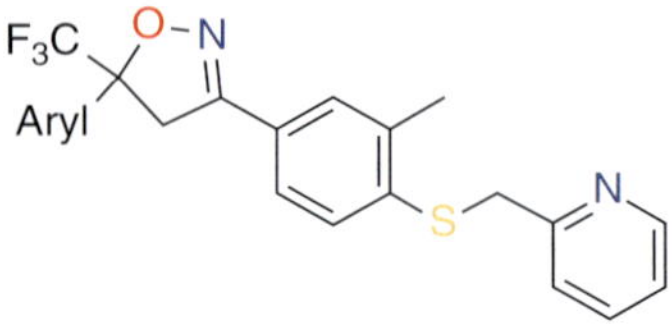

DuPont

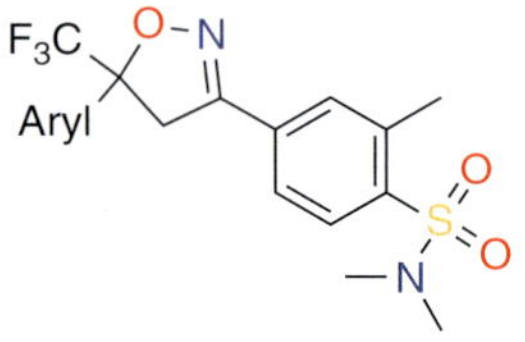

Nihon Nohyaku

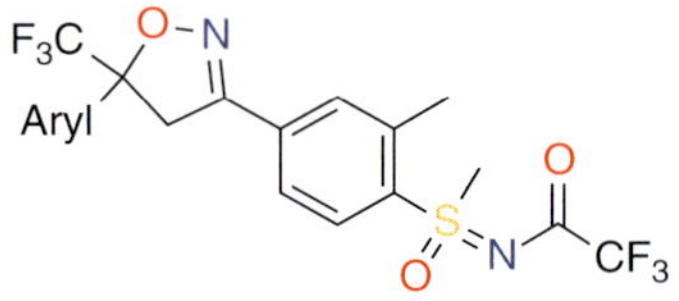

DOW

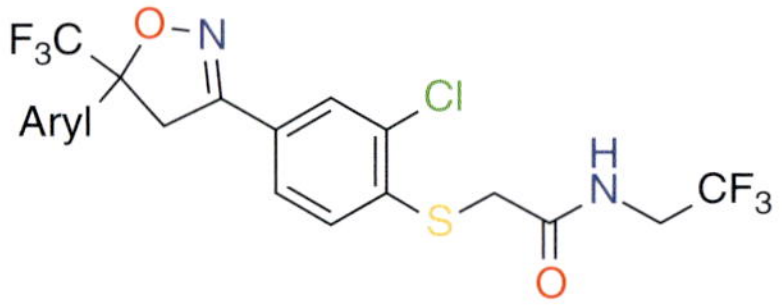

Syngenta

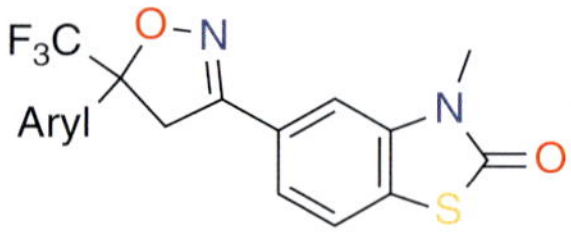

Bayer

Figure 16.14 Thio substitutions in lieu of an amide.

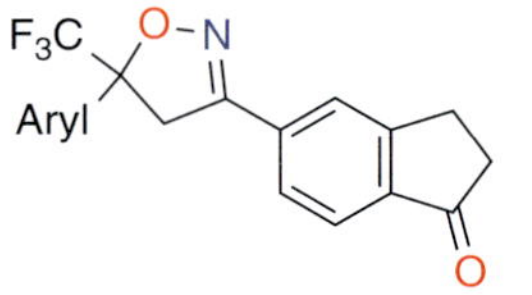

Bayer/Nippon soda

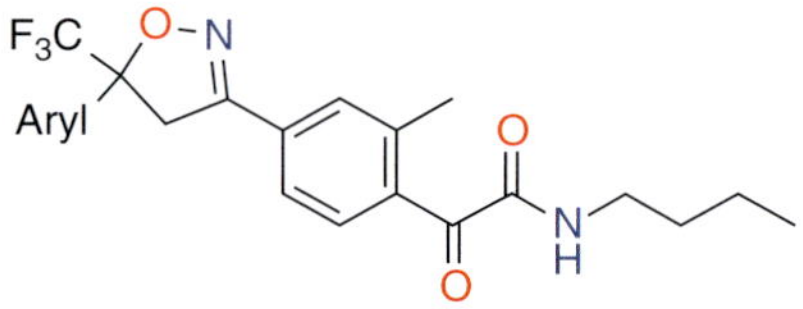

Syngenta

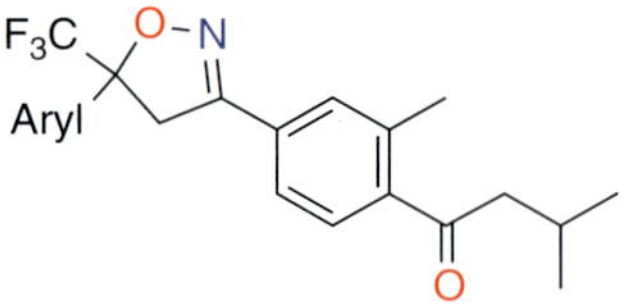

BASF

Figure 16.15 Ketone substitutions on the phenyl linker.

Isoxazolines as a Preferred Core

As noted in the introduction, all of the commercial compounds in this chemical class do, in fact, contain an isoxazoline ring as the central portion of the molecule. However, an industry-wide endeavor has sought to find suitable replacement groups for the isoxazoline: approximately one-third of the total number of patent applications regarding the isoxazoline class is focused on molecules containing other functional/linker groups as the central core. Loosely termed "scaffold-hopping," the intent is to find analogs of similar or equal potency and/or physico-chemical properties, while using molecules whose chemical structures exist outside the previously defined intellectual property space. This body of work has produced a number of analogs nearly equal to those for the isoxazolines themselves; these analogs can be placed into three generalized classes: five-membered ring groups, fused ring analogs, and acyclic derivatives.

A seemingly straightforward path to finding new, but related active compounds involved placing additional substitutions onto the five-membered isoxazoline ring. Alkylations, specifically methylations, or halogenations, specifically brominations and chlorinations, at the 4-position on the isoxazoline ring were first introduced by Nissan [59], with later examples to include the more challenging fluorinations by Syngenta [44] and BASF [22]. Sumitomo [60] demonstrated chemical routes generating analogs with nitro substitution at *C*-4. Also shown in Figure 16.16 is a more highly elaborated example by the Nagoya Institute of Technology [61] which includes a germinal fluorine and trifluoromethylsulfonyl substitution presumably meant to define and illustrate the synthetic route to the said analogs.

By its very nature, the 3,5,5-trisubstituted isoxazoline core leaves only the *C*-4 and *N*-2 positions available for direct alkylations. Syngenta appears to be the sole company to have published on *N*-2 alkylated isoxazoline derivatives, both in the saturated [44b, 62a] and the unsaturated [62b] variants (Figure 16.16). The isoxazoline core is also subject to oxidations, again at the *C*-4 and *N*-2 positions. Syngenta [63] has investigated the biological activity of *N*-2 oxidized analogs and again, researchers from the Nagoya Institute of Technology [27b] have demonstrated a manner of synthesis for these dipolar analogs. Syngenta [44c] has also disclosed products from a net oxidation at the *C*-4 position, namely, the ketone and subsequent oxime(s), also shown in Figure 16.16.

In a similar vein, a single atom replacement can lead to a variety of possibilities, as shown in Figure 16.17. A substitution of the isoxazoline oxygen atom for a methylene to provide dihydropyrroles (1-pyrrolines) was first disclosed by Nissan [64a] in 2007. Apparently a highly suitable bioisostere for the isoxazoline, the dihydropyrrole moiety is present in nearly 11 000 related compounds, described or mentioned, in approximately 30 patent applications coming from Nissan [64a–d], Nippon Soda [31c, 65], Bayer [55d, 66], Syngenta [32a–e, 44b, 67a–j], Zoetis [68], BASF [22, 69], and Sumitomo [33]. It is of interest to note that a synthetic route to access enantiomerically enriched dihydropyrroles was described by Bayer [66f]. The Bayer method differed considerably from the methods used to enrich the isoxazolines and involves a metal-catalyzed cycloaddition wherein one of the reaction partners contained a chiral ester auxiliary to facilitate, in principle, both a diastereomerically enriched set of cycloadducts and the separation of the said adducts. This method required a base-catalyzed isomerization and a decarboxylation to remove the chiral auxiliary and afforded the dihydropyrrole products which were

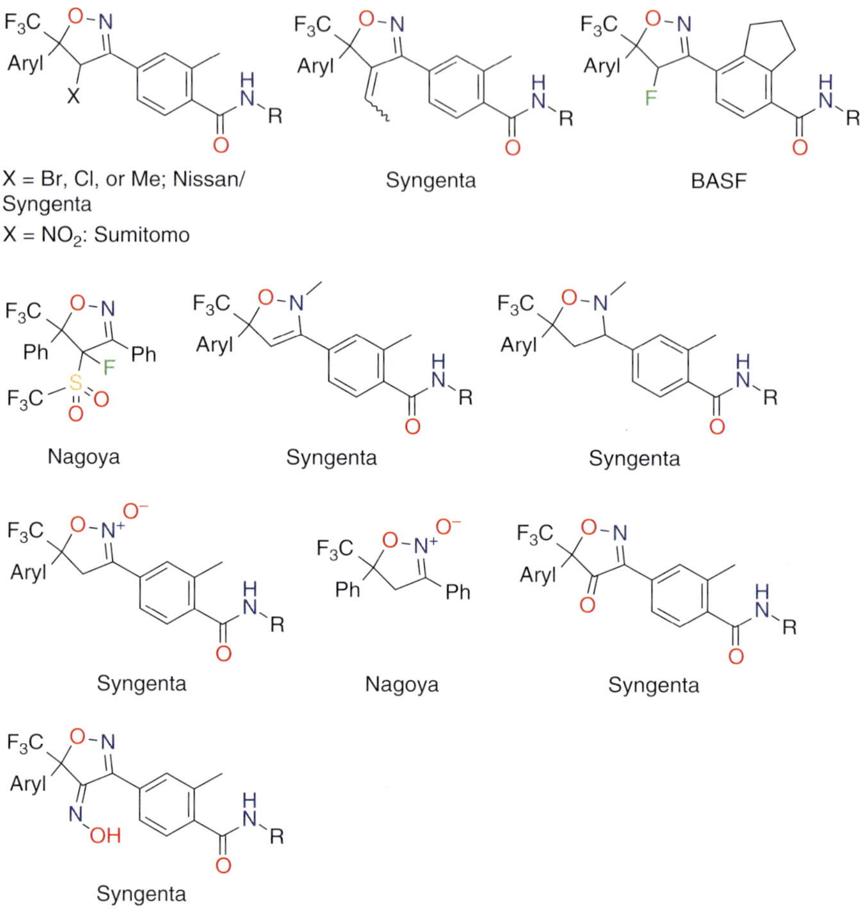

Figure 16.16 Substitutions on the isoxazoline linker.

then subjected to chiral HPLC (high-performance liquid chromatography) to obtain the enantiomerically pure final products.

Replacement of the isoxazoline nitrogen with a carbon to provide dihydrofurans (Figure 16.17) was first disclosed by Syngenta [32a, c, d, 70] in 2011, with additional entries coming later from Zoetis [68b, 71] and BASF [72]. Note that in several of these applications, derivatives wherein the alkenyl bond migrated to the 3,4-position were also synthesized and tested. Relative to the aforementioned dihydropyrroles, a far lesser number of the dihydrofurans were published, and this could be due to reduced potency or biological spectrum, chemical or metabolic stability, or a combination of factors.

Switching the carbon at the 4-position to an oxygen in order to yield a dioxazole (Figure 16.17) was first demonstrated by Nissan in the context of the ryanodine receptor agonists [73]. As a substitution pattern more directly applicable to the

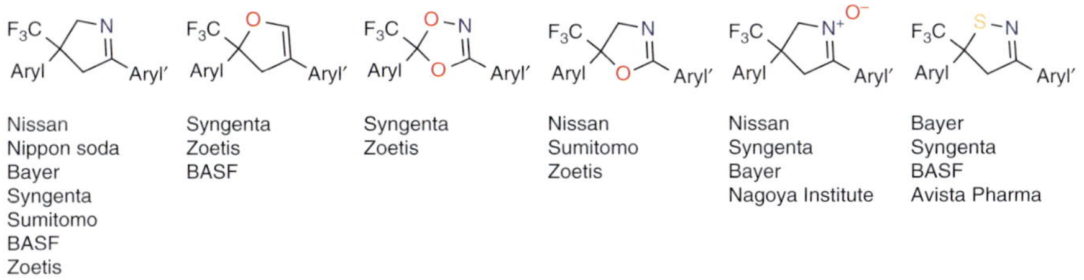

Figure 16.17 Single-atom variants of isoxazolines.

isoxazoline series, Syngenta [44b] was the first to utilize the attractive chemistry to access the dioxazole cores; two applications by Zoetis [68a, b] which focused on dihydropyrroles also disclosed the dioxazole compounds.

Migrating the oxygen from the 1-position in the isoxazoline to the "4"-position affords a dihydrooxazole (Figure 16.17). As was the case for the dioxazoles, these compounds first appeared in the synthetic programs at Nissan to find diamide ryanodine receptor agonists with insecticidal activity [73a, b, 74]. Nissan [64a] was also the first to use this moiety with the *para*-amidoaryl linkers similar to the general patterns observed in fluralaner (**4**), afoxolaner (**5**), and so on. Presumably due to the relative synthetic ease of accessing the dihydrooxazoles, approximately 5000 compounds have been disclosed in applications from Nissan [64a, 73a, b, 74], Sumitomo [33], and Zoetis [68a, b].

Replacing the isoxazoline oxygen at the 1-position with a carbon and then oxidizing the nitrogen at the 2-position yields a nitrone (Figure 16.17); overall this is equivalent to the *N*-oxidation of a dihydropyrrole. As a 1,3-dipole, the formal negative charge on the oxygen can also be depicted on the carbon at the 2-position (benzylic carbon in this case). This ability to delocalize the charge has implications for chemical reactivity and the physical chemical properties; the dipolar nitrone analogs are not as polar as molecules with a fixed charge separation (as the canonical structure would imply). In addition to the approximately 1300 analogs which have been described in patent applications from Nissan [64b], Syngenta [32a, 67a], and Bayer [75], researchers from the Nagoya Institute of Technology [76] have published a method for the asymmetric synthesis of the dihydropyrroles which lead to the nitrone following oxidation. The relative lack of nitrone analogs described is suggestive that this substitution pattern is lacking in the potency or the physicochemical properties needed for biological efficacy.

A direct bioisostere of the isoxazoline can also be obtained by exchanging the ring oxygen for a sulfur to afford a thioisoxazoline or isothiazoline (Figure 16.17). Chemists at Bayer [55d] were the first to publish a handful of these analogs in 2009; and several years later, scientists at Syngenta [77] and BASF [45c, 78] disclosed an additional approximately 1000 compounds. Finally, in 2016, Avista Pharma [79] published the thioisoxazoline version of sarolaner (7) and other related derivatives.

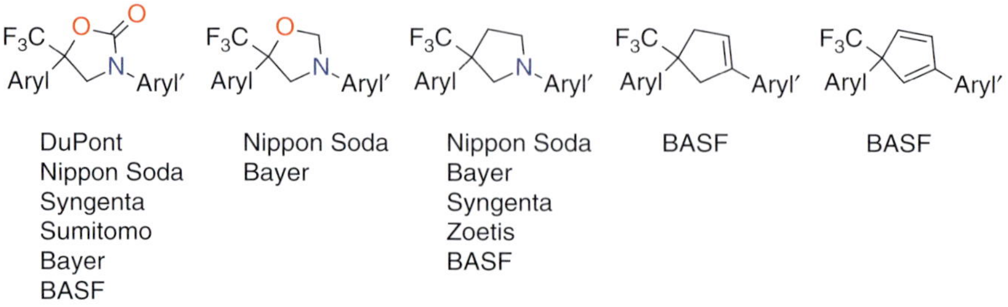

Figure 16.18 Additional isoxazoline replacement rings.

A number of additional five-membered ring variants have been investigated and representative structures are shown in Figure 16.18. A migration of the isoxazoline nitrogen into a direct bond with the aryl linker, along with a carbonyl bridging to the ring oxygen yields a 3-aryl oxazolidinone. Straightforward to access via synthesis from the requisite anilino-alcohol and an activated carbonyl source (e.g. carbonyl diimidazole), approximately 5K molecules have been described in the relevant patent literature. The first examples came from DuPont [80], with subsequent applications coming from Nippon Soda [81], Syngenta [44b], Sumitomo [33], Bayer [82], and BASF [83].

In practice, the oxazolidines (Figure 16.18) are often obtained from the chemical manipulation of the oxazolidinones via a ring opening to afford the anilino-alcohol and then reaction with formaldehyde or its equivalent. A far lesser number of these (formally reduced) analogs have been described relative to the oxazolidinone derivatives. The first examples of the oxazolidine-containing analogs appeared in 2008 from the Nippon Soda company [31c, 81]. In 2012, Bayer [82c, d] also published a set of analogs found active against mites, beetles, and cutworms.

Another group of *N*-aryl analogs meant to mimic the isoxazolines is the pyrrolidines (Figure 16.18). A large amount of effort has been expended on these compounds; over 8000 analogs were described and tested in 24 different patent applications published from 2008 to late 2016. The earliest examples came from Nippon Soda in 2008 [81], followed soon thereafter by 7 applications from Bayer [84] and 11 from Syngenta [32a, c, e, 67b–d, 85]. Zoetis [68a, b] and BASF [83] also contributed examples of pyrrolidine-linked pesticidal isoxazoline mimics.

Devoid of any polar atoms in/on the ring, the cyclopentenes and cyclopentadienes put forth by BASF [86] represent a large departure from the bioisosteres disclosed in that the central ring is now incapable of forming any hydrogen bonding interactions (Figure 16.18). Along with the additional bioisosteres shown, the large number of biologically active examples in the BASF application suggests the potency of the "isoxazoline class" may have less to do with the molecular target binding to/with the isoxazoline ring and more to do with the isoxazoline's use as a preferred linking group to provide molecules with an optimal combination of size, shape, and physicochemical properties for *in vivo* potency, as well as possessing attractive chemical syntheses.

A Coming Together of Groups

Several companies have explored an alternate means of combining the isoxazoline ring with the phenyl linking group; in one such arrangement, conceptually excising two atoms and then fusing the two groups together leads to the dihydrobenzofuran moiety containing the germinal trifluoromethyl and aryl groups (Figure 16.19). Process chemists at Merck [87] were the first to publish the synthesis of this chemotype in 2005, albeit in a different context and years before any connection to the isoxazoline parasiticidal class could be made. Syngenta was the first to realize the potential with respect to insecticidal usage [88]. Interestingly enough, the derivatives from Syngenta contained what could be considered an extra phenyl or extended alkenyl linking group, possibly suggestive that derivatives similar to the (simply contracted) Merck example have the incorrect length/shape for parasiticidal potency. Entries from Nippon Soda came later the next year, with two patent applications linking the dihydrobenzofuranyl moiety to their pyridinyl warheads via differing oximyl arrangements [31, 89]; due to the nature of the aryl group, it is the author's hypothesis that these particular analogs from Nippon Soda are likely parasiticidal due to an alternate mode of action relative to the dihydrobenzofuranyl compounds put forth by Syngenta.

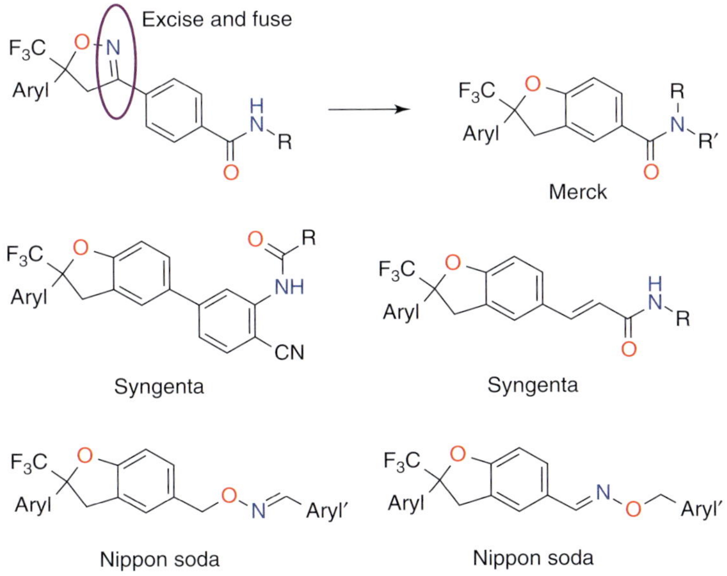

Figure 16.19 Fused ring isoxazoline replacements.

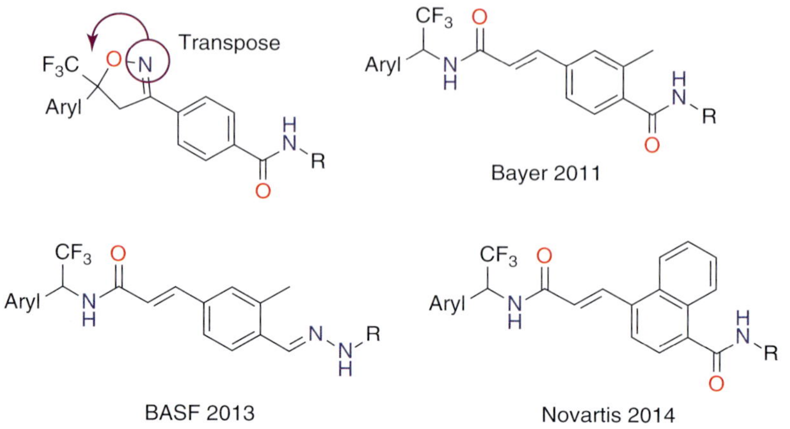

Figure 16.20 Acrylamide replacements for the isoxazoline ring.

An Open Future for Analogs

A somewhat opposite approach to the contraction of groups to provide fused-ring derivatives is to open up or displace the isoxazoline in search of new active analogs. In this vein, several distinctly different and clever angles have been pursued. The first general examples of the isoxazoline ring-opening approach (Figure 16.20) came from Bayer in 2011 and provided a set of acrylamide derivatives wherein the geminal trifluoromethyl/aryl substitution was maintained [90]. BASF also produced a set of acrylamides, varying the nature of the amide or related side chain bound to the central phenyl ring [91]. Novartis was the third company to publish a set of acrylamides based on the isoxazoline series, including the naphthyl, thiophenyl, and benzofuranyl linkers in lieu of the central phenyl ring [92].

The second ring-opening approach was one wherein chemists from Dow merely "removed" the nitrogen and oxygen to arrive at a propenyl linker in lieu of the isoxazoline (Figure 16.21). Their first publication in 2012 provided examples including triazole-substituted phenyl groups containing the *ortho*-cyano group and a racemic propenyl linker [93a]. Subsequent publications showed examples with the desired stereochemistry at the germinal trifluoromethyl-aryl center and further elaborations of the amide side chain [93b–h]. It should be noted that these examples represent another set of analogs devoid of the hydrogen bonding capability of the parent isoxazolines.

The third and final example of "opening up" the phenyl isoxazoline moiety presented herein is the set of vinylogous phenyl isoxazolines recently published by Syngenta [94]. Insertion of an alkenyl linker between the isoxazoline and the central phenyl ring afforded analogs with a slightly longer and more flexible structure (Figure 16.22). Of particular note is that the majority of the analogs disclosed in a 2016 application contain a *meta* arrangement between the phenyl ring and the inverse amide. One can surmise that the altered structure, relative to the parent

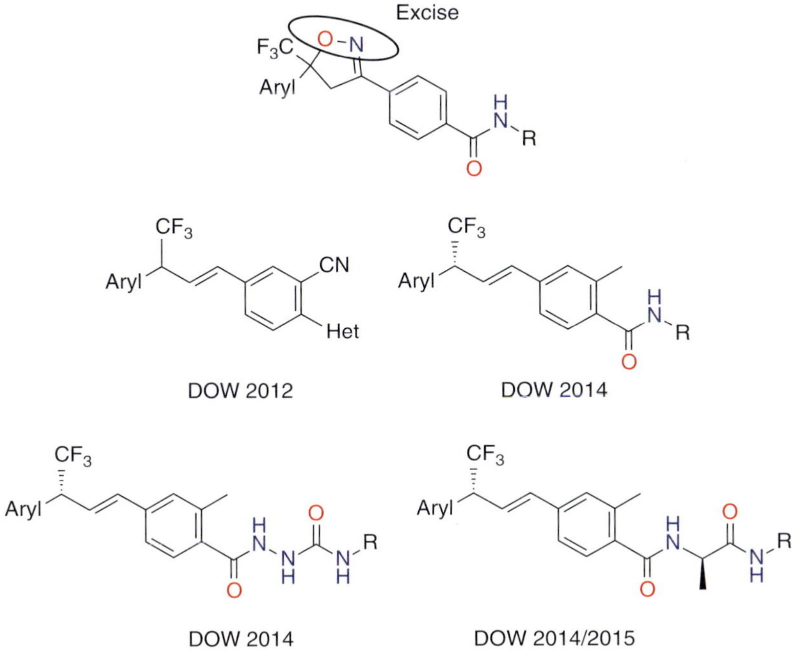

Figure 16.21 "Open ring" isoxazoline replacements from Dow Agrosciences.

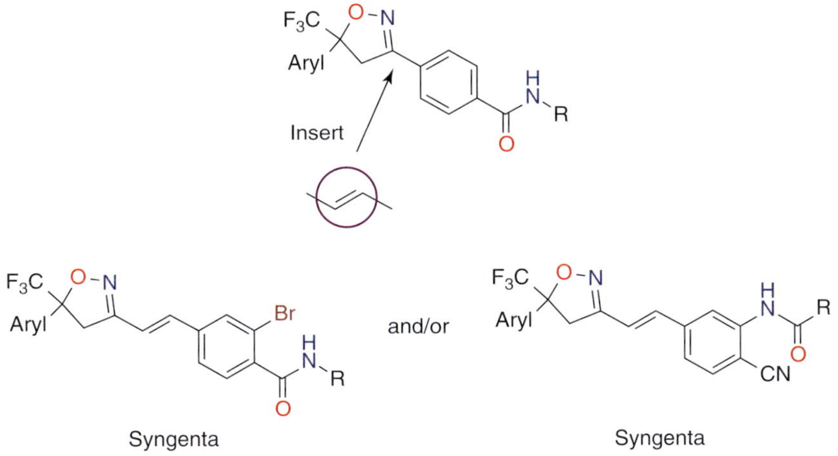

Figure 16.22 Vinyl-extended phenyl isoxazolines from Syngenta.

phenyl isoxazoline, confers a differing target-site requirement for the terminal amide side chains simply due to size and shape. It remains to be seen whether any of these "open-ring" or extended analogs possess the requisite binding affinities needed to achieve potency while maintaining efficacy in the *in vivo* applications.

Conclusion

This chapter began by illustrating the origins of the parasiticidal isoxazoline chemical class, apparently arising from systematic changes made to the diamide and anthranilimide compounds being developed at the turn of the century. Those early isoxazoline examples were then quickly developed into and commercialized as flea and tick control products for pets; a monumental accomplishment as they comprise the first new class of purely synthetic, *orally active* ectoparasiticides since the neonicotinoids and are also the first series to control ticks following oral administration. A brief discussion of the synthetic routes was provided, along with examples of the methods used to access the single enantiomers as the most recent commercial entries are presented in their optically active forms. A representative view of select physicochemical properties for the isoxazolines was given, along with commercial parasiticides from other chemotypes in order to provide context. The patents and patent applications for this burgeoning chemical class number in the hundreds, and an attempt was made to illustrate all of the companies and timelines associated with the significant discovery efforts to identify and develop compounds in the class with a variety of attributes and potential applications. Finally, a deconstructive method was used to examine the isoxazoline molecules and the many related analogs the class has spawned. With the tremendous body of work done to date, it will be interesting indeed to see what isoxazoline-inspired molecules and their applications may be described in the near future.

Disclaimer

This document is provided for scientific purposes only. Any reference to a brand or a trademark herein is for informational purposes only and is not intended for any commercial purposes or to dilute the rights of the respective owners of the brand(s) or trademark(s).

References

1. Lahm, G.P., Cordova, D., Barry, J.D., Pahutski, T.F. *et al.* (2013) 4-Azolylphenyl isoxazoline insecticides acting at the GABA gated chloride channel. *Bioorg. Med. Chem. Lett.*, **23** (10), 3001–3006.
2. Jeanguenat, A. (2013) The story of a new insecticidal chemistry class: the diamides. *Pest Manag. Sci.*, **69** (1), 7–14.
3. García-Reynaga, P., Zhao, C., Sarpong, R., and Casida, J.E. (2013) New GABA/glutamate receptor target for [^{3}H] isoxazoline insecticide. *Chem. Res. Toxicol.*, **26** (4), 514–516.
4. Selzer, P. and Weber, T. (2016) Isoxazolines: a novel chemotype highly effective on ectoparasites. *ChemMedChem*, **11** (3), 270–276.
5. E. I. du Pont de Nemours and Company (2009) Naphthalene isoxazoline invertebrate pest control agents. WO 2009002809 A2.
6. Nissan Chemical Industries, Ltd. (2005) Preparation of isoxazoline-substituted benzamide derivatives as insecticides, acaricides, and parasiticides. WO 2005085216 A1.
7. Pfizer, Inc. (2012) Preparation of spirocyclic isoxazoline derivatives as antiparasitic agents. WO 2012120399 A1.

8. Novartis AG (2010) Isoxazoline derivatives and their preparation and use for the control of parasites in and on warm-blooded animals. WO 2010070068 A2.
9. Sparks, T.C. (2013) Insecticide discovery: an evaluation and analysis. *Pestic. Biochem. Physiol.*, **107** (1), 8–17.
10. Nissan Chemical Industries, Ltd. (2007) Preparation of isoxazoline-substituted benzamide compounds as pesticides. WO 2007026965 A1.
11. Merial, Inc. (2016) Long-acting injectable formulations comprising an isoxazoline active agent, methods and uses thereof. WO 2016138339 A1.
12. Intervet International B.V. (2016) Isoxazoline solution containing vitamin E for use with sanitized drinking water. WO 2016207234 A1.
13. Novartis AG (2013) New use of compounds for controlling sea lice on fish. WO 2011157733 A2.
14. Novartis Tiergesundheidt AG (2016) Preparation of diarylisoxazoline for controlling parasites in and on warm-blooded animals or fish. WO 2016077158 A1.
15. Pfizer, Inc. (2012) Preparation of isoxazoline derivatives as antiparasitic agents for use in animals. WO 2012017359 A1.
16. Zoetis LLC (2014) Spirocyclic isoxazoline derivatives for treatment of sea lice in fish. WO 2014039422 A1.
17. (a) Matoba, K., Kawai, H., Furukawa, T., Shibata, N. *et al.* (2010) Enantioselective synthesis of trifluoromethyl-substituted 2-isoxazolines: asymmetric hydroxylamine/enone cascade reaction. *Angew. Chem. Int. Ed.*, **49** (33), 5762–5766; (b) Researchers from Shanghai University have also used the Shibata asymmetric catalysis to synthesize Fluralaner; their process begins with the 3,-5 dichlorinated β,β-difluorostyrene and uses silver(I) fluoride to provide the β-trifluoromethylketone; standard methods then furnish the α,β-unsaturated ketone: Gao, B., Zhao, Y., and Hu, J. (2015) AgF-mediated fluorinative cross-coupling of two olefins: facile access to α-CF_3 alkenes and β-CF_3 ketones. *Angew. Chem. Int. Ed.*, **54** (2), 638–642.
18. (a)Nissan Chemical Industries, Ltd. (2009) Method for catalytic asymmetric synthesis of optically active isoxazoline compound. WO 2009063910 A1;(b)Nissan Chemical Industries, Ltd. (2013) Method for asymmetric synthesis of optically active isoxazoline compounds. WO 2013069731 A1.
19. (a)Syngenta Participations AG and Syngenta LTD (2011) Preparation of isoxazoline derivatives having pesticidal activity via stereoselective synthesis using a quinine derivative as a chiral phase transfer catalyst. WO 2011104089 A1;(b)Syngenta Participations AG (2016) Process for the preparation of optically active isoxazoline compounds having a cycloserine substituent. WO 2016023787 A1.
20. (a)Zoetis LLC (2013) Process for the preparation of chiral isoxazoline azetidine derivatives as antiparasitic agents. WO 2013116236 A1;(b)Zoetis LLC (2014) Synthesis of spirocyclic isoxazoline derivatives. WO 2014081800 A1.
21. Sumitomo Chemical Company, Ltd. (2014) Method for producing optically active isoxazoline compound. WO 2014192841 A1.
22. BASF SE (2015) Preparation of azoline compounds for combating or controlling invertebrate pests. WO 2015128358 A1.
23. (a)Syngenta Participations AG (2014) Preparation of arylthiomethylsulfonamide derivatives as insecticides. WO 2014079937 A1;(b)Syngenta Participations AG (2014) Preparation of N-(arylsulfanylmethyl) carboxamide derivatives as insecticides. WO 2014079941 A1.
24. (a)Anacor Pharmaceuticals, Inc., USA and Eli Lilly and Company (2013) Boron-containing small molecules. US20130131016;(b)Anacor Pharmaceuticals, Inc., USA and Eli Lilly and Company (2013) Boron-containing small molecules. US20130131017.

25. Novartis AG (2014) Process for the enantioenrichment of diaryloxazoline derivatives. WO 2014090918 A1.

26. Lee, K., Silverio, D., Torker, S., Hoveyda, A. *et al.* (2016) Catalytic enantioselective addition of organoboron reagents to fluoroketones controlled by electrostatic interactions. *Nat. Chem.*, **8** (8), 768–777.

27. (a) Kawai, H., Okusu, S., Tokunaga, E., and Shibata, N. (2013) Enantioselective synthesis of 5-trifluoromethyl-2-isoxazolines and their *N*-oxides by [hydroxy(tosyloxy)iodo] benzene-mediated oxidative *N–O* coupling. *Eur. J. Org. Chem.*, **2013** (29), 6506–6509;(b)National University Corporation Nagoya Institute of Technology (2014) Preparation of 5-trifluoromethyl-2-isoxazoline-N-oxides and optically active 5-trifluoromethyl-2-isoxazolines. JP 2014009209 A.

28. Mugumbate, G. and Overington, J. (2015) The relationship between target-class and the physicochemical properties of antibacterial drugs. *Bioorg. Med. Chem.*, **23** (16), 5218–5224.

29. Lipinski, C., Lombardo, F., Dominy, B., and Feeney, P. (2001) Experimental and computational approaches to estimate solubility and permeability in drug discovery and development settings. *Adv. Drug Deliv. Rev.*, **46** (1–3), 3–26.

30. (a)Novartis AG (2012) Preparation of isoxazoline derivatives useful for controlling invertebrate pests. WO 2012107533 A1;(b)Novartis AG (2012) Preparation of isoxazole derivatives for controlling parasites. WO 2012120135A1.

31. (a)Nippon Soda Co., Ltd. (2015) Preparation of pyridine derivatives as fungicides, pesticides, insecticides or miticides. WO 2015199065 A1;(b)Nippon Soda Co., Ltd. (2015) Preparation of isoxazolin-5-carboxamide compounds as pesticides. WO 2015147022 A1;(c)Nippon Soda Co., Ltd. (2012) Ectoparasite control agent for animals comprising heterocyclic compounds. WO 2012060317 A1;(d)Nippon Soda Co., Ltd. (2010) Preparation of nitrogen-containing heterocyclic compounds as pesticides. WO 2010032437 A1;(e)Nippon Soda Co., Ltd. (2009) Preparation of isoxazoline compounds as pest control agents. WO 2009022746 A1.

32. (a)Syngenta Participations AG (2014) Insecticides and methods of soil pest control. WO 2014029639 A1;(b)Syngenta Participations AG (2014) Insecticidal compounds and methods of controlling insects in rice. WO 2014019951 A1;(c)Syngenta Participations AG (2013) Isoxazoline, pyrroline, and dihydrofuran derivative compounds and methods of soil dwelling pest control. WO 2013190046 A1;(d)Syngenta Participations AG (2013) Isoxazoline, pyrroline, and dihydrofuran derivative compounds and methods of controlling insects of the family Curculionidae, in particular Anthonomus species. WO 2013190050 A1;(e)Syngenta Participations AG (2012) Insecticidal compounds based on arylthioacetamide derivatives and their preparation. WO 2012156400 A1;(f)Syngenta Participations AG (2011) Isoxazoline derivatives as insecticides and their preparation and use. WO 2011067272 A1;(g)Syngenta Participations AG (2010) Preparation of phenyldihydroisoxazolylbenzo-thiadiazole carboxamide derivatives and analogs for use as insecticides. WO 2010084067 A2.

33. Sumitomo Chemical Company, Limited (2010) Preparation of azolyl aryl hydrazides as pesticides. WO 2010090344 A1.

34. Zoetis LLC (2014) Spirocyclylisoxazolines as antiparasitic agents and their preparation. WO 2014039484 A1.

35. Merial Limited (2011) Dihydroazole compounds as antiparasitic agents and their preparation and use in the treatment of parasitic infection or infestation in animals and as pesticides. WO 2011075591 A1.

36. Dow AgroSciences, LLC (2008) Preparation of diarylisoxazoline

derivatives as insecticides. WO 2008130651 A2.
37. (a)Bayer Cropscience AG (2008) Preparation of arylisoxazoline derivatives as insecticides and acaricides. WO 2008122375 A2;(b)Bayer Cropscience AG (2008) Preparation of diarylisoxazolines as insecticides and animal parasiticides. WO 2008019760 A1.
38. (a)Nissan Chemical Industries, Ltd. (2008) Preparation of isoxazoline-substituted benzamide compounds as pesticides. JP 2008239611 A;(b)Nissan Chemical Industries, Ltd. (2007) Preparation of isoxazoline compounds as pest control agents. WO 2007105814 A1;(c)Nissan Chemical Industries, Ltd. (2007) Preparation of isoxazoline compounds as pesticides. JP 2007016017 A.
39. Nihon Nohyaku Co., Ltd. (2007) Preparation of isoxazoline derivatives as pest-controlling agents. WO 2007125984 A1.
40. (a)E. I. du Pont de Nemours and Company (2007) Preparation of 5-arylisoxazolines as insecticides and acaricides. WO 2007075459 A2;(b)E. I. du Pont de Nemours and Company (2007) Preparation of isoxazolines for controlling invertebrate pests. WO 2007079162 A1.
41. Xu, M., Wagerle, T., Long, J.K., Lahm, G.P. *et al.* (2014) Insecticidal quinoline and isoquinoline isoxazolines. *Bioorg. Med. Chem. Lett.*, **24** (16), 4026–4030.
42. (a)BASF SE (2011) Preparation of azoline compounds for combating invertebrate pests. WO 2011073444 A2;(b)BASF SE (2010) Preparation of isoxazoline compounds for combating invertebrate pests. WO 2010112545 A1.
43. Norbrook Laboratories Ltd. (2015) Preparation of isoxazolines useful for the treatment of parasite infestation in animal. WO 2015132592 A1.
44. (a)Syngenta Participations AG (2010) Preparation of benzamide isoxazolines as insecticides. WO 2010020521 A1;(b)Syngenta Participations AG (2010) Preparation of tetracyclic derivatives as insecticides. WO 2010020522 A1;(c)Syngenta Participations AG (2010) Preparation of benzamide isoxazolines as insecticides. WO 2010086225 A1;(d)Syngenta Participations AG (2013) Preparation of insecticidal isoxazoline derivatives and use in pest control. WO 2013026933 A1.
45. (a)BASF SE (2013) Preparation of azoline compounds substituted by a carbocyclic condensed ring system useful for controlling invertebrate pests. WO 2016102488 A1;(b)BASF SE (2013) Preparation of azoline compounds substituted by a condensed ring system useful for controlling invertebrate pests. WO 2016102482 A1;(c)BASF SE (2014) Preparation of bicyclyl-substituted isothiazoline compounds for combating or controlling invertebrate pests. WO 2014206910 A1.
46. Nippon Soda Co. Ltd. (2009) Nitrogen-containing heterocyclic compound and pest control agent. WO 2009022746 A1.
47. Sumitomo Chemical Co. Ltd. (2012) Animal ectoparasite-controlling agent. EP2412238.
48. Sumitomo Chemical Co. Ltd. (2012) Animal ectoparasite control agents containing semicarbazide compounds and control of ectoparasites using the compounds. JP 2012031094 A.
49. BASF SE (2010) Preparation of substituted isoxazolylaryl amidine compounds for combating animal pests. WO 2010072602 A1.
50. BASF SE (2011) Process for producing imine compounds for combating invertebrate pests. WO 2011161130 A1.
51. BASF SE (2010) Imine compounds for combating invertebrate pests. WO 2010072781 A2.
52. Novartis AG (2010) Isoxazoline derivatives as pesticides and their preparation and use for the control of parasites in and on warm-blooded animals. WO 2010079077 A1.
53. (a)Anacor Pharmaceuticals, Inc. and Eli Lilly and Co. (2013) Boron-containing small molecules. US 20130131016 A1;(b)Anacor Pharmaceuticals, Inc. and Eli Lilly and Co. (2013) Boron-containing small molecules. US 20130131017 A1.

54. (a) Zhang, Y.K., Plattner, J., Easom, E., Zhou, Y. *et al.* (2015) Discovery of an orally bioavailable isoxazoline benzoxaborole (AN8030) as a long acting animal ectoparasiticide. *Bioorg. Med. Chem. Lett.*, **25** (23), 5589–5593;(b) Zhang, Y.K., Plattner, J., Easom, E., Zhou, Y. *et al.* (2016) Optimization of isoxazoline amide benzoxaboroles for identification of a development candidate as an oral long acting animal ectoparasiticide. *Bioorg. Med. Chem. Lett.*, **26** (13), 3182–3186.

55. (a)Nissan Chemical Industries, Ltd. (2007) Preparation of isoxazoline compounds as pesticides. JP 2007106756 A;(b)E. I. du Pont de Nemours and Company (2007) Preparation of pyrazoline derivative acaricides and insecticides. WO 2007123855 A2;(c)Dow Agrosciences LLC (2008) Preparation of diarylisoxazoline derivatives as insecticides. WO 2008130651 A1;(d)Bayer Cropscience AG (2009) Aryl-azole compounds as pesticides and their preparation. WO 2009112275 A1.

56. Nippon soda Co. Ltd. (2012) Preparation of 5-phenyl-3-(2,3-dihydro-1H-inden-5-yl)-4,5-dihydroisoxazole derivatives as pesticides. WO 2012026403 A1.

57. Syngenta Participations AG (2013) Preparation of insecticidal 2-arylacetamide compounds. WO 2013135674 A1.

58. BASF SE (2011) Preparation of substituted ketonic isoxazoline compounds and derivatives for combating animal pests. WO 2011092287 A1.

59. (a)Nissan Chemical Industries Ltd. (2009) Preparation of substituted isoxazoline or enone oxime compounds as pest control agents. WO 2009005015 A1;(b)Nissan Chemical Industries Ltd. (2010) Preparation of isoxazoline and enone oximes as pesticides. JP 2010168367 A.

60. Sumitomo Chemical Co., Ltd. (2012) Preparation of 5-trifluoromethyl-4-nitro-2-isoxazolines by direct trifluoromethylation of 4-nitro-2-isoxazolines. WO 2012060232 A1.

61. Nagoya Institute of Technology (2012) 5-Triflurommethyl-4-trifluoromethylsulfonyl-2-isoxazoline derivative, and method for the preparation thereof. WO 2014024752 A1.

62. (a)Syngenta Participations AG (2010) Preparation of phenylisoxazolidinylbenzamide derivatives for use as insecticides. WO 2010108733 A1;(b)Syngenta Participations AG (2011) Isoxazoline derivatives as insecticides and their preparation and use. WO 2011101402 A1.

63. Syngenta Participations AG (2013) Preparation of isoxazoline derivatives as insecticidal agents. WO 2013026931 A1.

64. (a)Nissan Chemical Industries, Ltd. (2007) Preparation of dihydroazole-substituted benzamides as pesticides. JP 2007091708 A;(b)Nissan Chemical Industries, Ltd. (2009) Preparation of substituted dihydroazole compounds as pest control agents. WO 2009072621 A1;(c)Nissan Chemical Industries, Ltd. (2012) Substituted dihydropyrrole parasitic control agents for animal. JP 2012031148 A;(d)Nissan Chemical Industries, Ltd. (2012) Substituted benzamide compounds as control agent against parasite and sanitary insect. WO 2012067235 A1.

65. Nippon Soda Co., Ltd. (2008) Nitrogen-containing heterocyclic compounds and pesticides containing them. JP 2008133273 A.

66. (a)Bayer Cropscience AG (2009) Arylpyrrolines and related compounds as insecticides and their preparation. WO 2009097992 A1;(b)Bayer Cropscience AG (2010) Preparation of arylpyrrolines and their use as insecticides. WO 2010133336 A1;(c)Bayer Cropscience AG (2011) Preparation of arylpyrrolines compounds. WO 2011054871 A1;(d)Bayer Cropscience AG (2011) Preparation of pyrroline derivatives for use as pesticides. WO 2011128299 A1;(e)Bayer Cropscience AG (2012) Preparation of pyrroline derivatives as pesticides. WO 2012004326 A1;(f)Bayer Cropscience AG (2011) Process for the preparation of 5-substituted 3-aryl-3-(trifluoromethyl)-3,4-dihydro-2H-pyrroles. WO 2011141414 A1;(g)Bayer Cropscience AG (2012) Pyrroline derivatives, and insecticides and animal parasite control agents containing them. JP 2012017289 A.

67. (a)Syngenta Participations AG (2011) A process for the preparation of dihydropyrrole derivatives for use as pesticides. WO 2011154555 A1;(b)Syngenta Participations AG (2012) Preparation of dichlorophenylpyrrolidine derivatives as pesticides. WO 2012163959 A1;(c)Syngenta Participations AG (2013) Preparation and use of insecticidal compounds for pest control in soybean, in particular stinkbugs. WO 2014019957 A2;(d)Syngenta Participations AG (2014) Preparation and use of insecticidal compounds for pest control in soybean, particularly stinkbugs. WO 2014019609 A1;(e)Syngenta Participations AG (2014) Insecticides and methods of controlling insects. WO 2014029640 A1;(f)Syngenta Participations AG (2014) Insecticide preparation and use in controlling insects, preferably Anthonomus. WO 2014029706 A2;(g)Syngenta Participations AG (2014) Insecticide preparation and use in controlling insects in rice. WO 2014029707 A1;(h)Syngenta Participations AG (2014) Insecticide preparation and use in controlling insects, preferably Plutella. WO 2014029708 A1;(i)Syngenta Participations AG (2014) Insecticide preparation and use in controlling insects, preferably soil pests such as corn rootworm. WO 2014029710 A2;(j)Syngenta Participations AG (2010) Preparation of dihydro-pyrrole derivatives as insecticidal compounds. WO 2010149506 A1.

68. (a)Zoetis LLC (2013) Preparation of azetidine derivatives as antiparasitic agents for use in veterinary medicine. WO 2013169622 A1;(b)Zoetis LLC (2014) Spirocyclic derivatives as antiparasitic agents and their preparation. WO 2014039489 A1.

69. BASF SE (2012) Preparation of imine substituted 2,4-diarylpyrroline derivatives as pesticides. WO 2012042007 A1.

70. (a)Syngenta Participations AG (2011) Preparation of dihydrofuran derivatives for use as insecticides. WO 2011101229 A1;(b)Syngenta Participations AG (2013) Preparation of dihydrofuran derivatives as insecticidal compounds. WO 2013026724 A1;(c)Syngenta Participations AG (2013) Preparation of dihydrofuran derivatives as insecticidal compounds. WO 2013026726 A1;(d)Syngenta Participations AG (2014) Insecticidal compounds and methods of pest control in soybean, particularly useful for control of stinkbugs. WO 2014019950 A1;(e)Syngenta Participations AG (2014) Insecticides and methods of controlling insects, especially boll weevil in cotton. WO 2014029709 A1.

71. Zoetis LLC (2013) Preparation of dihydrofuranazetidine derivatives for use as antiparasitic agents. WO 2013116230 A1.

72. BASF SE (2016) Preparation of cyclic compounds substituted by a condensed ring system useful for controlling invertebrate pests. WO 2016102490 A1.

73. (a)Nissan Chemical Industries, Ltd. (2004) Preparation of N-(4-heterocyclylphenyl)phthalic acid diamide compounds as pest control agents. WO 2004018410 A1;(b)Nissan Chemical Industries, Ltd. (2005) Preparation of cycloalkyl or heterocyclylbenzanilide compounds as pesticides. JP 2005272452 A.

74. Nissan Chemical Industries, Ltd. (2005) Compositions containing benzanilides and their application as pesticides. JP 2005272443 A.

75. (a)Bayer Cropscience AG (2012) Preparation of pyrroline N-oxide derivatives as pesticides. WO 2012034957 A1;(b)Bayer Cropscience AG (2012) Preparation of pyrroline N-oxide derivatives as pesticides. JP 2012062267 A.

76. Nagoya Institute of Technology (2014) Asymmetric conjugate addition reaction of nitromethane

to β-aryl-β-trifluoromethyl-enone, and enantioselective preparation of diarylpyrroline or N-oxide thereof. JP 2014122169 A.

77. (a)Syngenta Participations AG (2013) Preparation of isothiazoline derivatives as insecticides. WO 2013037626 A1;(b)Syngenta Participations AG (2014) Preparation of isothiazole derivatives as insecticidal compounds. WO 2014001120 A1;(c)Syngenta Participations AG (2014) Preparation of isothiazole derivatives as insecticidal compounds. WO 2014001121 A1;(d)Syngenta Participations AG (2014) Preparation of [4-(isothiazol-3-yl)arylthio]acetamide derivatives as insecticides. WO 2014079935 A1.

78. (a)BASF SE (2013) Preparation of isothiazoline compounds for combating invertebrate pests. WO 2013092943 A1;(b)BASF SE (2014) Isothiazoline compounds and methods for controlling pests in soybean. WO 2014202751 A1;(c)BASF SE (2014) Suspension concentrate composition comprising isothiazoline insecticide and activated charcoal. WO 2014202437 A1;(d)BASF SE (2014) Preparation of naphthylisothiazoline derivatives for use as pesticides. WO 2014206907 A1;(e)BASF SE (2014) Preparation of carbobicyclylisothiazoline derivatives for use as pesticides. WO 2014206908 A1;(f)BASF SE (2014) Thiophene- or furan-substituted isothiazoline compounds as pesticides. WO 2014206909 A1;(g)BASF SE (2014) Isothiazoline compounds for combating invertebrate pests. WO 2014206911 A1.

79. Avista Pharma Solutions (2016) Preparation of isothiazoline-substituted spirocyclic azetidinyl-isobenzofuran derivatives as antiparasitic agents. WO 2016115315 A1.

80. E. I. du Pont de Nemours and Company (2007) Preparation of five-membered heterocyclic invertebrate pest control agents. WO 2007123853 A2.

81. Nippon Soda Co., Ltd. (2008) Preparation of nitrogen-containing heterocyclic compounds and their use as pesticides. JP 2008110971 A.

82. (a)Bayer CropScience AG (2011) Pesticidal heterocyclic compounds and their preparation. WO 2011051455 A1;(b)Bayer CropScience AG (2011) Oxazolidinones, their insecticides and control agents for animal parasites. JP 2011093855 A;(c)Bayer Intellectual Property GmbH (2012) Preparation of pesticidal diaryl heterocyclyl derivatives. WO 2012146572 A1;(d)Bayer CropScience AG (2012) Preparation of pesticidal diaryl heterocyclyl derivatives. JP 2012232910 A.

83. BASF SE (2012) Preparation of imine compounds as pesticides. WO 2012042006 A1.

84. (a)Bayer # Bayer CropScience AG (2008) Preparation of aryl pyrrolidines as insecticides and acaricides. WO 2008128711 A1;(b)Bayer CropScience AG (2010) Preparation of arylpyrrolidine derivatives and analogs for use as pesticides. WO 2010043315 A1;(c)Bayer CropScience AG (2011) Preparation of arylpyrrolidines. WO 2011080211 A1;(d)Bayer CropScience AG (2011) Pesticidal arylpyrrolidines and their insecticides and parasiticides. JP 2011136928 A;(e)Bayer CropScience AG (2012) Preparation of pesticidal arylpyrrolidines and their use as animal parasite-controlling agents. WO 2012035011 A1;(f)Bayer Intellectual Property GmbH (2013) Preparation of pesticidal arylpyrrolidines. WO 2013135724 A1;(g)Bayer CropScience Aktiengesellschaft (2016) Insecticidal arylpyrrolidines, method for synthesizing same, and use thereof as agents for controlling animal pests. WO 2016180802 A1.

85. (a)Syngenta Participations AG (2012) Preparation of pyrrolidinylpyridinylcarboxamide derivatives and analogs for use as insecticides. WO 2012045700 A1;(b)Syngenta Participations AG (2013) Enantioselective processes to insecticidal 3-aryl-3-trifluoromethyl-substituted pyrrolidines. WO 2013050301 A1.

86. BASF SE (2015) Preparation of cyclopentene and cyclopentadiene

compounds for controlling invertebrate pests. WO 2015114157 A1.
87. Kuethe, J.T., Wong, A., Journet, M., and Davies, I.W. (2005) A rapid synthesis of 2-aryl-5-substituted-2,3-dihydrobenzofurans. *J. Org. Chem.*, **70** (9), 3727–3729.
88. (a)Syngenta Participations AG (2014) Preparation of dihydrobenzofuran derivatives as insecticidal compounds. WO 2014135095 A1;(b)Syngenta Participations AG (2014) Dihydrobenzofuran derivatives as insecticidal compounds and their preparation. WO 2014172871A1.
89. Nippon Soda Co., Ltd. (2016) Preparation of pyridine derivatives as fungicides, pesticides, insecticides and miticides. WO 2016039048 A1.
90. Bayer CropScience AG (2011) Haloalkyl-substituted amides as insecticides and acaricides and their preparation. WO 2011054436 A2.
91. BASF SE (2013) Preparation of acrylamide compounds for combating invertebrate pests. WO 2013167633 A1.
92. Novartis AG (2014) (Hetero) arylacrylamides for the control of ectoparasites. WO 2014096381 A1.
93. (a)Dow Agrosciences LLC (2012) Preparation of substituted (3-phenyl-4,4,4-trifluorobut-1-en-1-yl) benzene compounds useful in pesticidal compositions. WO 2012177813 A1;(b)Dow Agrosciences LLC (2014) Preparation of diarylpropene derivatives as pesticides and their pesticidal compositions. WO 2014100163 A1;(c)Dow Agrosciences LLC (2014) Preparation of diarylpropene derivatives as pesticides and their pesticidal compositions. WO 2014100166 A1;(d)Dow Agrosciences LLC (2014) Preparation of diarylpropene derivatives as pesticides and their pesticidal compositions. WO 2014100170 A1;(e)Dow Agrosciences LLC (2014) Preparation of diarylpropene derivatives as pesticides and their pesticidal compositions. WO 2014100190 A1;(f)Dow Agrosciences LLC (2014) Preparation of diarylpropene derivatives as pesticides and their pesticidal compositions. WO 2014100206 A1;(g)Dow Agrosciences LLC (2014) Preparation of diarylpropene derivatives as pesticides and their pesticidal compositions. WO 2014120355 A1;(h)Dow Agrosciences LLC (2015) Methods for the preparation of diarylpropene amide derivatives as pesticidal agents. WO 2015191430 A1.
94. Syngenta Participations AG (2016) Preparation of isoxazoline-styrene derivatives as insecticides. WO 2016155831 A1.

Index

Ectoparasites: Drug Discovery Against Moving Targets, First Edition.
Edited by Charles Q. Meng and Ann E. Sluder